U.S. Marine Corps Wilderness Medicine Survival Course

Plus

Marine Rifle Marksmanship

Department of Defense

Pentagon
Publishing

UNITED STATES MARINE CORPS
Mountain Warfare Training Center
Bridgeport, California 93517-5001

WILDERNESS MEDICINE COURSE STUDENT HANDBOOK

TABLE OF CONTENTS

TOC

TABLE OF CONTENTS (CONT.)

UNITED STATES MARINE CORPS
Mountain Warfare Training Center
Bridgeport, California 93517-5001

FMST.07.18
04/02/02

STUDENT HANDOUT

MOUNTAIN SAFETY (SUMMER)

TERMINAL LEARNING OBJECTIVE: Given a unit in a wilderness environment and necessary equipment and supplies, apply the principles of mountain safety to prevent death or injury per the reference. (FMST.07.18)

ENABLING LEARNING OBJECTIVE: Without the aid of references and given the acronym "BE SAFE MARINE", list in writing the 12 principles of mountain safety, in accordance with the references. (FMST.07.18a)

OUTLINE

1. **PLANNING AND PREPARATION:** (FMST.07.18a) As in any military operation, planning and preparation constitute the keys to success. The following principles will help the leader conduct a safe and efficient operation in any type of mountainous environment. We find this principle in the acronym "BE SAFE MARINE". Remember the key, think about what each letter means and apply this in any type of environment.

B - Be aware of the group's ability.

E - Evaluate terrain and weather constantly.

S - Stay as a group.

A - Appreciate time requirements.

F - Find shelter before storms if required.

E - Eat plenty and drink lots of liquids.

M - Maintain proper clothing and equipment.

1

A - Ask locals about conditions.

R - Remember to keep calm and think.

I - Insist on emergency rations and kits.

N - Never forget accident procedures.

E - Energy is saved when warm and dry.

a. Be Aware of the Group's Ability. It is essential that the leader evaluates the individual abilities of his men and uses this as the basis for his planning. In his evaluation, the leader must include the group's overall physical conditioning, and the consideration of change in climate and how long the unit has had to acclimatize.

(1) Mental attitude of your group. Is morale high? How much tactical training has the group had in a particular type of terrain?

(2) Technical aspect of your group. Have they been on skis, snowshoes, etc.?

(3) Individual skills. At this point, you must choose who is most proficient at the individual skills that will be required for your mission, navigation techniques, security, call for fire, rope installations, track plans, bivouac site selection, skijoring, etc.

b. Evaluate Terrain and Weather Constantly.

(1) Terrain. During the planning stages of your mission, the leader must absorb as much information as possible on the surrounding terrain and key terrain features involved in your area of operation. Considerations to any obstacles must be clearly planned for. Will you need such things as fixed ropes, rope bridges, climbing gear, etc?

(a) Stress careful movement in particularly dangerous areas, such as loose rock and steep terrain.

(b) Always know your position. Knowing where you are on your planned route is important.

(2) Weather. Mountain weather can be severe and variable. Drastic weather changes can occur in the space of a few hours with the onset of violent storms, reduced visibility, and extreme changes. In addition to obtaining current weather data, the leader must plan for the unexpected "worst case". During an operation he must diagnose weather signs continually to be able to foresee possible weather changes.

(a) Constantly evaluate the conditions. Under certain conditions it may be advisable to reevaluate your capabilities. Pushing ahead with a closed mind could spell disaster for the mission and the unit.

c. Stay as a Group. Individuals acting on their own are at a great disadvantage in this environment.

(1) Give the unit adequate rest halts based upon the terrain and elevation, physical abilities of the unit, combat load and mission requirements.

(2) Remember to use the buddy system in your group.

(3) Maintain a steady pace so that it will allow accomplishment of the mission when all members of the unit reach the objective area.

d. Appreciate Time Requirements. Efficient use of available time is vital. The leader must make an accurate estimate of the time required for his operation based on terrain, weather, unit size, abilities, and on the enemy situation. This estimate must take into account the possibility of unexpected emergencies and allow sufficient leeway to make unplanned bivouacs in severe conditions.

(1) Time-Distance Formula (TDF). This formula is designed to be a guideline and should not be considered as the exact amount of time required for your movement. Furthermore, this formula is for use in ideal conditions:

3 km/mph + 1 hour for every 300 meters ascent; and/or + 1 hour for every 800 meters descent.

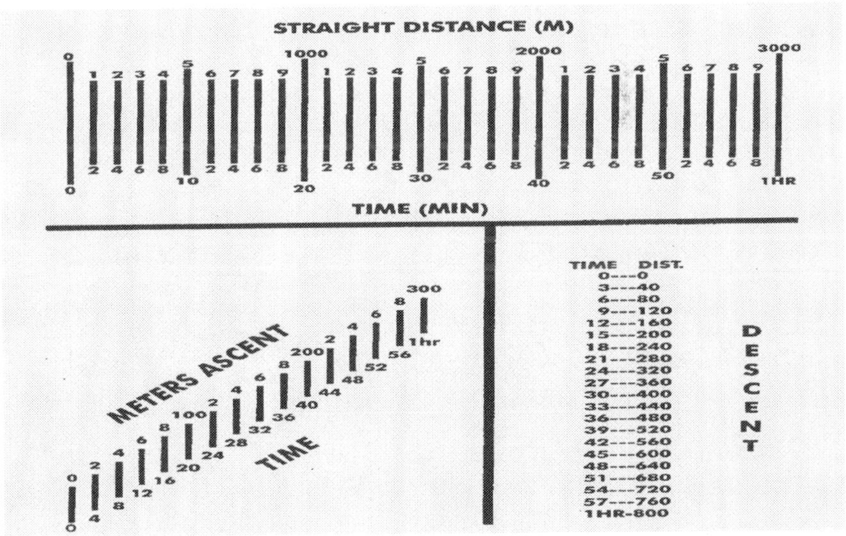

(2) Route Planning. Route cards are not to be used in place of an overlay, but as a tool to be used in route planning. Overlays/Route cards should contain the following information at the minimum:

Unit Designation:
Unit Commander.
Number of personnel.
Inclusive dates and times of movement.
Grid coordinates of each checkpoint and bivouac.
Map references.
Azimuth and distances for each leg.
Elevation gain/loss per leg.
Description of the ground.
ETA and ETD.

ROUTE CARD

UNIT I.D.		UNIT COMMANDER		NUMBER OF PERSONNEL		DATE AND TIME	MAP REFERENCE
LEG	AZM	DIST	GRID	ETA	ETD	ELEVATION	DESCRIPTION

(3) As in any military operation, route planning and execution are of vital importance. Prior to departure, the unit commander must submit a route card or patrol overlay to his higher headquarters and keeps a duplicate copy for himself. This preplanned route should be followed as closely as possible, taking into account changes based on the tactical situation. In non-tactical situations, the preplanned route should be followed to reduce search and rescue time in an emergency situation.

e. Find Shelter before Storms if required. Under certain conditions, inclement weather can provide tactical advantages to the thinking unit commander, but by the same token it can reduce the efficiency of a unit to nil if an incorrect evaluation of the situation is made.

(1) Bivouac. If the group decision is to bivouac, then it's vital that we know the principles for an unplanned bivouac.

(a) Unplanned bivouac. The principles and techniques discussed here apply both to unplanned and tolerated bivouacs. In any survival situation, especially in a mountainous environment, the most immediate danger is from exposure to the elements. Being lost will not directly kill an individual. Starvation takes time, but hypothermia can manifest itself in a matter of hours resulting in death. Adhering to the following principles will give an individual the best chance to spend a relatively safe bivouac with the prospect of continued effort toward mission accomplishment.

1. Make shelter. The requirements for expedient shelters and the building procedures will be covered in another section. The basic requirement for protection from the elements is essential.

2. Keep warm. The retention of body heat is of vital importance; any action in which body heat is lost should be avoided. The following points should be considered:

a. Adequate shelter.

b.. Insulation from the ground using branches, a rucksack, etc.

c. Wear extra clothing.

d. Use extra equipment for insulation.

e. Produce external heat while trying to conserve fuel for future use.

3. Keep dry. Being wet causes the loss of body heat 24 times faster than when dry. Adequate protection from the elements is of prime importance to prevent the onset of hypothermia.

f. Eat Properly and Drink Plenty of Fluids

(1) Food. The human body can be compared to a furnace, which runs on food to produce energy (warmth). By planning the consumption of food to suit the specific situation, adequate nutrition and extra warmth can be supplied.

(2) Water. The intake of adequate amounts of water will maintain the body in proper working order. Danger from dehydration is as high in mountain regions as in hot dry areas. Loss of liquids is easily seen and felt in hot climates; whereas in the

mountains, the loss of body fluids is much less noticeable. High water intake, at least 6 quarts per day when in bivouac, 8 quarts per day when active, will help to prevent dehydration.

g. Maintain Proper Clothing and Equipment

(1) Clothing

(a) Our clothing has to perform an important function in our mission; therefore, when choosing our clothing we have to take into account some essential requirements.

1. Protection against wind and rain.

2. Layered and easily adjustable.

3. Lightweight and durable.

(b) To help us remember how to maintain and wear our clothing, we use the acronym, "COLD".

C - Keep clothing Clean.

O - Avoid Overheating.

L - Wear clothing loose and in Layers.

D - Keep clothing Dry.

NOTE: In the mountains a man should never be separated from his gear. Here are some basic and essential items that should be considered during your planning stage.

(2) Required equipment:

(a) A daypack should always be carried per squad and one Marine should carry a combat load.

(b) Map and compass. Every individual in a leadership position and his assistant should carry a map and compass. The maps should be weatherproofed and extra maps should be distributed throughout the unit.

(c) Repair kit. This kit should include those items necessary to do emergency repairs on your equipment.

(d) Survival Kit. Always carried on your person. The contents of a survival kit will be covered in another period of instruction dealing specifically with survival kits.

h. Ask Locals About Conditions. An often-overlooked source of information is the indigenous population of an area. Local weather patterns, rockslide areas, watering points, and normal routes can all be obtained by careful questioning. The leader must obtain current information of the actual conditions along his intended route. Of particular importance are recent precipitation and enemy sightings.

i. Remember to Keep Calm and Think.

(1) Emergency situation. Having recognized that you are lost and that an emergency situation exists, the following principles should be followed:

(a) Keep calm and do not panic. At this point you must make every effort to conserve body heat and energy.

(b) Think. When an individual is cold, tired, hungry or frightened he must force himself to organize his thoughts into a logical sequence.

(c) The group must try to help itself by either finding the way back to safety or by preparing shelters and procuring food.

(d) Above all else, the group must act as a tight-knit unit. In emergency situations, individual dissension can cause a total loss of control and unit strength.

(2) If the decision is reached that the group should seek its way back to safety, several possibilities exist. In most situations, the safest approach will be to retrace the route to the last known point and continue from there. The other course of action is to get a group consensus on the present location and send out a small search party to locate a known point. This party must ensure that they mark their trail adequately to return to the group. If all attempts at finding a way back to known terrain fails, a definite survival situation exists and actions discussed later in this section must be instituted.

j. Insist on Emergency Rations and Kits. Just like what was covered in the SUMMER MOUNTAIN WARFIGHTING LOAD REQUIREMENTS class, survival rations and a survival kit should always be carried.

k. Never Forget Accident/Emergency Procedures.

(1) Causes of accidents. The general procedures used to handle accidents differ little in this environment, but several distinct points should be kept in mind. The most frequent causes of accidents are as follows:

(a) Overestimation of physical and technical abilities.

(b) Carelessness.

(c) General lack of observation of one's surroundings.

(d) Lack of knowledge and experience by leaders.

(e) The failure to act as a group.

(f) Underestimation of time requirements to move through mountainous terrain and underestimation of the terrain itself.

(2) Preventive measures. The only truly effective preventive measures for the above lie in the education and experience of leaders at all levels. Too often, leaders sit by watching during training and as a result have no concept of the requirements involved in the mountainous environment. Only by active involvement can a leader gain the knowledge and experience needed to effectively lead in this environment.

(3) General procedures for handling an accident. These require only a good dose of common sense as outlined below.

(a) Perform basic first aid.

(b) Protect the patient from the elements to include insulation on top and bottom.

(c) Evacuate if necessary.

(d) Send for help if required. If possible, never send a man for help alone.

(e) Send the following information regarding the accident:

1. Time of accident.
2. Nature and location of accident.
3. Number injured.
4. Best approach route to accident scene.

(4) If one man of a two-man team is injured, the injured man must be given all available aid prior to going for help. If the injured man is unconscious, he should be placed in all available clothing and sleeping gear and anchored if on steep terrain. A note explaining the circumstances, and reassuring him, should be left in a conspicuous spot. This note must also contain the following information:

(a) When you expect to return.

(b) Where you went.

(c) What you did before you left (medication, etc.).

(5) International distress signal (whistle):

(a) Six short blasts in 1 minute from person requesting help.

(b) The return signal is three blasts in 1 minute from the respondent.

(6) Other methods if help is required:

(a) Red pyrotechnics.

(b) SOS, (... --- . . .).

(c) "Mayday" by voice communications.

l. Energy is saved when warm and Dry. With the previous 11 principles in mind this one should fall right into place. Save your heat and energy by following these steps:

(1) Dress properly.

(2) Eat properly.

(3) Drink properly.

(4) Ensure shelter meets criteria.

(5) Produce external heat (fires, stove, extra clothing, etc.) to save body heat and energy for future use.

(6) Don't lose body heat by getting wet.

UNITED STATES MARINE CORPS
Mountain Warfare Training Center
Bridgeport, California 93517-5001

FMST.07.01
03/08/02

STUDENT HANDOUT

NUTRITION

TERMINAL LEARNING OBJECTIVE: Given a unit in a mountainous environment, and the necessary equipment and supplies, apply the principles of nutrition in a mountainous environment to prevent death or injury per the reference. (FMST.07.01)

ENABLING LEARNING OBJECTIVES:

(1) Without the aid of references and from a given list, choose the correct caloric intake needed per day in a mountainous environment, in accordance with the references. (FMST.07.01a)

(2) Without the aid of references and from a given list, choose which type of diet increases your tolerance to cold, in accordance with the references. (FMST.07.01b)

(3) Without the aid of references and from a given list, choose one of the preventive measures of excessive water loss in a mountainous terrain, in accordance with the references. (FMST.07.01c)

OUTLINE.

1. **STARVATION FACTORS.** Factors that combine to result in "Near Starvation" during military operations in any mountainous environment:

a. Difficulty in carrying sufficient food.

b. Lack of time for preparing meals and beverages.

c. Loss of appetite and low food palatability resulting from hypoxia/hypothermia.

d. Fat intolerance resulting from altered metabolism.

2

2. **PHYSIOLOGICAL ASPECT TO COLD STRESS.**

a. Man's adaptation to cold is essentially behavioral. Shelters and clothing are utilized to provide a micro-environment, which will allow survival. Body composition adjustments such as increases in body weight and fat are also examples of the body's adaptation to the cold. The winter months lend themselves to greater periods of inactivity, leading to these changes.

b. In the Falkland War, constipation was a major concern. Dehydration was found to be the culprit in the majority of cases, but diet had a significant role as well. MRE's are low in fiber, containing less than 10 grams. Man needs at least 31 grams of fiber per day to ease passage of stool. Constipation can be a debilitating condition, affecting overall unit readiness.

(1) Low fiber diets can lead to diverticulosis, diverticulitis and Irritable Bowel Syndrome (IBS). IBS flare-ups tend to occur during upsetting situations and emotional stress. Educate your troops on the importance of fiber, especially in a cold weather environment.

(2) Fiber is supplied by foods derived from plants. The best sources are:

(a) Whole grain cereals.

(b) Whole wheat breads.

(c) Fresh and dried fruits.

NOTE: Vegetables also help prevent constipation, but the fiber from whole grain cereals and whole wheat bread is more effective than vegetables.

3. **CALORIC REQUIREMENTS.** (FMST.07.01a) Recommended caloric intake (Naval Medical Command Instruction 10110.1) for troops operating in a cold weather environment is 4,500 calories per day.

a. Energy Expenditure. Military operations in a cold weather environment usually involve strenuous activity. Walking on snow expends roughly twice the energy that walking on hard ground does. If troops are unfamiliar with personal over-snow mobility equipment, i.e. snowshoes and skis, awkward movement and thus wasted energy will result.

(1) The heavy clothing necessary for functioning in the cold increases energy expenditures by 5-15%. You will have the opportunity to experience this effect in the upcoming field evolution.

(2) Properly clothed, experienced individuals should have negligible heat loss from radiation, conduction, and convection. However, losses associated with respiration and evaporation can be significant.

(3) In intense cold, particularly with high winds, heat loss to the environment may be inevitable.

4. **NUTRITION IN A HIGH ALTITUDE ENVIRONMENT.** A diet high in carbohydrates and high in fat is the most beneficial, while a high protein diet is generally the least desirable.

a. The marked weight loss seen in individuals at high altitude is largely due to increased metabolic demands and anorexia secondary to hypoxia. Total caloric intake commonly does not meet energy demands.

b. Altered metabolic processes that have been reported at high altitude include delayed gastric emptying, intestinal malabsorption and decreased gastric enzyme secretion.

5. **BEST DIET FOR COLD WEATHER**. (FMST.07.01b)

a. High carbohydrate and high fat foods should be included rather than high protein foods.

b. Small meals (snacks) spaced at time intervals of two hours or less should be consumed rather than the usual three meals-a-day.

c. During a typical eight-hour day of cold exposure, three high-fat (unsaturated fat) meals providing 60% of the day's calories, coupled with two normally scheduled meals (breakfast and dinner) should provide good cold tolerance.

6. **BASIC CONSTITUENTS OF FOOD**. The three basic constituents of food are carbohydrates, fats and proteins.

a. Carbohydrates. Also known as the quick energy foods that provide energy to produce heat with the byproducts of carbon dioxide and water when oxidized by the body.

(1) Carbohydrates are present in food mainly as sugars and starches and are broken down during digestion into simple sugars that are converted into glucose.

(2) Carbohydrates are stored as glycogen in the liver and muscles and can be broken down quickly into glucose to provide a rapid source of energy. However, these stores are not large and are markedly depleted by fasting for as little as 24 hours.

(3) Eating a high carbohydrate diet for several days will double one's glycogen stores and can increase endurance by up to three times compared to the ordinary balanced diet.

(4) The main carbohydrate food sources are: CARBOHYDRATE SOURCES

(a) Fruits.

(b) Vegetables.

(c) Cereals.

(d) Sugar.

b. Fats: In the body, fat serves as the main storage form of energy. One gram of fat has the equivalent of 9.3 Cal/gm as opposed to 4.1 Cal/gm for carbohydrates and proteins. They also produce energy, heat, CO_2 and H_20.

(1) In cases of starvation, body fat tends to be broken down into acidic compounds. If these compounds accumulate in the blood faster than they can be burned, they cause the body tissues and blood to become excessively acidic (acidosis).

(2) During brief physical activity, energy is derived equally from fat and carbohydrates, but as the duration of the activity lengthens, the percent of energy supplied by fat increases as carbohydrate energy is quickly depleted.

(a) Fat sources:

1. Butter.
2. Lard.
3. Cooking oil.
4. Mayonnaise.
5. Ice cream.

FAT SOURCES

(4) Fat is also found in lesser amounts in:

(a) Dairy products. (d) Nuts.

(b) Meat. (e) Vegetables.

(c) Eggs. (f) Cereals.

c. Proteins: Protein is a reparative food of complicated molecules composed of chains of amino acids.

(1) The protein structure is made up of 25 different amino acids. Eleven of these are important for growth while eight are essential for normal tissue maintenance.

(2) Common protein food sources are:

PROTEIN SOURCES

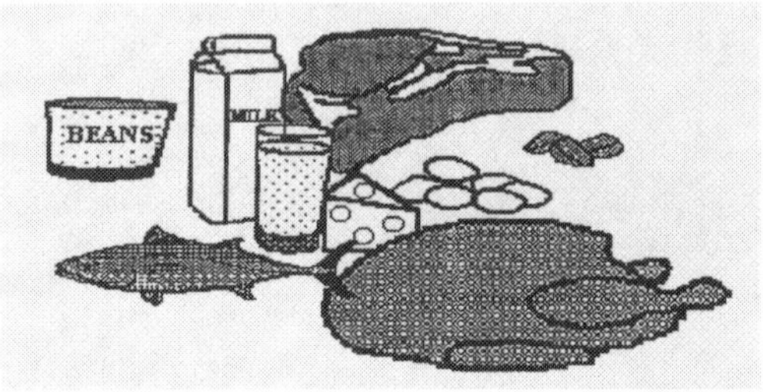

(a) Eggs.

(b) Dairy products.

(c) Meat.

(d) Poultry.

(e) Fish.

(f) Legumes (peas and beans, nuts and cereals with animal products, especially eggs, being better sources of complete protein).

(3) A pure protein diet may cause fatalities in 3-8 weeks from "rabbit starvation". This refers to those who have tried to live on relatively fat-free rabbit meat only. Fat must be a part of a normal diet.

7. **PERCENTAGE OF FATS, PROTEINS AND CARBOHYDRATES IN A SUMMER MOUNTAINOUS ENVIRONMENT**.

a. Requirements:

(1) 20% Protein.

(2) 35% Fats.

(3) 45% Carbohydrates.

b. When troops move to higher altitudes and/or are doing strenuous activity, carbohydrates should definitely be increased to at least 70 - 80% of energy intake. (Experienced mountaineers find carbohydrates more palatable at altitude).

8. **VITAMINS.** Vitamins are essential to the metabolic functioning of the body and are even more important in a cold weather environment because of an increase in metabolism and stress. Because the body cannot make vitamins, they are supplied in the food we consume. A daily multivitamin tablet supplement is recommended in a cold weather environment, but mega doses (more than 10 times the RDA) may be harmful.

a. Water-soluble vitamins are not stored in the body and need to be consumed daily. They are not stable to cooking and oxidation.

(1) B complex, Niacin, Pantothenic acid, Biotin, Choline, Folic Acid and C.
(Note: Vitamin C facilitates absorption of iron from food.)

b. Fat-soluble vitamins can be stored in the body and are more stable to cooking and oxidation.

(1) A, D, E and K.

9. **FOODS THAT MAKE YOU FEEL WARM**.

a. Hot Beverages. Increases peripheral vasodilatation.

b. Adequate Dietary Iron. Mild iron deficiency increases chilling, thus decreasing cold tolerance.

c. Fat in the Diet. Fat slows gastric emptying and anecdotal evidence suggests less chilling and more restful sleep at night.

d. Carbohydrates. Keeps energy levels up. "Tired" is associated with "Cold".

e. Spices. Red pepper (Tabasco sauce) and monosodium glutamate increase peripheral vasodilatation, while capsicum and capsaicin, which are found in Cayenne, produces the same result when applied externally to the feet.

f. Water. Dehydration leads to lethargy, which tends to reduce food intake resulting in less heat production.

10. **INCREASING COLD TOLERANCE**.

a. Frequent Feedings. A high carbohydrate diet is superior to a high protein diet in increasing tolerance to cold.

(1) Diets deficient in both calories and protein reduce physical work capacity in the cold. The provision of both components is required to maintain work capacity.

11. **HIGH FAT DIETS AND SUSCEPTIBILITY TO FROSTBITE**.

a. High fat diets consumed at frequent (2 hour) intervals, maintain body temperatures better than high carbohydrate diets, but not if normal (4 hour) intervals are observed between meals.

b. Fat enriched diets lead to a propensity toward increased arterial thrombosis and increased platelet aggregation, which translates into increased potential for thrombosis.

c. Frostbite injuries are associated with disruption of the microcirculation, progressive micro vascular thrombosis, alteration of platelet function, and fibrin deposition in affected areas.

d. Therefore, diets high in (saturated) fat may predispose to frostbite.

CAUTION! A high-fat diet is important in a high-altitude, cold-weather environment. The benefits associated with the high-energy source called Fat far outweigh the risks of arteriosclerosis from the diet itself. A high-fat diet is appropriate in this specific environment during the time frame of cold, high-altitude exposure. In garrison, at sea level, a high fat diet should be avoided in order to maintain a healthy lifestyle.

12. **IRON DEFICIENCY AND THERMOREGULATION**.

a. Iron deficiency results in 9% less heat production.

b. Consuming only 1/3 of the RDA for iron results in a 29% greater heat loss during cold exposure. "The furnace is going, but the windows are open" is a phrase that helps describe this effect.

c. Iron absorption is increased by consuming iron rich foods:

(1) Lean meat.

(2) Poultry.

(3) Beets.

(4) Beans.

(5) Green leafy vegetables.

(6) Fish.

(7) Vitamin C.

d. Bottom line, IRON improves the body's response to cold.

13. **WATER REQUIREMENTS**.

a. On average, water makes up about 60% of the one's entire body weight. The average sedentary person excretes about 2.7 quarts of water each day. Of this total, 1.3 quarts are lost in the urine, 1.1 quarts through the skin and lungs and 10 ounces in the stool.

b. At high altitude, and during strenuous exercise, the amount of water lost through the skin and lungs due to the cold, dry air increases greatly.

c. Cold weather decreases the sense of thirst, which may lead to a state of chronic, mild dehydration. At temperatures below freezing and at elevations above the snow line, the lack of liquid water and the time and effort required to melt snow compound the problem.

d. Prevention of excessive water loss: (FMST.07.01c)

(1) Force fluids. Drink 6-8 quarts of water daily.

(2) Observe for dark urine and other symptoms of dehydration. (Headache, nausea/vomiting, dizziness, constipation, etc.)

(3) Limit caffeine to 1-2 cups per day.

(4) Do not rely on thirst as an indicator of hydration status.

NOTE: Eating snow or ice will cause extreme inflammation of the mucous membranes of the mouth after two days. Warm canteens of water placed in your sleeping bag will give you water to drink in the morning. They will initially act as hot water bottles. (Place them upright, as they tend to leak ...keep bag dry.)

e. Health Problems Associated with Excessive Water Loss:

(1) Dehydration.

(2) Constipation.

(3) Hypothermia.

(4) Frostbite.

(5) Heat Exhaustion.

f. The importance of hydration is essential while in a cold weather environment. If these guidelines are followed then complications arising from poor hydration should be prevented.

14. **SPECIAL PURPOSE MILITARY RATIONS**.

RATION	APPLICATION
Ration, Cold Weather, RCW.	Cold weather operations where weight and freezing are a problem.
MRE Supplement Packet.	Cold weather calorie supplements for the MRE.
Ration, Light Weight, RLW.	Long range patrols, reconnaissance, and assault.
Nutrition Sustainment Module, NSM.	Calorie dense food, bar ration for mission lasting 3-5 days where soldier must carry his own food.

Family of Operational Rations, FOR.	Modular rations that allow an optimum mix of meals for tactical operations under all climatic conditions.

15. **FIELD STRIPPING OF MRE'S.**

a. Tips on how to conserve space and weight:

(1) Remove the entire contents from the plastic wrapper.

(2) Remove the cardboard, and all materials that you will not use.

(3) Be sure to place the toilet paper and other materials from the accessory pack either back in your pocket or into the brown packaging.

(4) Carefully repack the stripped meal back into the wrapper and seal with tape.

(5) Your meal will be half its original size and two-thirds its original weight.

UNITED STATES MARINE CORPS
Mountain Warfare Training Center
Bridgeport, California 93517-5001

FMST.07.09
03/31/02

WILDERNESS PATIENT ASSESSMENT

TERMINAL LEARNING OBJECTIVE: Given a casualty in a cold weather environment and the necessary equipment and supplies, perform a patient assessment in a mountainous environment to determine the nature and extent of injury per the reference. (FMST.07.09)

ENABLING LEARNING OBJECTIVES:

(1) Without the aid of references, from a given list, choose when CPR may be legally discontinued, in accordance with the references. (FMST.07.09a)

(2) Without the aid of references and from a given list, choose which situations CPR may not be started in a remote wilderness environment, in accordance with the references. (FMST.07.09b)

(3) Without the aid of references, in the proper sequence, correctly perform a patient assessment in the proper sequence in accordance with the references. (FMST.07.09c)

OUTLINE

1. **INTRODUCTION.**

   a. An organized prioritized assessment of a wilderness trauma victim is essential to proper care. A modified Advanced Trauma Life Support (ATLS) assessment system will be introduced in this chapter.

2. **TRIAGE.**

   a. Two situations of triage may exist in a wilderness environment.

(1) The number of patients and the severity of their injuries do not exceed the ability of the care-giver and his/her equipment to render care. In this situation, treat life-threatening problems first.

3

(2) The number of patients and the severity of their injuries exceed the capacity of the care-giver and his/her equipment to render care. In this situation, those with the greatest chance of survival are treated first

3. **CPR: URBAN vs. WILDERNESS SPECIFICS.**

a. In a remote wilderness situation where advanced life support assistance may be hours or days away, CPR probably has no chance of success. In addition, administering CPR under wilderness conditions may put group members in serious danger, due to physical hazards and the effects of exhaustion.

b. Based on the above, several suggestions regarding CPR in remote wilderness areas can be made. Even though the following suggestions seem reasonable, they should be viewed as recommendations, without legal force, and not doctrine. At the present time, a rescuer who begins CPR is legally obligated to continue unless or until one of the following conditions is fulfilled:

(1) CPR may be discontinued legally when: (FMST.07.09a)

(a) The patient revives.

(b) ACLS is started by higher authority.

(c) Relieved by another rescuer.

(d) Too exhausted to continue.

(e) Pronounced dead by proper authority.

(2) In a remote wilderness environment, CPR may not be started under one of the following situations: (FMST.07.09b)

(a) The patient is in cardiac arrest caused by trauma.

(b) The patient is a drowning victim who has been immersed for over an hour.

(c) The patient is in cardiac arrest and advanced life support is more than an hour away, especially if the patient must be carried out.

(d) The patient's cardiac arrest was unwitnessed and the time of onset is unknown.

(e) The patient appears to be dead, based on rigor mortis, lethal injuries, or a body temperature below 60°F.

(f) Giving CPR would be hazardous to rescuers.

(3) After 30 minutes of CPR with no signs of life, further CPR is probably useless and may reasonably be discontinued. Administering CPR to a victim who is being evacuated by litter or sled is very difficult, if not impossible. Unless an ambulance or helicopter can be brought in rapidly, chances of survival are slim.

(4) Exceptions include patients in cardiac arrest caused by hypothermia; patients with another illness or injury complicated by hypothermia (avalanche burial and near drowning in cold water), and patients in cardiac or respiratory arrest caused by lightning injury. In these cases, the outlook is probably more favorable and CPR should be aggressively administered.

4. **PATIENT ASSESSMENT** (FMST.07.09c). There are different considerations to think about in a wilderness setting. The environment plays a major role in treatment, care and even the extraction of the patient. Not only do we have to be mindful of the enemy situation, by keeping tactical. We also have to take into consideration the elements. Snow, rain ice, and the terrain can and will pose a great danger to the patient and the rescuers. We must protect the patient from the elements. We can do this by placing an insulating barrier between the patient and the ground. Also exposing one body part at a time and recovering as you move on will help to fight hypothermia.

A) SCENE SIZE-UP.

1) Determine scene safety:

a. Rescuers: Your safety is priority. You are no good to the team as a casualty.

b. Patient: Protect the patient from further injuries, enemy, and the elements.

c. Bystanders: They can cause many accidents. Limit the personnel on scene.

d. Tactical: Have the Marines maintain security and find cover for the rescuer and the patient.

e. Fire: Forest fires can pose a formidable threat.

f. Electrical: Make sure to shut off electricity first. Rushing in injures many rescuers.

g. Special Rescues: Swift water and high angle rescues are very dangerous and only trained professionals should attempt the rescue.

2) Determine the mechanism of injury / nature of illness:

a. Obtain SAMPLE and Mechanism of Injury (MOI) history from patient, witnesses, or other caregivers.

S = Signs and/or Symptoms.
A = Allergies.
M = Medications.
P = Past Illnesses.
L = Last Meal.
E = Events of Injury.

MOI = What, when, where, and how injury occurred.

b. Environment: Survey the area to get clues. (i.e. falling rocks, next to a cliff, broken branches, etc…)

c. Bystanders: Ask them what they saw.

3) Determine number of patients:

a. Survey area for number of patients.

b. Look for the hidden patients. Patients could be buried in rock, sand or snow. They also could be spread out in bushes and trees. One of the most over looked places is in the bystanders. Be sure to have people search and identify all patients.

4) Request additional help / Organize help if needed:

a. Medical assistance: Due to the lack of medical personnel and equipment, you might need to request additional help.

b. Technical assistance: When the scope of care is over your ability or higher than the casevac corpsman can provide, you should request higher trained personnel on the casevac bird.

c. Organize help at the scene: Use the people around you to help out.

5) Consider stabilization of the spine:

a. If indicated from: Falls of 10 feet or greater, high speed collisions, unwitnessed loss of consciousness, etc…

b. Designate an assistant: Use the bystanders / marines. Tell them exactly what you want them to do and show them how to do it.

c. Improvise control: If you are alone you can use the patient's boots, sand bags, rocks, etc as forms of temporary control.

B) INITIAL ASSESSMENT.

1) Make general impressions of the patient: What can you obtain from the patient's condition as you walk up? Is the patient talking, or unconscious?

2) Determine responsiveness / level of consciousness.

a) Use AVPU.

b) Assess pupils for size, equality, and reactivity.

3) Determine chief complaint / Apparent life threats.

4) Assess Airway and breathing.

a) Assessment of airway.

- Perform chin lift or jaw thrust maneuver, with C-Spine precautions.
- Clear the airway of foreign bodies.
- Insert an oropharyngeal or nasopharyngeal airway.
- Use of other airways as needed, and as equipment and training allow.

b) Assessment of breathing.

- Expose the neck and chest, assure immobilization of the head and neck.
- Determine the rate and depth of respirations.
- Auscultate the chest bilaterally. (Ear to chest wall, if necessary)
- Check for symmetry. (Rise and fall of the chest)

c) Management.

- Ventilate. (Face-mask, bag-valve-mask, if available)
- Alleviate tension pneumothorax.
- Seal open pneumothorax.
- Administer high concentration oxygen, if available.

5) Assess Circulation.

a) Assess and control major bleeding.

- Identify and control major bleeding.
- Pulse: Quality, rate, and regularity.
- Pulse Pressure: Check radial, then carotid.
- Skin color: cyanosis, pallor and temperature.

b) Management.

- Insert two large bore IV's with saline locks. IV fluid replacement, if indicated.
- Place in shock position.
- Prevent hypothermia.

6) Identify priority patients and make transportation decisions.

   a) Triage patients at this point.

   b) Make transportation decisions. In this type of environment it is possible that a casevac vehicle might not gain access to your position. At this point you will have to organize how the patient will be transported out. Also, you might have to set up relay points and rewarming stations.

C) DETAILED PHYSICAL EXAM: This is the head to toe assessment. We will find and treat all secondary injuries at this time. When performing this exam in the wilderness, you will need to expose one body part at a time and cover it back up as you go. When assessing the patient, you are looking for deformities, contusions, abrasions, penetrations/punctures, bruising, tenderness, lacerations, and swelling (DCAP-BTLS).

1) Head.

   a. Assessment.

   - Inspect and palpate entire head and face for lacerations, contusions, and fractures.
   - Re-evaluate pupils.
   - Re-evaluate level of consciousness.
   - Assess eyes for retinal hemorrhage, optic disc bulging, visual acuity disturbances, and contact lenses.
   - Evaluate cranial nerve function.
   - Inspect ears and nose for CSF leakage, hemorrhage
   - Inspect mouth, tongue, and teeth.

   b. Management.

   - Maintain airway, continue ventilation, and oxygenate as indicated.
   - Control hemorrhage.
   - Prevent secondary brain injury.
   - Remove Contact Lenses.

2) Assess the neck.

a. Assessment.

- Inspect for signs of blunt trauma, penetrating injuries, tracheal deviation, and use of accessory muscles during respiration, and jugular vein distention.
- Palpate for tenderness, deformity, swelling, or subcutaneous emphysema.

b. Management.

- Maintain open airway.
- Maintain immobilization and protection of cervical spine.
- Mark trachea.

3) Assess the chest.

3

a) Assessment.

- Inspect chest wall for signs of blunt trauma, penetrating injuries, and use of accessory muscles during respiration.

- Auscultate the chest wall and posterior bases for breath sounds.
- Auscultate for heart sounds.
- Palpate the entire chest wall for evidence of blunt trauma, penetrating injuries, flail chest, tenderness, crepitus, or subcutaneous emphysema.
- Percuss for hyperresonance or dullness.

c. Management.

- Needle thoracenthesis, if tension pneumothorax is suspected and proper training and equipment are present.
- Three-sided dressing for open chest wounds.
- Pericardiocentesis, if cardiac tamponade is suspected and proper training and equipment are present.

4) Abdomen.

a. Assessment.

- Inspect abdomen for signs of blunt trauma, penetrating injuries, and internal bleeding.
- Auscultate for the presence / absence of bowel sounds.
- Palpate the abdomen to elicit subtle tenderness, rebound, or guarding.

- Check pelvis for stability.

b b. Management.

c

- Correctly dress wounds.
- Keep patient NPO until definitive care is delivered.

5) Perineum / Rectum / Penis / Vagina.

d a) Assessment.

e

(1) Perineum.

- Contusions / hematomas.
- Lacerations.

(2) Rectum.

- Blood. Sphincter tone.
- High-riding prostate.

(3) Penis.

- Urethral bleeding.

(4) Vagina.

- Bleeding.
- Lacerations.

6) Musculoskeletal.

a) Assessment.

- Inspect extremities for evidence of blunt / penetrating injury. (deformities, altered range of motion).
- Palpate extremities for tenderness, crepitus, and abnormal movements.
- Palpate all pulses.
- Assess pelvis for stability. (Fractures) - Palpate spine for tenderness, crepitus, step-off's, depressions, and abnormal movements.

b) Management.

- Splint all suspected fractures in the position of function or comfort.

- Reduce dislocations as soon as possible, depending on training.
- Maintain immobilization and alignment of cervical, thoracic, and lumbar spine.

7) Reassess the vitals signs.

a) Assessment.

- Assess the vitals, pulse rate, breathing rate, pulse pressure.
- Re-evaluate the pupils and level of consciousness (GCS).
- Evaluate extremities for motor and sensory deficits.

b) Management.

- Maintain adequate immobilization of entire patient, as indicated.

5. **PATIENT PACKAGING AND EVACUATION.**

a. See the Mountain Casualty Evacuation chapter.

UNITED STATES MARINE CORPS
Mountain Warfare Training Center
Bridgeport, California 93517-5001

FMST.07.13
03/30/02

STUDENT HANDOUT

HIGH-ALTITUDE ILLNESSES

TERMINAL LEARNING OBJECTIVE: Given a simulated casualty, in a high altitude wilderness environment, treat high altitude health problems, in accordance with the references. (FMST.07.13)

ENABLING LEARNING OBJECTIVES:

(1) Without the aid of references, select from a given list the definition of acute mountain sickness, in accordance with the references. (FMST.07.13a)

(2) Without the aid of references, select from a given list the four main mechanisms of acute mountain sickness, in accordance with the references. (FMST.07.13b)

(3) Without the aid of references, select from a given list the most common and prominent symptom of acute mountain sickness, in accordance with the references. (FMST.07.13c)

(4) Without the aid of references, select from a given list the field management of acute mountain sickness, in accordance with the references. (FMST.07.13d)

(5) Without the aid of references, select from a given list the seven high altitude health preventive measures, in accordance with the references. (FMST.07.13e)

(6) Without the aid of references, select from a given list the correct definition of high altitude cerebral edema, in accordance with the references. (FMST.07.13f)

(7) Without the aid of references, select from a given list the pathophysiology of high altitude cerebral edema, in accordance with the references. (WMC.6.4g)

4

(8) Without the aid of references, select from a given list five of the symptoms of high altitude cerebral edema, in accordance with the references. (FMST.07.13h)

(9) Without the aid of references, select from a given list the field management of high altitude cerebral edema, in accordance with the references. (FMST.07.13i)

(10) Without the aid of references, select from a given list the correct definition of high altitude pulmonary edema, in accordance with the references. (FMST.07.13j)

(11) Without the aid of references, select from a given list three of the nine symptoms of high altitude pulmonary edema, in accordance with the references. (FMST.07.13k)

(12) Without the aid of references, select from a given list the field management of high altitude pulmonary edema, in accordance with the references. (FMST.07.13l)

(13) Without the aid of references, select from a given list choose the most important treatment for both high altitude cerebral edema and high altitude pulmonary edema, in accordance with the references. (FMST.07.13m)

OUTLINE

1. **ACCLIMATIZATION**.

a. Successful acclimatization depends on three factors:

(1) Degree of hypoxic stress (i.e. altitude)

(2) Rate of onset of hypoxic stress (i.e. ascent rate)

(3) Individual Physiology (i.e. genetic differences between individuals)

b. Physiologic Changes:

(1) Respiratory - Hypoxic Ventilatory Response (HVR)

- Hypobaric hypoxia
- Triggers carotid body
- Central respiratory center stimulated
- Ventilatory rate increases

HVR can be measured:

- "High" leads to successful acclimatization and superior performance.
- "Low" tends to be found in victims of altitude illnesses.
- HVR can be influenced by depressants (ETOH) and stimulants. (caffeine), and progesterone.
- **May** be negatively influenced by physical conditioning.

(2) Respiratory – Fast response: minutes to hours

- HVR leads to increased Ventilatory rate.
- Increased ventilations "blows off' C02."
- Low C02 levels yield a respiratory alkalosis and rising pH.
- Elevation of pH feeds back to depress the central respiratory center.
- No further immediate rise in ventilatory rate (a brake is placed on any further increase in respiration.)

(3) Renal – Slower response: kidneys respond over the next 24 - 48 hours:

- Kidneys respond to alkalosis by inducing a bicarbonate diuresis.
- PH slowly declines - brake on central respiratory center eases.
- Respiratory rate gradually rises over 4 - 7 days to new baseline.
- Final respiratory rate is altitude dependent.

(4) Cardiovascular - changes are a result of increased sympathetic tone (due to hypoxia).

- Increased: Blood Pressure, venous tone and Cardiac output (increased heart rate and contractility.)
- Stroke Volume declines in first 1 - 3 days due to bicarbonate diuresis.
- Left ventricular pressures are normal, usually; therefore, wedge pressures are usually normal.

(5) Pulmonary Circulation - critical in understanding the pathophysiology of HAPE.

- Ascent results in global Hypoxic Pulmonary Vasoconstriction.
- Elevated Pulmonary Artery Pressure (PAP) occurs.
- Exertion at altitude worsens this Pulmonary Hypertension.
- Damage to pulmonary vascular endothelium leading to leaky capillaries.

(6) Cerebral Circulation - critical in understanding pathophysiology of HACE

- Hypoxia -----------Vasodilatation
- Hypocapnia ----------Vasoconstriction
- Hypoxia is the dominant influence and vasodilatation results.
- On average, Cerebral Blood Flow (CBF) increases 20-40% on ascent to 12.000ft.

(7) Hemopoietic Response

- Erythropoietin levels increase within hours of ascent.

- New immature Red Blood Cells (RBC) seen within days.
- New mature RBC's present within 4-10 days
- RBC mass increase seen over weeks to months (altitude dependent).
- Early changes due to hemoconcentration.
- Polycythemia is a potential danger – this can lead to a condition known as CMS, or Chronic Mountain sickness.

(8) Muscle Tissue

- Intramuscular capillary and mitochondrial density increases after three weeks exposure to altitude.

(9) Sleep Pattern

- Stage 1 increase, Stage 2 unaffected
- Stage 3 & 4 decreases
- Rapid Eye Movement (REM) decreases
- Arousals increase
- Periodic Breathing (Cheyne-Stokes).
  - Increased respiratory rate leads to respiratory alkalosis
  - Elevated pH depresses central respiratory center
  - Apnea
  - 02 saturation declines, C02 levels rise
  - Carotid body stimulated
  - Ventilatory rate rises

(10)Summary: Physiologic changes seen at altitude result from the body's responses to hypo baric hypoxia. It has been speculated that man would be limited to an 8000-foot elevation limit without supplemental oxygen if not for the acclimatization process. As Messner and Haebler demonstrated on Mt. Everest, the process of acclimatization enables man to explore the entire planet.

2. **<u>ACUTE MOUNTAIN SICKNESS (AMS)</u>**.

a. Definition - AMS is an acute, self-limiting illness, which results when an unacclimatized individual ascends rapidly to high altitude. (FMST.07.13a)

(1) It is rare below 8,000 feet but will occur in approximately 42% of persons rapidly exposed to altitudes greater than 10,000 feet in elevation.

(2) Aerobic fitness is no predictor for risk of AMS with exposure to altitude. There is some evidence that prior aerobic fitness leads to increased AMS incidence, most likely due to excessively rapid ascent or over-exertion at altitude. Elite athletes may also be at increased risk due to depressed cardiovascular and Ventilatory responses to exertion.

(3) Prior ascents to altitude without symptoms of AMS are no guarantee against having symptoms of AMS with future ascents. However, those with past AMS are at an increased risk of AMS with repeat exposure to high altitude. Generally, there is no correlation between the severity of the illness and increasing altitudes. The incidence is slightly increased in younger persons.

(4) There is still much we do not know about altitude-induced physiologic changes. Current explanations in the literature center around genetic differences in hypoxic ventilatory response and vascular endothelium release of inflammatory mediators and Nitric Oxide.

b. Pathophysiology of AMS: The physiologic changes associated with the symptom complex known as AMS are linked to: (FMST.07.13b).

(1) Poor ventilatory response: An individual's Hypoxic Ventilatory Response (HVR) dictates how well he/she will respond to the hypobaric hypoxia of altitude. Those with a brisk HVR will do well, whereas those with a blunted response tend to suffer the symptoms of AMS. A poor ventilatory response leads to hypoxia and hypercapnia (elevated C02 levels). These two combine to augment cerebral blood flow by inducing cerebral vasodilatation.

(2) Fluid retention: The increased sympathetic tone associated with hypobaric hypoxia explains many of the changes seen in the renal system at altitude. Low Renal Blood Flow, increased anti-diuretic hormones, low Glomerular Filtration Rates, and decreased Urine Output all contribute to a Net Fluid Gain. This retained fluid redistributes within the body's fluid spaces. As a result, intravacular and extravascular fluid shifts occur, especially in the brain, lung, and peripheral interstitial tissues.

(3) Increased Intra-cranial Pressure: Increased cerebral blood flow coupled with CNS inflammation resulting in increased vascular permeability leads to increased intra-cranial pressure and associated symptoms.

(4) Altered Pulmonary Mechanics. On ascent to altitude, there is an increase in extravascular lung water. This is thought to be mediated, in part, by a phenomenon known as hypoxic pulmonary vasoconstriction. Blood vessels in the lungs constrict when they are exposed to low oxygen levels, this global vasoconstriction leads to increased right-sided heart pressures. The result of increased right-sided heart pressures causes fluid to leak from the vascular space into the alveoli, which results in impaired gas exchange. This has been demonstrated via Pulmonary Function Testing bronchoalveolar lavage, right sided-heart catheterization and arterial blood gas analysis showing decreased Vital Capacity, decreased Peak Expiratory Flows, pulmonary exudates rich in proteins and Increased Alveolar-Arterial oxygen Gradients. One study showed a 54% increase in estimated lung mass. It is not uncommon to hear crackles in all lung fields when auscultating asymptomatic individuals. The presence of intra-alveolar fluid inhibits the diffusion of oxygen

across the alveolar membrane, thereby reducing oxygenation across the pulmonary capillary membrane.

Signs and Symptoms of AMS. Symptoms tend to occur within the first 1- 3 days of exposure to altitude. Symptoms should resolve spontaneously within 15 - 24 hours, if ascent is arrested. The three most prominent symptoms are Headache, Nausea and Vomiting. (FMST.07.13c) Although nonspecific, headache must be present to make a diagnosis of AMS. 95% of AMS victims will complain of one or more of these three. Other common complaints include dizziness, anorexia, drowsiness, malaise, weakness, and insomnia. It is possible that findings of dyspnea on exertion and mental status changes along with ataxic gait may accompany the classic AMS signs and symptoms. These are clear indications for descent, as these are the hallmark signs of progression to HACE and the development of HAPE. Again, headache is the most common and prominent symptom of AMS. The Lake Louise Consensus criteria was created to give clinicians formal guidance in diagnosing AMS. The criteria require the presence of headache plus one or more symptoms from the above list. (i.e. nausea, vomiting, anorexia, insomnia, dizziness, fatigue, and dyspnea on exertion.)

c. Field Management (FMST.07.13d)

(1) Stop any further ascent. Light duty. No tobacco.

(2) Symptomatic Treatment. Mild analgesics such as Aspirin or Tylenol for the headache, Compazine or Phenergan for nausea/vomiting.

(3) Other Meds: Acetazolamide (Diamox), Dexamethasone (Decadron), and oxygen. The effects of Diamox and Decadron are additive when used together and are a very effective means for treating AMS.

(4) Hyperventilate: It is unrealistic to try to consciously control one's respiratory rate over any appreciable period of time. It is nearly impossible to maintain conscious hyperventilation while functioning in the field.

(5) Descend. If none of these methods help, then individuals should descend 1,000-3,000 feet. This will usually result in marked relief of symptoms. Most cases, however, improve in 1-2 days with symptomatic treatment.

(6) ALL PATIENTS WITH AMS MUST BE EVALUATED FOR HAPE AND HACE.

HOW?

- Field sobriety test.
- Mini-mental status exam.
- Auscultation of lung fields.

d. Preventive Measures: (FMST.07.13e)

(1) The best approach to high-altitude travel is staged, gradual ascent, when moving at altitudes greater than 3000 meters. Gradual ascent means no faster than 3000 ft/day up to 14,000 and no greater than 1,000 ft/day over 14,000 ft.

(2) Avoid any alcohol and/or sedatives during the first two nights.

(3) Maintain adequate fluid intake.

(4) Work high - Sleep low.

(5) High Carbohydrate Diet (approximately 70% of diet).

(6) Avoid overexertion, but remember that mild exercise assists acclimatization.

(7) No smoking.

(8) Pharmacological options: Prevention of AMS is possible with Acetazolamide (Diamox). It dramatically lowers the incidence of AMS when taken prophylactically. The usual dose is 125 mg PO BID starting 24-48 hours before ascent and should be continued for 3-4 days after the start of the mission. It is also effective when used as a treatment especially when combined with Decadron. In individuals with contraindications or allergies to Diamox use Decadron (dosed PO/IM/IV 4 mg Q6).

3. <u>HIGH ALTITUDE CEREBRAL EDEMA (HACE)</u>.

a. Definition. HACE is a high-altitude illness characterized by swelling of the brain. (FMST.07.13f)

(1) HACE can occur as low as 8,000 feet, but typically occurs at more than 12,000 feet.

(2) HACE, considered by most, to be a progression of AMS, occurs as a failure to heed the warnings of AMS signs and symptoms and continue ascent.

b. Pathophysiology. Increased cerebral blood flow along with the fluid shifts mentioned above lead to vasogenic edema (leaky blood vessels) inside the confined space of the skull. It is also thought that inflammation plays a key role in the development of HACE and occurs as follows. Hypoxia triggers the release of inflammatory mediators (likely Transforming Growth Factor Beta, TGF-β and Vascular Endothelial Growth Factor, VEGF.) These mediators cause increased vascular permeability that result in fluid accumulation within the brain – edema and increased intracranial pressure. Once this pressure nears arterial pressure, cerebral blood flow will be impaired. Eventually, intra-cranial pressure will rise further with a concomitant increase in the risk of brain herniation leading to marked alterations in mental status, ataxia and even cranial nerve palsies. Increased CSF production further compounds the problem. (FMST.07.13g)

c. Differential Diagnosis: Other illnesses are possible in individuals ascending to altitude. Your differential should include the following when neurologic deficits are involved:

(1) Cerebrovascular spasm and Transient Ischemic Attacks.

(2) Cerebral vascular thromboembolic events.

(3) Intra-cranial hemorrhage (aneurysm, AVM).

(3) Hypothermia.

(4) Meningitis

(5) Encephalitis

(6) "Hangover"

d. Signs and Symptoms:

(1) Early signs and symptoms are those found in AMS. (i.e. H/A, NV)

(2) When the above warning signs go unheeded the true signs and symptoms of HACE can quickly follow: (FMST.07.13h)

(a) Ataxia (loss of muscle coordination leading to difficulty maintaining balance), especially prominent in the victim's gait.

(b) Mental Status Changes, poor judgement, personality change which may progress to stupor, coma and death relatively rapidly without proper treatment.

(c) Lassitude, confusion, hallucinations, convulsions, behavioral aberrations.

(d) Paralysis of one or more extremities (see Differential Diagnosis).

(3) The most important impediment to early recognition of HACE is its insidious onset. Early signs and symptoms frequently go unrecognized or are ignored by patients, as well as their companions, who also may be suffering to some degree from the effects of altitude.

e. Field Treatment (FMST.07.13i)

(1) Early recognition is the KEY. Look for Ataxia and Mental Status Changes.

(2) Treatment should be immediate since fatalities can occur within a few hours in severe cases.

(3) Once diagnosed, the patient should be placed in the most comfortable position possible, descent should be directed immediately and administration of high concentration O2 should be given, in route, if available.

(4) Decadron (early), 10mg PO/IM/IV followed by 4 mg Q6. Diamox can also be used 125-250 mg BID.

(5) Consider diuretics – use carefully.

(6) If loss of consciousness, think airway management and ABC's.

(7) Casevac to an advanced medical facility ASAP!

(8) A Gamow Bag can be a lifesaver if decent is not immediately possible. Perform 4-6 compressions per day each treatment should last 40-60 minutes.

NOTE: The Gamow Bag is a one-man portable hyperbaric chamber. It weighs 14.5 lbs., is operated by a foot-pump, and is made of nylon with an airtight zipper. With the HAPE or HACE victim inside, the bag is pressurized with the foot-pump to an internal pressure of 2 psi. This is the equivalent to a 5000 - 6000-foot descent in regards to the partial pressure of oxygen inside the bag. Air is circulated within the bag via the foot-pump. This can be a life-saving resource in the event a victim is caught at altitude and cannot descend due to weather or other uncontrollable variables.

f. Prevention

(1) Preventive measures are the same as those discussed for AMS.

(2) Decadron can be used as a prophylactic medication along with Diamox, in cases of unavoidable rapid ascent to extreme altitude. However one must continue this regimen, while at altitude, or face rapid onset of symptoms once the medication is stopped.

4. **<u>HIGH ALTITUDE PULMONARY EDEMA (HAPE)</u>**.

a. Definition. HAPE is a high-altitude illness that is characterized by filling of the lungs with edema fluid. (FMST.07.13j)

(1) HAPE rarely occurs below 12,000 feet.

(2) HAPE victims in the past were commonly described, after autopsy, as victims of "pneumonia" and "Inflammation of the lungs". Since 1960, medical science has had the technology to investigate the complexities of the physiological changes associated with hypobaric hypoxia.

(3) Incidence: Occurs in 1-2 % of people brought rapidly to 12,000ft.

(4) Commonly occurs within 2-4 days of ascent.

(5) Risk for developing HAPE remain the same as above:

(a) Rate of ascent
(b) Sleeping altitude
(c) Level of exertion
(d) Gender (M>F)
(f) Individual physiology

b. Pathophysiology of HAPE:

(1) Hypoxic Pulmonary Vasoconstriction

(a) Normal part of the acclimatization process (elevated PAP).
(b) Adaptive at sea level (i.e. lobar pneumonia).
(c) Maladaptive at altitude (global hypoxia).
(d) Tends to be uneven, leading to high pressure/high flow area.

(2) Overperfusion

(a) Localized areas exposed to high pressure and high flow.
(b) Shearing forces and stress failure at the microvascular level.
(c) Endothelial damage to capillaries leads to inflammatory response and increased vascular permeability.
(d) Elevated hydrostatic pressure combined with increased vascular permeability leads to leakage of protein rich fluids into the alveoli and decreased oxygenation

(3) Nitric Oxide deficiency

Individuals prone to HAPE may have a deficiency in the amount of Nitric Oxide produced in the vascular endothelium resulting in more profound hypoxic pulmonary vasoconstriction.

c. Pathologic Findings of HAPE: Postmortem studies of victims of HAPE have yielded some interesting results:

(1) Grossly, the lungs are congested and swollen, as seen with other causes of pulmonary edema. Average lung weight on autopsy is 2-4 times normal. Histologically, one sees pulmonary edema with a protein rich exudate filling the alveoli. It is also common to find the alveolar spaces filled with hyaline membranes.

(2) The left heart is completely normal in appearance. However, the right ventricle, right atrium, and pulmonary arteries are distended and dilated.

(3) Greater than 50% of HAPE victims are found to have evidence of HACE.

d. Signs and Symptoms. These tend to occur within 2-4 days of arrival at altitude. Usually the symptoms of AMS are present before or occur with the symptoms of HAPE. (FMST.07.13k)

(1) Early signs:

(a) Dry cough, frequently occurring at night.

(b) Dyspnea on Exertion (DOE), especially with ambulation uphill.

(c) Mild chest pain - usually perceived as an ache beneath the sternum.

(d) Decreased work performance and increased recovery time between events.

(e) Peripheral or Central Cyanosis.

(2) Later signs:

(a) Dyspnea at Rest.

(b) Productive cough which yields large amounts of pink, frothy sputum.

(c) Rapid pulse and respiratory rates.

(d) Audible crackles on auscultation.

(e) **Mental status changes**, **Ataxia**, Loss of Consciousness.

(f) CXR findings:

- Prominent pulmonary arteries.
- Normal heart size.
- Patchy infiltrates.

ECHO findings:

-Non-cardiogenic pulmonary edema -increased PAP
-Tricuspid regurgitation.
-Normal left ventricle
-Prominent pulmonary arteries -enlarged right atrium
-Enlarged right ventricle.

NOTE: Crackles heard on lung exam, as an isolated finding is NOT an indication for descent or grounds for the diagnosis of HAPE. It is common (up to 15% of climbers on Mt. Rainier) to find asymptomatic rales in climbers at altitudes of 10,000-12,000ft. Concern is heightened when crackles occur in constellation with other signs and symptoms.

e. Field Treatment (FMST.07.13l)

(1) The most important emergency care measure is immediate descent to a lower altitude, since fatalities can occur within 6-12 hours in severe cases.

(a) A descent of at least 2,000-3,000 feet can be a definitive treatment if the condition is caught early enough.

(2) The patient should be placed in the most comfortable position (usually sitting) and given **high-flow O2 if available**. It is important to minimize activity, as exertion increases pulmonary artery pressures. Remember oxygen is Lasix for the lungs.

(3) Medication regimens include:

(a) Nifedipine (10mg PO initially then 30mg SR Q12) – calcium channel blocker - pulmonary artery/arteriole dilator. SR=sustained relief.

(b) Diamox and Decadron may be appropriate regimens, depending on your medication availability.

(c) Morphine (5mg IV initially decreased PAP) as a last resort.

(4) Oxygen is always appropriate, however, it is not always available.

(5) Gamow Bag can be a lifesaver if decent is not immediately possible.

f. Prevention Measures. Preventive measures are the same as those discussed for AMS.

(1) Acetazolamide (Diamox) has been proven to be successful in the prevention of HAPE. 125 - 250 mg PO bid helps prevent HAPE in individuals with recurrent episodes.

(2) Nifedipine (30 mg SRq12h) - also effective in individuals with recurrent HAPE.

(3) It should be noted that a history of HAPE is NOT a contraindication to the individual participating in high-altitude activities. Each case should be evaluated individually.

NOTE: Remember decent is the most important treatment for both HACE and HAPE. (FMST.07.13m)

5. **SYSTEMIC EDEMA AT HIGH ALTITUDES**.

Occasionally, certain individuals, especially women, will develop edema of the face, hands, and feet at altitude. This edema usually tends to occur in the absence of other symptoms. In susceptible persons, repeat episodes are common. The edema goes away upon return to lower altitude. There is no need for descent in these cases because the patient is usually able to tolerate the edema and continues to function well. However, they should be examined carefully to rule out the presence of HAPE or HACE. Diuretics are usually not warranted. It is also probably reasonable to restrict salt intake. The only indication for descent is if the edema is causing functional disability, i.e. inability to don one's boots, or facial edema severe enough to restrict vision.

6. **SICKLE CRISIS**.

Persons with sickle cell trait are at risk for developing sickling attacks when exposed to hypoxic stress, as might occur at high altitudes. However, sickling can occur at lower elevations. Individuals with Mediterranean family ties should be considered at risk for having sickle cell trait. The trait is more likely in black individuals as well. Any patient complaining of Left Upper Quadrant pain, SOB, and/or arthralgia at altitude should have sickling crisis included in the differential diagnosis regardless of race. It has been known to occur in white females as well. It has been recommended by some that individuals with sickle cell trait should be restricted to an altitude of 8,000 feet and below. However, more liberal recommendations exist, suggesting that the individual and their symptoms dictate ultimate altitude.

7. **ALTITUDE THROAT**.

At higher altitudes, mouth breathing is often increased during exertion. Inhalation of this cold, dry air can result in drying of the mucous membranes in the throat and upper airway. This condition responds to humidified air, salt-water gargles or lozenges. If exudate or edema is present, throat culture and/or antibiotics may be warranted.

8. **HIGH ALTITUDE FLATUS EXPULSION HAFE).**

Ascent to altitude can result in rapid expansion of intestinal gas, due to the decreased pressure of altitude. This produces abdominal discomfort and the passage of colonic gas. This condition is not associated with serious consequences; however, it may create tension amongst tent-team members. Treatment consists of oral administration of anti-flatulence such as simethicone, Mylanta II, or Gaviscon. Know and avoid your gas producing foods!

9. **HIGH ALTITUDE RETINAL HEMORRHAGE.**

Up to 60% of individuals ascending to altitudes greater than 18,000 feet will develop asymptomatic retinal hemorrhages. On fundoscopic exam, there will be hyperemia of the disc and dilation and tortuosity of the vessels. "Flame" hemorrhages and "cotton wool spots" can be seen on exam. The only indication for descent in an individual with retinal hemorrhages is loss of visual acuity. This usually results when a large hemorrhage involves the macular region of the retina. Remember that Carbon Monoxide poisoning can present with retinal hemorrhage as well.

10. **THROMBOEMBOLIC EVENTS.**

Thromboembolic disease occurs with increased frequency at higher altitudes. It is believed that the common circumstances of volume depletion due to dehydration and Polycythemia due to altitude exposure predispose individuals to coagulopathic events. In addition, extended periods of inactivity due to "going to ground" during storms contribute to this predisposition. There does not seem to be any direct influence of hypobaric hypoxia on the body's normal coagulation system.

11. **ACETAZOLAMIDE.**

Acetazolamide (Diamox) is a Carbonic Anhydrate Inhibitor, which is especially active in the brain, the lung, and the kidney.

a. Kidney:

(1) Uptake of bicarbonate ion is inhibited.

(2) Bicarbonate diuresis is induced.

(3) Metabolic acidosis ensues within one hour of medication.

(4) Sequence mimics and accelerates the natural acclimatization process.

(5) Enhanced HVR is the end result.

(6) Diuresis addresses problems associated with fluid retention.

b. Brain:

(1) Direct effects at the level of the Central Respiratory Center.

(2) Diminishes periodic breathing during sleep.

(3) Fewer episodes of apnea and extreme hypoxemia.

(4) CSF production is inhibited and CSF absorption is enhanced.

c. Side Effects:

(1) Polyuria.

(2) Flattens taste of carbonated beverages.

(3) Peripheral paraesthesias, myopia, impotence, nausea, and drowsiness.

(4) Contraindicated for Sulfa Allergic and G6PD-deficient individuals (reversible when medication is discontinued.)

UNITED STATES MARINE CORPS
Mountain Warfare Training Center
Bridgeport, California 93517-5001

FMST.07.35
03/13/02

STUDENT HANDOUT

HEAT ILLNESSES

TERMINAL LEARNING OBJECTIVE: Given a casualty in a wilderness environment, and necessary equipment and supplies, manage common heat related injuries to prevent death or further injury per the reference (FMST.07.35)

ENABLING LEARNING OBJECTIVES:

(1) Without the aid of references, given a list of heat illnesses match the definition to the correct condition, in accordance with the references. (FMST.07.35a)

(2) Without the aid of references, select from a given list the correct field treatment for heat cramps, in accordance with the references. (FMST.07.35b)

(3) Without the aid of references, select from a given list the correct field treatment for heat exhaustion, in accordance with the references. (FMST.07.35c)

(4) Without the aid of references, select from a given list the correct field treatment for heat stroke, in accordance with the references. (FMST.07.35d)

1. **CAUSES OF HEAT ILLNESSES**. Heat illness is caused by one of or a combination of the following factors:

a. Excessive environmental heat and humidity.

b. Faulty heat dissipation mechanisms within the body.

c. Stressed heat dissipation mechanisms.

d. Prolonged strenuous physical activity with inadequate fluid and electrolyte replacement.

2. **THERMOREGULATION**. The body is constantly adjusting itself in order to maintain its optimal temperature. This is called thermoregulation. A basic physics principle applies, in that heat is transferred from a higher to a lower temperature. What this means is that when the environmental temperature is higher than the skin temperature, the body gains heat, but when the environmental temperature is lower, the body loses heat. As a result, elevated environmental temperature adds to the heat burden of the body and can interfere with heat dissipation.

a. Heat Transfer Mechanisms. Keeping in mind the above principle we will now discuss the four methods of heat transfer.

1) Conduction: This is heat exchange between two surfaces in direct contact. An example would be lying on a hot or cold surface.

2) Convection: This is heat exchange from a surface to a gas or liquid, usually air or water. The amount of heat transferred depends on many variables, such as density, flow rate and surface area exposed.

3) Radiation: This is heat transfer between the body and its surroundings. All matter either absorbs heat or emits thermal radiation. The sun emits heat and the body absorbs it. A interesting fact is that highly pigmented skin while protective against UV radiation absorbs about 20% more heat than lightly pigmented skin.

4) Evaporation: This is the heat flow from the body surface to the environment by sweating.

b. If the body is unable to maintain thermal equilibrium by radiation, convection, and conduction and core temperature rises, sweating must occur to permit heat loss by vaporization of water. Since gastric emptying of water is about 1L/hr and sweat rates can exceed 1L/hr, a healthy diet along with constant replenishment of water lost is imperative to prevent illness.

3. **PREDISPOSING FACTORS**.

a. An underlying illness such as infection.

b. The temperature and humidity of previous days effect the likelihood of becoming a heat casualty.

c. Most heat illnesses occur during the morning hours.

d. Thirst is not an accurate indication of hydration; therefore you must drink even when you don't feel thirsty.

e. People usually do not voluntarily drink as much water as they lose.

f. All persons in the heat should be considered dehydrated, unless they have been forced to drink more than they desired.

4. **PREVENTIVE MEASURES**.

a. Drink 6-8 qts every day to replace fluids lost by sweating or from other bodily functions.

b. Consumption of a solute rich meal or a fluid replacement beverage about 2 hours before exercise can potentially counteract the sweat/salt losses.

5. **EARLY SIGNS OF HEAT RELATED ILLENSSES**. Changes in cognitive functions appear to occur frequently prior to the physical symptoms of heat illnesses. It is therefore imperative that medical personnel and leaders watch for confusion and impaired ability to function or work in the members of their group.

6. **SPECIFIC HEAT RELATED ILLNESSES**.

a. Heat Induced Syncope. (FMST.07.35a) This event is caused by the gravitational displacement of blood into the legs of an upright and stationary person combined with hemodynamic displacement of blood into peripheral circulation to support heat transfer at the body surface.

1) Prodromal symptoms.

(A) Restlessness

(B) Sighing

(C) Dysphoria

(D) Nausea

(E) Yawning

2) Treatment.

(A) The fainting is usually brief and responds to horizontal positioning.

(B) Rest in a cool, shaded area.

(C) Oral fluids.

(D) Persons should be cautioned against standing in hot environments and advised to flex leg muscles repeatedly while standing still to enhance venous return and warned to sit or lie down if warning signs reappear.

b. Heat Cramps. These are painful spasms of muscles, caused by sweating resulting in excessive loss of body salt. Generally, the spasms involve voluntary skeletal muscle subjected to prior exertion. This can occur even if a person is replacing water loss, but not salt losses.

1) Precipitating Factors: Cramps are preceded by several hours of sustained effort, combining heavy sweating in hot surroundings, with the ingestion of large volumes of water.

2) Symptoms: Painful muscle cramps in the arms, legs, and abdomen.

3) Treatment: (FMST.07.35b)

(A) Rest in a cool, shaded area.

(B) Gentle massage of affected muscles (may relieve spasms) or place the affected muscle in a stretch.

(C) Generally, heat cramps respond quickly to salt solutions.

1.) 1/4 to 1/2 teaspoon of salt in a quart of water (sip).

2.) In severe cases IV isotonic saline 0.9% NaCL.

NOTE: Salt tablets are not recommended due to the gastric irritation they cause.

c. Heat Exhaustion. This occurs when body salt losses and dehydration from sweating are so severe that a person can no longer maintain adequate blood pressure. It tends to develop over several days, that present ample opportunity exists for electrolyte and water imbalances to occur. Some 20% of cases are found to also have some form of gastroenteritis. Additionally, 60% of heat exhaustion casualties will also have symptoms of heat cramps.

1) Problems in recognition.

(A) Early signs may go unrecognized or be mistaken for malingering.

(B) Headache and visual disturbances may mimic prodromal of migraine headache.

(C) Dizziness and syncope may lead to an evaluation for arrythmia.

(D) Chills, ^ temperature, and sweating may lead to an evaluation for infection.

(E) Altered mental status or ataxia may lead to an evaluation for Meningitis or high altitude medical problem.

2. Symptoms:

(A) Headache

(B) Fatigue

(C) Piloerection

(D) Nausea/Vomiting

(E) Profuse sweating

(F) Dizziness

(G) Hyperirritability, Anxiety

(H) Chills

(I) Heat cramps

3. Clinical Findings:

(A) Tachycardia

(B) Hypotension

(C) Core temperature may be normal or elevated. Spontaneous cooling may occur, usually with profuse sweating. This will not happen with Heat stroke.

(D) Hyperventilation

(E) Syncope

(4) Treatment is based on symptoms and focuses on reduction of body temperature and rehydration. (FMST.07.35c)

(A) Lay down in a cool, shaded area.

(B) If body temperature is elevated, lower by any means available.

(C) Sprinkling with water and fanning.

(D) Ice bags over superficial great vessels and fanning.

(E) Periodic toweling of the skin renews the evaporative process.

(F) If patient is alert, orally rehydrate with cool water or commercially available rehydration mixture. Goal is 2L over 1st 2 hours.

(G) If patient is not alert, rehydrate via IV with LR.

(H) Victim must rest and continue to be rehydrated over next 24 to 36 hours.

d. Heatstroke (Hyperthermia). " A true medical emergency". This is a failure of the body's cooling mechanisms that rid the body of excessive heat build-up. Prior to 1950 the mortality rate was 40 to 75%. Long term survival of victims is directly related to the speed of effective treatment measures. Traditionally, the diagnosis of Heatstroke required the following 3 signs: a core temperature > 104°F(41°C), altered mental status, and cessation of sweating. Although, this symptom complex does represent full-blown Heatstroke adherence to this strict criteria may delay critical interventions. Unless an alternative etiology is obvious, the previously healthy person who collapses after physical exertion in hot weather should be considered to have exertional Heatstroke. Treat for Heatstroke whenever you suspect it and always suspect it in the heat.

1. Symptoms:

   (A) Headache
   (B) Shortness of breath
   (C) Nausea/ vomiting
   (D) Low BP
   (E) Skin may be wet or dry
   (F) Constricted pupils
   (G) Seizures
   (H) Dry mouth
   (I) Core temp > 104°F (41°C)
   (J) Weakness/ dizziness
   (K) Hot, red skin
   (L) Tachycardia
   (M) Confusion, Aggression, Combative
   (N) Coma

2. Clinical Findings.

   (A) CNS: altered mental status, agitation, ataxia, delirium, hallucinations, convulsions, seizures, coma, pupillary constriction.

   (B) Cardiovascular: hypotension (a late and ominous finding), shock, sinus tachycardia with ST segment & T wave abnormalities.

   (C) Pulmonary: hyperventilation, pulmonary edema.

   (D) Renal: acute renal failure, hematuria, pyuria, proteinuria, UA casts.

   (E) Gastrointestinal: nausea, vomiting, +/- disseminated intravascular coagulation, hematemesis, melena.

   (F) Hematologic: WBC 20-30 thousand, < platelets, < clotting factors.

   (G) Hepatic: increased liver function tests.

3. Emergency Treatment. (FMST.07.35d)

(A) Rapid reduction of body core temperature by any means available until core temperature is 102°F.

1) Ice water bath (controversial).
2) Cool water bath.
3) Pouring or hosing cool water over victim and fanning.
4) Placing ice bags at neck, armpits, groin, forehead, chest etc..

(B) Maintain ABC's.

(C) Monitor core temperature.

(D) IV access infuses RL or normal saline 500cc/hr x 4hrs. (watch for pulmonary edema).

(E) Control seizures with valium.

(F) Transport STAT. Cool enroute.

UNITED STATES MARINE CORPS
Mountain Warfare Training Center
Bridgeport, California 93517-5001

FMST.07.15
04/01/02

STUDENT HANDOUT

COMBAT CASUALTY CARE

TERMINAL LEARNING OBJECTIVE: Given a simulated casualty in a simulated combat environment, treat the casualty in accordance with the references. (FMST.07.15)

ENABLING LEARNING OBJECTIVES:

(1) Without the aid of references, choose from a given list the three correct phases of care of tactical casualty management, in accordance with the references. (FMST.07.15a)

(2) Without the aid of references, given a simulated casualty and the necessary equipment, conduct combat casualty care, in accordance with the references. (FMST.07.15b)

OUTLINE

1. Summary of Findings and Recommendations.

   a. Pre-hospital care of combat casualties is divided into three phases: (FMST.07.15a)
      1. Care under Fire.
      2. Tactical Field Care
      3. Combat Casualty Evacuation (CASEVAC) Care.

   b. Suppression of hostile fire may take temporary precedence over the rendering of care in the Care under Fire phase.

   c. Cervical spine immobilization is not required for victims of penetrating head or neck trauma.

   d. CPR should not be attempted on the battlefield for victims of blast or penetrating trauma who have no pulse, respirations, or other signs of life.

   e. The use of tourniquets is encouraged in the Care under Fire phase. The tourniquet is removed and bleeding controlled with direct pressure in subsequent phases when feasible.

6

f. The nasopharyngeal airway is the airway of first choice for unconscious patients until the CASEVAC phase.

g. For patients with airway obstruction in the Tactical Field Care phase unrelieved by a nasopharyneal airway, the next airway maneuver is cricothyroidotomy.

h. Endotracheal intubation is not the standard of care until the CASEVAC phase.

i. Evaluation of the laryngeal mask airway and the Combitube for use by SOF corpsmen and medics is recommended.

j. Progressive, severe respiratory distress in the setting of unilateral blunt or penetrating chest trauma on the battlefield should result in a presumptive diagnosis of tension pneumothorax and decompression of that hemithorax with needle thoracostomy.

k. Chest tube insertion is not indicated until the CASEVAC phase of care.

l. Oxygen is not expected to be carried onto the battlefield, but CASEVAC assets should have it.

m. No IV fluids are to be given until the CASEVAC phase unless the patient is in hypovolemic shock from a bleeding site which has now been controlled. Fluid resuscitation is not necessary for patients who are not in shock and not appropriate for patients in hypovolemic shock from uncontrolled hemorrhage (penetrating wounds of the chest and abdomen).

n. Saline locks may be used instead of IV's if fluid resuscitation is not required.

o. 18 gauge catheters may be used instead of 14 or 16 gauge.

p. Hespan (1000cc) is the fluid of choice for initial resuscitation of patients with controlled hemorrhagic shock.

q. Morphine is to be used IV (5mg) instead of IM.

r. IV cefoxitin is to be used as soon as possible for patients with penetrating abdominal trauma, grossly contaminated wounds, massive soft tissue trauma, open fractures, or any patient in whom a long delay until definitive treatment is expected.

s. Casualties should not be completely undressed for a secondary survey in the field. Removal of clothing should be limited to that necessary to expose known or suspected wounds.

t. More specific combat casualty care planning should be carried out and based on specific mission-oriented scenarios.

u. The establishment of Combat Casualty Transport Teams and their use on CASEVAC assets is recommended.

v. Electronic monitoring should be routinely used on CASEVAC assets unless conditions prohibit.

w. MAST trousers are not recommended as a standard of care in any phase.

2. Phases of Care:

a. Phase One: Care Under Fire

1. Return fire as directed or required.

2. Keep yourself from getting shot.

3. Try to keep the casualty from sustaining additional wounds.

4. Stop any massive external hemorrhage with a tourniquet.

5. Take the casualty with you when you leave.

b. Phase Two: Tactical Field Care

1. Airway management

- Unconscious casualty without airway obstruction: > Nasopharyngeal airway.

- Unconscious casualty with airway obstruction: > Cricothyroidotomy.

- Cervical spine immobilization is not necessary for casualties with penetrating head or neck trauma.

2. Breathing

- Consider tension pneumothorax and decompress with needle thoracostomy if a casualty has unilateral penetrating chest trauma and progressive respiratory disstress.

3. Bleeding

- Control any remaining bleeding with a tourniquet or direct pressure.

4. IV

- Start an 18 gauge IV or saline lock.

5. Fluid Resuscitation

- Controlled hemorrhage without shock: > No fluids necessary.

- Controlled hemorrhage with shock: > Hespan 1000cc

- Uncontrolled (infra-abdominal or thoracic) hemorrhage: > No IV fluid resuscitation.

6. Inspect and dress wound.

7. Check for additional wounds.

8. Analgesia as necessary: > Morphine (5mg) IV, wait 5minutes, repeat as necessary.

9. Splint fractures and recheck pulses.

10. Antibiotics

Cefoxitin 2gm slow IV push (over 3-5 minutes) for penetrating abdominal trauma, grossly contaminated wounds, massive soft tissue trauma, open fractures, or any patient in whom a long delay until CASEVAC or definitive treatment is expected.

11. Cardiopulmonary Resuscitation

CPR on the battlefield for victims of blast or penetrating trauma who have no pulse, no respirations, and no other signs of life will not be successful if attempted.

c. Phase Three: Combat Casualty Evacuation (CASEVAC) Care

1. Airway management

Unconscious casualty without airway obstruction: > Nasopharyngeal airway, endotracheal intubation, Combitube or laryngeal mask airway.

Unconscious casualty with airway obstruction: > Cricothyroidotomy if endotracheal intubation and/or other airway devices are unsuccessful.

2. Breathing

Consider tension pneumothorax or hemothorax and decompress with needle thoracostomy if a casualty has unilateral penetrating chest trauma and progressive respiratory distress.

Consider chest tube insertion if a suspected tension pneumothorax is not relieved by needle thoracostomy

Oxygen.

3. Bleeding

Consider removing tourniquets and using direct pressure to control bleeding if possible.

4. IV

Start an 18 gauge IV or saline lock if not already done.

5. Fluid resuscitation

No hemorrhage or controlled hemorrhage without shock: > Lactated Ringer's at 250cc/hr.

Controlled hemorrhage with shock: > Hespan 1000cc initially.

Uncontrolled (infra-abdominal or thoracic) hemorrhage: > No IV fluid resuscitation.

Head Wound patient: > Hespan at KVO unless there is concurrent controlled hemorrhagic shock.

UNITED STATES MARINE CORPS
Mountain Warfare Training Center
Bridgeport, California 93517-5001

FMST.07.37
04/01/02

STUDENT HANDOUT

BURN MANAGEMENT

TERMINAL LEARNING OBJECTIVE. Given a casualty in a wilderness environment, and necessary equipment and supplies, treat burn injuries to prevent death or further injury per the reference. (FMST.07.37)

ENABLING LEARNING OBJECTIVES

1. Without the aid of references, given a list of the depths and appearances of burns, match them to the type of burn, in accordance with the references. (FMST.07.37a)

2. Without the aid of references, given a list choose the major burns that require rapid casualty evacuation, in accordance with the references. (FMST.07.37b)

3. Without the aid of references, given a list choose the first step in the treatment of burn injuries, in accordance with the references. (FMST.07.37c)

OUTLINE

1. **GENERAL:** Each year there are about 2 million individuals burned severely enough to seek medical attention. Around 70,000 require hospitalization. Deaths are on the decrease largely due to the increase in the use of smoke detectors. It has been estimated that 90% of all burn injuries are preventable. Children and young adults are frequently affected by burn injuries.

2. **PHYSIOLOGY:** For non-chemical burns, the primary events of injury occur during the time of heat contact. Coagulation necrosis takes place within cells and denaturation of collagen in the dermis. Blood vessels are completely destroyed or endothelium damage is severe enough to cause clotting which leads to ischemic necrosis of remaining viable cells. Burn wounds are not static. Surrounding the "coagulation zone" is a zone of capillary and small vessel stasis. Circulation becomes stagnant from a strip formed by red blood cells, and aggregation of platelets and white blood cells. The fate of the burn wound depends on the progression of this zone of stasis.

7

3. In patients with burns less than 10% of the total body surface area (TBSA), the inflammatory process is generally limited to the burn site itself. As burns approach 20% TBSA, the local

response becomes systemic. Burns of 30% TBSA or greater, often lead to multiple organ system failure, which in a patient with a severe burn almost invariably leads to a fatal outcome, often due to acute renal failure.

Cardiovascular changes begin immediately after a burn. The degree of these changes depends initially on the size of the burn and to a lesser extent the depth of the burn. Burns less than 15% TBSA can be fluid resuscitated with crystalloid fluids. As burns pass 20% TBSA, massive shifts of fluid and electrolytes occurs. This results in massive fluid shift from the intravascular to the extravascular space. Unless intravascular volume is repleted, classic hypovolemic shock occurs.

3. TYPES OF BURNS

a. Scald Burns: Scalding is the most common cause of burns, usually resulting from hot water. Water at 140°F creates a deep partial-thickness or full-thickness burn in three seconds. At 156°F the same burn occurs in one second. Freshly brewed coffee is generally about 180°F. Boiling water always causes a deep burn. Soups and sauces are thicker, remaining in contact with the skin longer. Exposed areas tend to burn less deeply than areas covered with a thin layer of clothes. Clothing retains heat and keeps liquid in contact with the skin longer. Immersion scalds are deep and cause severe burns resulting from longer contact time. These burns are more common in children and elderly patients because of their thin skin. Burns from hot oils are generally deep partial-thickness or full-thickness. Temperatures for cooking oils reach around 400°F, and tar can be up to 500°F.

b. Flame Burns: The next most common burn injuries are flame burns. Although household injuries have gone down with the advent of smoke detectors, a high number of burn injuries still occur from careless smoking, improper use of flammable liquids, automobile accidents, and clothing ignited from stoves or space heaters.

c. Flash Burns: Flash burns result from intense heat for a brief time caused by explosions of natural gas, propane, gasoline, and other flammable liquids.

d. Contact Burns: These burns usually result from direct contact of hot metals, plastics, glass, or coals. Contact burns are usually limited in size, but are deep wounds. With the increase of wood stove usage, this type of burn injury is on the rise in children. Contact burns, dealing with an unconscious patient or involving molten materials, are usually fourth degree. The most common contact burn in the wilderness is from hot coals.

e. Electrical Burns: Electrical burns are actually thermal burns from very high-intensity heat. When electricity encounters resistance from the body it turns into heat. Severity of the burn depends on the portion of the body affected relative to the amperage. The smaller the body part, the more intense the heat is, and the part is less able to dissipate the heat. Consequently the fingers and toes are almost always destroyed with severe damage to the forearms. Larger areas can dissipate the current enough to prevent extensive damage. Arc burns take the most direct route, not the path of least resistance. These are deep and destuctive burns which mostly occur at the joints.

Electrical burns may have other associated injuries that should not be overlooked on the initial assessment. Fall related injuries are common. The intense muscle contraction resulting from electrocution can cause fractures of the lumbar vertebrae, humerus, or femur, and also may cause dislocation of the hips and shoulders. The casualty may have cardiac symptoms of a myocardial contusion or infarction. There can be a conduction system failure or an actual rupture of the heart wall. If there are no signs present of cardiac problems intially after a shock of 110 or 220 volts there is a minute chance they will appear later. Nervous system damage is possible whenever a current passes from one side of the body to the other more than likely affecting the spinal cord.

f. Chemical Burns: Usually caused by strong acids or alkalis and are most often the result of industrial accidents, home use of drain cleaners, assault, and improper use of harsh solvents. Chemical burns continue to burn until they are inactivated by reaction with tissue or dilution by flushing with water. Acids tend to tan the skin, creating an impermeable barrier that limits futher penetration. Alkalis combine with cutaneous lipids and saponify[1] the skin until they are neutralized. The appearance of the wound can be misleading. Unless the observer can be 100% sure, all chemical burns should be considered full thickness.

g. NBC Burns: Nuclear weapons cause burns in two ways: by direct absorption of thermal radiation through exposed surfaces (flash burns); or by the indirect action of fires caused by the weapon (flash burns). The first step in treatment is to prevent contamination to the rescuers and other patients.Vesicants such as mustard gas cause skin burns with edema and blister formation. Burns resulting from vesicants should be flushed with copious amounts of water. Absorbent powder (flour, talcum powder, Fuller's earth) can be used if water is scarce. The powders should be wiped away with a moist towel. The military uses M258A1 kits for skin decontamination. White Phosphorus ignites spontaneously when exposed to air. Remove the air with an air-tight seal; mud is optional. A brief rinse with 1% copper sulfate[2] will impead further oxidation.

4. **CLINICAL PRESENTATION:** The severity of the burn injury is related to the size of the burn, the depth of the burn, and the part of the body that is burned.

a. Burn Size: Burns are the only quantifiable form of trauma. The burn size is the single most important factor in predicting mortality, need for specialized care, and the complications expected from the burn. We can measure burn size with the "rule of nines". The "rule of nines" is where areas of the body represent approximately 9% of a person's total body surface area (TBSA), each leg 18% (the front of the leg 9%, and the back of the leg 9%), the front of the torso represents 18%, the back of the trunk is 18%, the head represents 9%, and the perineum 1%. For infants and small children under four years of age, the head represents a larger percentage 18%, and the legs a smaller percentage 13%, each arm 9%, front troso 13%, back troso 13%, and the perineum 1%. The percentage of the burn area and that of the unburned area need to add up to equal 100%. For smaller burn areas, use the Rule of Palmar Surface: the patient's palmar surface equals about 1% TBSA.

[1] Saponify: to convert fat into soap by treating it with an alkali.
[2] Navy uses sodium bicarbonate instead of copper sulfate

b. Burn Depth: The depth of the burn is either described in degree: first, second, third or fourth, or by the depth of the injury. The following describes burns in terms of partial-thickness or full-thickness. While these descriptions appear to separate burns into defined categories, many burns have a mixture of characteristics making a precise diagnosis difficult.

(1) First-Degree: These burns involve the epidermis only and do not blister. They are most commonly caused by ultraviolet light. An example would be a sunburn. The burned skin is painful and red. It should heal in 7 days without scarring.

(2) Superficial Second-Degree, or Superficial Partial-Thickness: These burns include the epidermis and upper layers of the dermis. They characteristically form blisters, under which the skin is red and moist, these burns are painful to the touch. Blisters can take up to two days to appear. These burns are usually caused by hot liquids. Wounds should heal in 14 to 21 days; they may scar depending on the extent of the burn.

(3) Deep Second-Degree, or Deep Partial-Thickness: Burns that involve deeper layers of the dermis. There is damage to hair follicles and sweat glands. They are characterized with a molted pink and white color with blisters forming immediately. These wounds may be less sensitive to touch than the surrounding normal skin, or maybe tender to touch. The patient often complains of discomfort rather than pain. Capillary refill may be slow or absent, when pressure is applied to the wound. These burns are caused by hot liquids, oil, steam, or flame. They may be difficult to distinguish from Third-Degree/Full-Thickness burns. Healing takes 3-9 weeks. Scarring is probable, the degree is related to the amount and the depth of dermal injury. Surgical grafting may be required.

(4) Third-Degree, or Full-Thickness: Involves the entire thickness of the skin, epidermis through the dermis, down to subcutaneous fat. All structures of the epidermis and dermis are destroyed. The wound is classically described as leathery, firm, and depressed when compared to normal skin. The tissue is charred, pale, and insensitive to light and touch. It is often misdiagnosed as a Deep Second-Degree/Deep Partial-Thickness due to the similar clinical findings. This burn will not spontaneously heal. Surgical repair and/or skin grafts are necessary. There is significant scarring.

(5) Fourth-Degree: These burns involve all layers of the skin, subcutaneous fat, muscle, and bone. These are devastating, life threatening injuries. They almost always have a charred appearance, and often only the cause of the burn gives a clue to the amount of underlying tissue destruction. (FMST.07.37a)

c. Burn Classification: The American Burn Association has devised a classification of burns, dividing them into major, moderate, and minor burns. Patients are placed in groups related to their risk. Low-risk patients are between the ages of 10 and 50 years old. High-risk patients are less than 10 and greater than 50 years old. Poor-risk are patients with underlying medical illnesses such as heart disease, diabetes, or chronic pulmonary problems.

(1) Minor Burns: Imply outpatient treatment. Minor burns involve TBSA of less than 15% in the low-risk group, or less than 10% TBSA in the high-risk group; full-thickness burns of less than 2% TBSA in anyone; and no other injuries.

(2) Moderate Burns: Are partial-thickness burns of 15-25% TBSA in the low-risk group, 10-20% TBSA in the high-risk group; full-thickness burns of equal to or less than 10% TBSA in anyone; burns not involving the hands, face, feet, perineum, or circumferential limbs.

(3) Major Burns: Defined as partial-thickness burns greater than 25% TBSA in the low-risk group, or greater than 20% TBSA in the high risk group; full-thickness burns of greater than 10% TBSA in anyone; burns involving the hands, face, feet, or perineum; burns crossing major joints; circumferential limb burns; burns complicated by inhalation injury; electrical burns; burns complicated by fractures or other forms of trauma; and burns in poor-risk patients. These burns require rapid casualty evacuation. (FMST.07.37b)

5. **TREATMENT:**

a. Initial Care:

(1) The first step in the treatment of burn injuries is to stop the burning process. (FMST.07.37c) Remove patients from heat, remove all their clothes that are affected by hot liquids, shut off electricity (very important), remove chemicals by flushing the affected area for 10 to 15 minutes, and decontamination of NBC. Heat can continue to injure tissue for a surprisingly long time. No first aid will be effective until the burning process has stopped.

(2) Manage the ABC's. Once an airway has been established give oxygen. If qualified, consider intubation earlier rather than later.

(3) Assess for associated injuries such as fractures or lacerations and inhalation injury.

(4) Remove clothing and jewelry from the burn area. This will help prevent constriction from swelling of the burn tissue. Do not try to remove anything that has adhered to the wound.

(5) Evaluate the Burn (depth, extent, pain).

b. General Treatment for the Patient:

(1) Gently wash the burn with tepid water and mild soap to remove any debris and to clean the skin surface around the burn site. Pat dry. Remove the skin from blisters that have popped open (do not open blisters unless necessary for function of hands or feet).

(2) Dress burn with a thin layer of antibiotic ointment.

(3) Cover the burn with Second Skin (if the burn is small enough) or cover with a thin layer of gauze, or clean, dry clothing. Burns on the face, neck and hands may be left open to the air after applying silvadene ointment. Covering the wound reduces pain and evaporative losses, but do not use an occlusive dressing.

(4) When evacuation is imminent, do not redress or reexamine the injury. If evacuation is prolonged, redress once a day. Remove old dressings, reclean (removing the old ointment), and apply fresh ointment and a clean dry covering. (Note: if stuck, soak off old dressings with clean, tepid water.)

(5) Do not pack wounds, which are larger than 20% of the body surface area, in ice. Do not leave wet coverings on burns for more than two hours at a time to reduce the risk of hypothermia.

(6) Stabilize the body temperature. When skin is lost, so is the patient's ability to thermoregulate.

(7) Have the patient drink as much fluid as he or she can tolerate without vomiting. Include some salt in the oral fluids, but do not make these solutions stronger than . 9%. This is the equivalent of a pinch of salt per 8 ounce glass.

(8) IV therapy as follows, Baxters (Parkland) Formula: In the first 24 hours give Ringers Lactate, 4ml times the weight in kilograms times the percent TBSA burned. Half of the solution is given in the first eight hours. The second half is given over the next 16 hours. The second 24 hours should be given fluids to maintain blood pressure, colloids (usually in the form of albumin) are the preferred choice.

4 x wt (kg) x TBSA= amount LR given in the first 24 hours

Example a 70kg patient with 50% TBSA this is the formula for IV therapy for the first 24 hours.

4 x 70kg x 50% = 14,000ml in 1ST 24 hours
7,000ml hours 1-8
3,500ml hours 9-16
3,500ml hours 17-24

c. General Management of the Burn: Caring for the wound itself is often the least important aspect of burn care. All burn wounds are sterile for the first 24 to 48 hours. Burn management is aimed at keeping the wound clean and reducing the pain.

(1) Elevate burned extremities to minimize swelling. Swelling retards healing and encourages infection. Have the patient gently and regularly move burned areas as much as possible.

(2) Burns are tetanus-prone wounds. Check on the last known tetanus shot. The patient may need a renewal.

(3) Ibuprofen is probably the best over the counter analgesic for burn pain (including sunburn).

(4) If you have no ointment or dressings, leave the burn alone. The burn's surface will dry into a scab like covering that provides a significant amount of protection.

6. **INHALATION INJURIES:** Half of all fire-related deaths are from smoke inhalation. Smoke inhalation doubles the mortality rate for any burn. Smoke inhalation is a general term for carbon monoxide poisoning, thermal airway injury, and smoke poisoning.

a. Carbon Monoxide Poisoning: Carbon monoxide (CO) is a colorless, odorless, tasteless gas that has an affinity for hemoglobin 200 times greater than that of oxygen. Carboxyhemoglobin (COHb) levels are measured as the percentage of hemoglobin bound to carbon monoxide. The following COHb levels correlate with clinical symptoms:

<10%	No symptoms
20%	Headache, nausea, vomiting, loss of dexterity
30%	Confusion, lethargy, ECG ST depression
40%	Coma
50%	Death

The half-life of COHb is 4 to 5 hours at room air, 90 minutes with 100% oxygen, and 20 to 25 minutes at 3 ata in a hyperbaric chamber. Treatment for less than 40% COHb is a 100% oxygen. At 40% or greater the hyperbaric chamber should be used. If patient is still in a coma after COHb levels are normal, prognosis is poor, and it has been experience that they rarely awaken.

b. Thermal Airway Injury: Injury to the respiratory tract from inhalation of hot gas or steam. The lower respiratory tract is usually not damaged, unless there was an inhalation of flame or steam. Any patients who were in an explosion, with burns of the hands, face, and upper torso are at risk. Signs of carbon such as soot in the pharnyx should alert the caregiver to airway injury. Maintaining the airway is the important issue here.

c. Smoke Poisoning: Is the inhalation of noxious gases that are the products of combustion. Cyanide poisoning is known to occur in victims of smoke inhalation. Other toxic gases released are sulfur dioxide, hydrogen chloride, phosgene, and ammonia. These gases cause damage to the respiratory tract similar to chemical burns. Physical signs include facial burns, intraoral or pharyngeal burns, singed nasal hairs, soot in mouth or nose, hoarseness, carbonaceous sputum, and expiratory wheezing.

UNITED STATES MARINE CORPS
Mountain Warfare Training Center
Bridgeport, California 93517-5001

WMC
04/09/02

LESSON PLAN

NOMENCLATURE AND CARE OF MOUNTAINEERING EQUIPMENT

LESSON PURPOSE. The purpose of this period of instruction is to introduce you to the types of equipment used here and how to care for it so as to prevent its untimely failure. This lesson relates to all climbing and installation work that you perform here.

1. **ROPES.** All ropes used in the military must meet UIAA standards or U. S. Federal Test Standard 191A. Most ropes have a 5-year shelf life and maximum 2-year service life.

   a. Static. Black in color.

      (1) Construction. Kernmantle

      (2) Minimum tensile strength. 7500 lbs.

      (3) Maximum elongation. 1.5%

      (4) Diameter. 11mm

      (5) Sizes.

         (a) 165 ft ± 5 ft

         (b) 300 ft ± 10 ft

      (6) Usage. Rescue operations and bridging where a low amount of elongation is desirable under a working load.

   b. Dynamic. Olive Drab in color.

      (1) Construction. Water-resistant treated Kernmantle to reduce friction.

      (2) Minimum tensile strength. 6500 lbs.

      (3) Maximum elongation. 6%

C-2

(4) Diameter. 10.5mm and 11mm

(5) Size. 165 ft ± 5 ft.

(6) Usage. For lead climbing/party climbing.

c. Maxim Dry Rope. Olive Drab or Multi-Colored.

(1) Construction. Water-repellent treated Kernmantle.

(2) Minimum tensile strength. 3472 lbs.

(3) Maximum elongation. 6%

(4) Diameter. 9mm

(5) Size. 150 ft ± 5 ft

(6) Usage. For glacier travel/ice climbing

d. Gold Line II.

(1) Construction. Eight strand braided nylon plymor.

(2) Minimum tensile strength. 4500 lbs.

(3) Maximum elongation. 20%

(4) Diameter. 11 mm

(5) Size. 300 ft or 600 ft spools

(6) Usage. Sling Ropes and litters only.

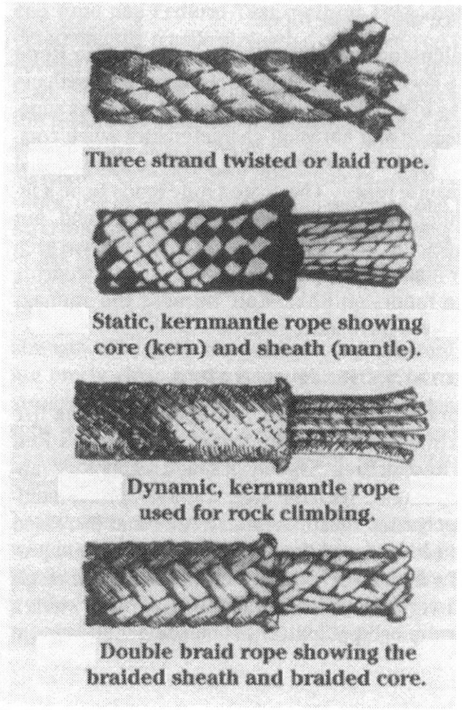

INSIDES AND OUTSIDE SHEATHS OF VARIOUS TYPES OF ROPES

NOTE: Sling ropes are made from 15 foot lengths of plymor or dynamic rope ONLY. Twenty-five foot practice coils should be constructed with static rope, but dynamic rope can be used.

2. (5 Min) **ADVANTAGES/DISADVANTAGES**

a. Advantages of Nylon Rope·

(1) High strength to weight ratio.

(2) Good energy absorption in dynamic ropes.

(3) Flexible.

(4) Rot resistant, not affected by frost.

b. Disadvantages of Nylon Rope

(1) Low melting point. Nylon fuses at 400°F and melts at 480°F.

(2) Susceptible to abrasions and cuts.

(3) Affected by chemicals and light.

c. Advantages of Manila Rope

(1) Easily gripped.

(2) Hard wearing.

(3) Does not deteriorate in heat.

d. Disadvantages of Manila Rode

(1) Heavy, kinks, especially when wet. Absorbs water and swells.

(2) Burns at +300°F.

(3) Edible by rodents.

3. (3 Min) **GENERAL INFORMATION**

a. Nylon rope stretches under tension and will rupture at between 30% and 70% elongation, depending on construction.

b. Nylon rope loses as much as 30% strength when wet.

c. Temperatures as low as 250°F will damage a nylon rope.

4. (5 Min) **NYLON WEBBING**

a. The type of nylon webbing available is tubular. Tubular nylon webbing is very strong and flexible. All rules that apply to nylon rope apply to tubular nylon webbing. The size of nylon webbing used is:

(1) 1 inch tubular nylon. Tensile strength approximately 4,000 - 4,500 lbs., depending on the manufacturer.

b. Pre-sewn Spectra Runners. Tensile strength approximately 5,500 lbs.

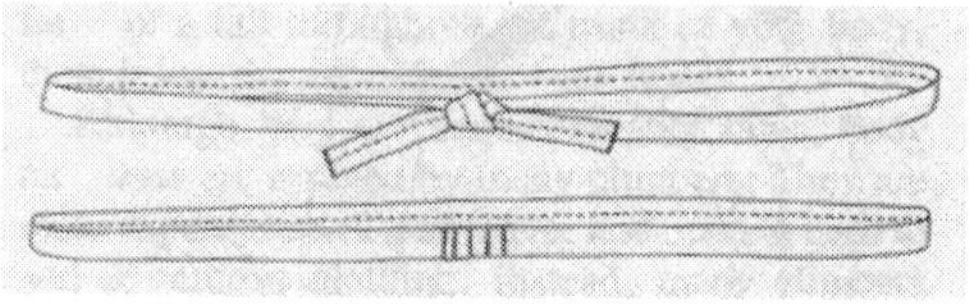

NYLON WEBBING

NOTE: These are minimum strengths. Some manufactures make even stronger webbing.

TRANSITION: We have just discussed general information in nylon webbing, are there any questions? Not only do we use ropes, but we also use carabiners in our installations, we will discuss the types of carabiners used:

5. (5 Min) **CARABINERS**. Also commonly known as snaplinks. Both locking and non-locking are used.

a. Purpose. Carabiners are used for the following purposes:

(1) To attach ropes or runners to pieces of protection.

(2) To attach the rappel rope to the rappel seat for seat-hip rappels or for crossing rope bridges.

(3) To attach the individuals safety rope to a safety line on a rope installation.

(4) To form field expedient pulley systems.

b. Nomenclature of a non-locking carabiner

(1) Gate

(2) Gate pivot pin.

(3) Locking pin.

(4) Body.

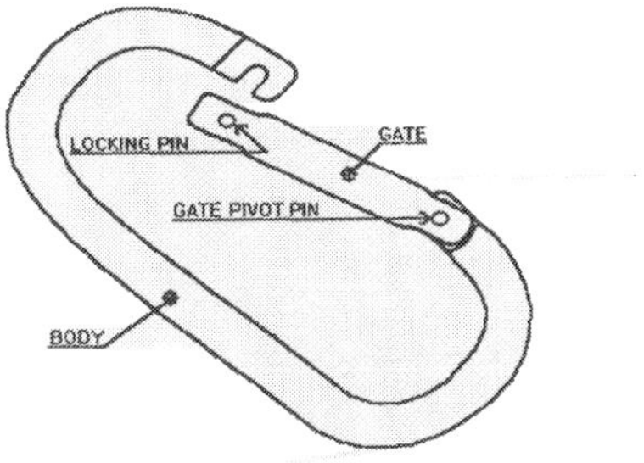

NON LOCKING ALUMINUM CARABINER

c. <u>Nomenclature of a locking carabiner</u>

(1) Gate.

(2) Gate pivot pin.

(3) Locking notch.

(4) Locking nut.

(5) Body.

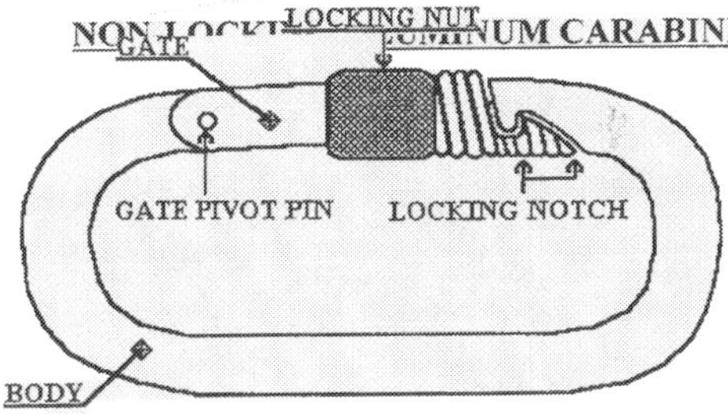

STEEL LOCKING CARABINER

d. There are two types of carabiners used. The two types and their characteristics are:

(1) <u>Steel locking carabiners</u>

(a) Large steel locking "D" (various manufacturers): Minimum tensile strength of 5,500 lbs.

(b) Steel-locking oval Stubai 82 is not in the MAC Kit and obsolete. However, it is being used at MWTC to save money, even though they are beyond the service life. Tensile strength of only 3,300 lbs.

(2) <u>Aluminum non-locking carabiners</u>

(a) Aluminum non-locking oval (various manufacturers): minimum tensile strength of 4,200 lbs.

e. Serviceability Check for a Carabiner. The following steps are used for you to check a carabiner for serviceability:

(1) The gate snaps shut with no friction and with no gap between the locking pin and locking notch.

(2) There is no excessive side to side movement of the gate.

(3) The pivot pin is tight.

(4) The locking pin is tight.

(5) The locking nut travels freely and locks securely.

(6) There are no cracks or flaws in the metal.

NOTE: The weakest part of a carabiner is the gate. If an engraver is used to mark a carabiner, it should be applied to the gate and not the load bearing side.

f. Preventive Maintenance for a Carabiner.

(1) Remove all dirt, moisture and grime.

(2) Lubricate with tri-flow graphite and clean off thoroughly.

NOTE: Whenever you use a locking carabiner ensure that the locking nut is always locked down (tightened).

6. (2 Min) **CARE OF THE CARABINER**. Do not drop the carabiner as this may result in either actual damage to the carabiner or in dirt getting into the workings of the carabiner and damaging it.

UNITED STATES MARINE CORPS
Mountain Warfare Training Center
Bridgeport, California 93517-5001

FMST.07.11
05/01/01

STUDENT HANDOUT

HYPOTHERMIA/REWARMING

TERMINAL LEARNING OBJECTIVE: Given a hypothermia casualty in a cold weather environment and necessary equipment and supplies, manage a hypothermia casualty to prevent death or further injury per the reference. (FMST.07.11)

ENABLING LEARNING OBJECTIVES:

(1) Without the aid of references, select from a given list the definition of hypothermia, in accordance with the references. (FMST.07.11a)

(2) Without the aid of references, select from a given list three of the five mechanisms by which the body loses heat, in accordance with the references. (FMST.07.11b)

(3) Without the aid of references, select from a given list the reason why gentle handling of a hypothermia victim is important, in accordance with the references. (FMST.07.11c)

(4) Without the aid of references, select from a given list the four rewarming techniques used to rewarm a hypothermia victim, in accordance with the references. (FMST.07.11d)

(5) Without the aid of references, select from a given list the proper steps of the sleeping bag rewarming method of rewarming a hypothermia victim, in accordance with the references. (FMST.07.11e)

(6) Without the aid of references, select from a given list the temperature of the water when rewarming a hypothermia casualty in a warm water bath, in accordance with the references. (FMST.07.11f)

8

1. **HYPOTHERMIA**:

a. Definition: A fall in core body temperature to 95°F (35°C) or less. (FMST.07.11a)

b. Hypothermia is potentially lethal. The mortality rate is greater than 50% in severe cases and cases complicated by trauma where the mortality can approach 100%. The most common cause of death in hypothermia is ventricular fibrillation.

c. When speaking of hypothermia, core temperature values must be used. In a controlled environment of an MTF or ER the gold standard is a core temperature taken by rectal or esophageal thermometry. The term "exposure" has often been used to speak of hypothermia but this term is inexact and has numerous other connotations. It should not be used as a synonym for hypothermia.

d. Incidence.

(1) Classically, hypothermia has been thought to occur in extremely cold environments. However, it can occur in temperate climates and at Southern latitudes. In fact, greater than 50% of cases of hypothermia occur at ambient temperatures of 50°F or greater. Contrary to popular belief, the highest incidence of hypothermia is seen in urban areas.

(2) Hypothermia tends tends to occur more frequently in males with a male to female ratio of 2.8:1.0. It's also more common in the elderly, those with drug and alcohol dependency and those with psychiatric diagnoses.

Many standard clinical thermometers don't register a low enough temperature to accurately diagnose hypothermia. Currently, a low-reading rectal thermometer is available through the Federal Stock System (NSN 6515-00-139-4593). All medical personnel attached to cold weather units or deploying to cold weather environments should carry this thermometer. In a high altitude cold weather environment medical providers should be adept at correlating signs and symptoms with specific temperature ranges so as not to unduly expose the patients to the elements. The more accurate, and invasive temperature determination should only be performed when the patient can be adequately protected from the elements.

2. **PATHOPHYSIOLOGY**:

a. Heat Conservation/Loss.

(1) The body is a furnace that generates heat. The ability of the body to thermo regulate depends on the ratio of heat production to heat loss. An imbalance of these two processes can lead to hypothermia. About 75% of food energy consumed goes toward generating body heat while the other 25% goes toward maintaining basic metabolic function.

(2) Thermogenesis can be divided into two main categories:

(A) Shivering thermogenesis creates heat by muscle activity with subsequent breakdown of glycogen. Shivering usually begins at a core temperature of around 97°F (36°C) and finally ceases at core temperatures of 85°- 90°F (30°- 34°C).

(B) Non-shivering thermogenesis creates heat by hormonal interaction with target organs such as the heart. Epinephrine, norepinephrine, growth hormone and thyroid hormone have all been implicated in playing a role in non-shivering thermogenesis.

(3) Changes in hormone secretion patterns are extremely important in acclimatization to cold weather.

(4) Once heat is generated it is retained by three main mechanisms: Peripheral vasoconstriction, abolition of sweating and piloerection. Peripheral vasoconstriction is by far the most important and efficient mechanism of heat conservation. The main internal temperature sensors in the body are located in the hypothalamus in the brain. When these areas sense a drop in core temperature they initiate peripheral vasoconstriction at the surface of the skin to prevent a loss of core body heat.

(5) The body loses heat by five main mechanisms: (FMST.07.11b)

(A) **Radiation** is heat transferred to the surrounding environment via infrared radiation.

(B) **Convection** is the loss of heat via air/water current.

(C) **Conduction** is the direct contact-transfer of heat from a hot to a cold object.

(D) **Evaporation** is the conversion of water from a liquid phase to a gaseous phase (perspiration).

(E) **Respiratory heat loss** takes place when warm humidified air from the lungs is expelled into the atmosphere.

(F) The mechanism accounting for the greatest percentage of heat loss is highly variable and depends on ambient conditions, i.e., a nude man standing in a cold room will lose approximately 65% of his heat by radiation. The same man standing in calm cold water will lose most of his heat by conduction and at a rate 25 times faster.

**NOTE: The liver and brain are the most thermogenic organs on a per weight basis. However, due to their greater mass, skeletal muscle and skin produce the largest percentage of body heat.

(6) Numerous factors predispose to the development of hypothermia. These include but are not limited to:

(A) Malnutrition.
(B) Lack of adaptation.
(C) Inactivity.
(D) Sedatives.
(E) Physical exhaustion.
(F) Burns.
(G) Extremes of age.
(G) Dehydration.
(H) Comorbid illness.

3. **SIGNS AND SYMPTOMS**:

a. Clinical manifestations of hypothermia reflect CNS, and cardiorespiratory involvement.

(1) 98°-95°F (37°-35.6°C): Mild shivering, cold sensation and impaired fine motor coordination.

(2) 95°- 92°F (35°-32.8°C): Violent shivering, difficulty speaking, sluggish thinking, and amnesia. Large muscle in coordination.

(3) 92°-86°F (32.2°-30°C): Shivering is replaced by muscular rigidity. Exposed skin is blue or puffy. Movements are jerky. Dulled sensorium, but victim may be able to maintain posture and the appearance of being in contact with surroundings. Possible atrial arrhythmias.

(4) 85°-81°F (29.4°-27.2°C): Coma, lack of reflexes, possible ventricular arrhythmias. Cessation of shivering below 86oF.

(5) Below 78°F (25.6°C): Failure of cardiac and respiratory centers, pulmonary edema, ventricular fibrillation, and asystole.

b. Some patients with hypothermia will exhibit a phenomenon that is called, "paradoxical undressing."

(1) In these instances the hypothermia victim is found nude with their cloths beside them. If the victims are women they are often misdiagnosed as sexual assault cases. This phenomena results from loss of hypothalamic control over peripheral vasoconstriction which leads to flushing of the periphery with relatively warm blood causing the individual to feel hot. Subsequently the individual undresses to cool down.

c. Again it must be stated that the only way to correctly diagnose hypothermia is with a core temperature reading.

4. **FIELD TREATMENT**:

a. ABC's, C-Spine precautions, remove wet clothing, replace with dry clothes and insulate with a vapor barrier system. (Wrapping the victim in two plastic garbage bags or bubble wrap is ideal. Further insulation may be necessary). Handle the victim carefully to prevent ventricular fibrillation caused by unnecessary jarring and rough handling. (FMST.07.11c) CASEVAC.

b. Victims with altered consciousness should be NPO. Under no circumstances should ETOH be given. This is a peripheral vasodilator and will only enhance heat loss.

c. If CASEVAC is not feasible for a long period of time, or if the victim is only mildly hypothermic, it may be necessary to rewarm in the field by using a proper sleeping bag rewarming technique. Additionally, insulated hot packs placed at the high heat loss areas will help.

d. The hypothermic victim often has no discernible vital signs.

(1) Before beginning CPR thoroughly assess pulses for 30-45 seconds to detect viable cardiovascular activity.
(2) Severely hypothermic victims in ventricular fibrillation and asystole have been successfully resuscitated even after periods as long as 4 hours.

5. **REWARMING**: The re-warming of a hypothermia casualty in a BAS setting is less than optimal since the AMAL for a BAS does not include monitoring devices, laboratory facilities, etc., but rewarming of a casualty can be successfully conducted even in these fairly primitive conditions. The following are three basic techniques of re-warming:

- Passive external re-warming: a sleeping bag or blanket that is not pre-warmed.
- Active external re-warming: a pre-warmed sleeping bag, bear hugger or warm water bath.
- Active internal or core re-warming: IV fluids, warmed/humidified O2, lavage, extracorporal re-warming.

a. Four ideal re-warming strategies that may be available in the field include: (FMST.07.11d)

(1) Sleeping bag rewarming.

(2) Water bath/counter Current rewarming.

(3) Warm/humidified airway rewarming.

(3) Warmed I.V. solutions. NS is preferable as a cold liver does not like to metabolize Lactate.

6. **REWARMING TECHNIQUES (BAS)**:

a. Sleeping Bag Rewarming. (FMST.07.11e) This is the simplest, but least effective method.

(1) Make the diagnosis. Take a rectal temperature to determine if the casualty is clinically hypothermic (95°F or less). Use this temperature as a baseline to determine stabilization.

(2) Warm sleeping bag (pre-warm sleeping bag with two volunteers). Placing a hypothermic casualty into a cold sleeping bag will cause further heat loss by conduction.

(3) Strip casualty. Remove all wet clothing avoiding unnecessary handling. Muscular movement will pump cold blood to the core.

(a) Place the casualty between the two volunteers, if space permits. Their body heat is transferred from the two volunteers to the cold casualty.

(4) Monitor core temperature frequently throughout transport to ensure that the casualty's core temperature is not continuing to drop.

(5) Adequate insulation is required to prevent further heat loss to the environment.

(6) Augment heat by placing insulated heating pads in the high heat loss areas: head, axilla, groin, popliteal region and antecubital fossa. (Be careful not to cause burns)

**NOTE: It should be noted that an evacuation bag can be used; this will provide more room for the casualty and volunteers.

b. Water Bath Re-warming. (FMST.07.11f) This method has a long history of success in rewarming even severely hypothermic casualties. The casualty is placed in a portable/field expedient bathtub or a life raft. The extremities and head are kept out of the water to avoid significant afterdrop. The water bath temperature should be 104°-108°F. An anesthesia temperature probe should be used to constantly monitor the core temperature. Since the casualty will cool the bath water around him, it will be necessary to continue to add warm water to maintain the proper water bath temperature. Keep in mind however that this technique has definite drawbacks. In the event of ventricular fibrillation caused by re-warming shock, unsynchronized cardioversion can be very dangerous in a wet environment. Furthermore, if the casualty in the water is not on a firm surface CPR, if needed, can be very difficult to perform ineffective.

**NOTE: Individual in heat distress can be rapidly cooled using this method with "cool" water instead.

c. Warm Airway Re-warming. A warmed and humidified air/oxygen mixture is used to provide a warming media within the lungs. These devices consist of a method of generating warm humidified air and are usually portable. The warm humidified air modestly increases the amount of heat that can be delivered to core. There should be a thermometer in the airway tubing to monitor the temperature. To prevent injury to the bronchi, the temperature should not exceed 115°F (108° to 115°F is ideal). Warm airway rewarming should be used in conjunction with other rewarming methods. This method, by itself, will probably not deliver sufficient heat to rewarm a severely hypothermic casualty by itself and may in fact blunt the shivering response robbing the body of an important endogenous mechanism for producing heat.

Caveat: When possible, active external re-warming should always be combined with active core re-warming techniques so as to minimize the effects of afterdrop. This occurs as a result of peripheral vasodilatation causing warm core blood to move to the surface while relatively cold surface blood moves to the core and may lower the fibrillation threshold.

**NOTE: Hypothermic submersion incident casualties cannot tolerate humidified air for any length of time.

d. Warmed I.V. Solutions. Hypothermia is a common response to I.V. therapy. Solutions that have been prewarmed have been shown to prevent this complication. There are various methods for warming I.V. solutions:

(1) Crystalloid solutions can be warmed by warm water bath and microwaved with no adverse changes to its integrity.

(2) Fresh Frozen Plasma (FFP; -20°C): Studies have shown that microwaving for 30 seconds followed by gentle manipulation of the bag for 10 seconds, repeated five times will result in only 2.6% cell destruction.

(3) Packed Red Blood Cells (PRBC 4°C): Using conventional rewarming methods, studies have shown that the outer five millimeters of the bag can have isolated areas of hemolysis. Until a device is made that can uniformly shake up PRBC during warming, we do not recommend microwaving of blood. Rewarming of blood by dilution with warm, calcium-free crystalloid solution is useful.

WIND CHILL CHART

WIND	AMBIENT TEMPERATURE (FAHRENHEIT/CENTIGRADE)													
CALM F	35	30	25	20	15	10	5	0	-5	-10	-15	-20	-25	-30
CALM C	27	-1	-4	-7	-9	-12	-15	-18	-21	-23	-26	-29	-32	-34
EQUIVALENT WIND CHILL FACTOR (FAHRENHEIT/CENTIGRADE)														
5 MPH F	33	27	21	16	12	7	1	-6	-11	-15	-20	-26	-31	-35
8 KPH C	1	-3	-6	-9	-11	-14	-17	-21	-24	-26	-29	-32	-35	-37
10 MPH F	21	16	9	2	-2	-9	-15	-22	-27	-31	-38	-45	-52	-58
16 KPH C	-6	-9	-13	-17	-19	-23	-26	-30	-33	-35	-39	-43	-47	-50
15 MPH F	16	11	1	-6	-11	-18	-25	-33	-40	-45	-51	-60	-65	-70
23 KPH C	-9	-12	-17	-21	-24	-28	-32	-36	-40	-43	-46	-51	-54	-57
20 MPH F	12	3	-4	-9	-17	-24	-32	-40	-46	-52	-60	-68	-76	-81

32 KPH C	-11	-16	-20	-23	-27	-31	-36	-40	-43	-47	-51	-56	-60	-63
25 MPH F	7	0	-7	-15	-22	-29	-37	-45	-52	-58	-67	-75	-83	-89
40 KPH C	-14	-18	-22	-26	-30	-34	-38	-43	-47	-50	-55	-59	-64	-67
30 MPH F	5	-2	-11	-18	-26	-33	-41	-49	-56	-63	-70	-78	-87	-94
48 KPH C	-15	-19	-24	-28	-32	-36	-41	-45	-51	-53	-57	-61	-66	-70
35 MPH F	3	-4	-13	-20	-27	-35	-43	-52	-60	-67	-72	-83	-90	-98
56 KPH C	-16	-20	-25	-31	-33	-37	-42	-47	-51	-55	-58	-64	-68	-72
40 MPH F	1	-4	-15	-22	-29	-36	-45	-54	-62	-69	-76	-87	-94	-101
64 KPH C	-17	-20	-26	-30	-34	-38	-43	-48	-52	-56	-60	-66	-70	-74

7 REWARMING TECHNIQUES (HOSPITAL):

a. Optimal re-warming techniques for higher echelon care facilities still tend to employ rather simple therapeutic modalities. The three safest and most effective are listed below:

- Warmed IVF.
- Warmed/humidified oxygen.
- Bear Hugger.

b. Many texts advocate the use of extracorporal blood re-warming. However, this technique has some disadvantages. It requires a high level of technical and medical support, which are often not available. Since patients undergoing bypass must be anti-coagulated, patients with multiple traumatic injuries may not be good candidates. Lastly, research indicates that anticoagulation with heparin may actually worsen the severity of frostbite that may occur in patients with hypothermia.

c. Lavage: peritoneal, thoracic and mediastinal lavage have all been suggested for use in severe hypothermia. However, the use of these techniques in inexperienced hands tends to carry a much higher complication rate. If performed correctly, peritoneal lavage can be very beneficial with a lower complication rate than thoracic lavage. Gastric, colonic and bladder lavage are indeed safer but are of limited efficacy, as they do not offer access to areas with significant surface area.

d. ACLS: At core temperatures below 86°F attempt three progressive unsynchronized shocks at 200J, 300J, and 360Js. If no response continue CPR and REWARM THE VICTIM. Re-attempt counter-shocks when core temperature rises above 86°F. Above 86°F, perform counter shocks in 2° increments above 86°F, i.e.88°F, 90°F and 92°F. Employ ACLS drug protocols above 93°F along with conventional normothermic ACLS protocol.

UNITED STATES MARINE CORPS
Mountain Warfare Training Center
Bridgeport, California 93517-5001

FMST.07.12
03/21/02

STUDENT HANDOUT

SUBMERSION INCIDENTS

TERMINAL LEARNING OBJECTIVE: Given a casualty in a cold weather environment and the necessary equipment and supplies, manage a submersion incident casualty to prevent death or further injury per the reference. (FMST.07.12)

ENABLING LEARNING OBJECTIVES:

(1) Without the aid of references, given a list of the events, which take place to a submersion victim, select the correct sequence of events, in accordance with the references. (FMST.07.12a)

(2) Without the aid of references, select from a given list, three of the five methods that can be used to increase survival time during cold water immersion, in accordance with the references. (FMST.07.12b)

(3) Without the aid of references, select from a given list, three of the six favorable prognostic signs post submersion incident, in accordance with the references. (FMST.07.12c)

(4) Without the aid of references, select from a given list, three of the four unfavorable prognostic signs post submersion incident, in accordance with the references. (FMST.07.12d)

(5) Without the aid of references, select from a given list the correct field management priorities of a submersion victim, in accordance with the references. (FMST.07.12e)

10

OUTLINE.

1. **TERMINOLOGY**.

a. Definition: A submersion incident is that medical problem that occurs after a casualty has been submersed under water.

(1) Drowning refers to death by suffocation following submersion in water.

(2) Near drowning refers to submersion in water with at least temporary survival (greater than 24 hours).

NOTE: An effort is being made to standardize the terminology used regarding submersion incidents, in order to clarify communication between care-givers. Terms like secondary drowning and other definitions are being discouraged.

2. **PATHOPHYSIOLOGY**.

a. Sequence of events: (FMST.07.12a)

(1) Panic initially, followed by a violent struggle.

(2) Gulping and swallowing air and water to avoid aspiration.

(3) Breath holding until hypoxia leads to unconsciousness.

(4) Once consciousness is lost, the gag reflex relaxes and passive influx of water into the lungs occurs. A small percentage of victims (10-15%) have significant laryngospasm, which prevents any appreciable volume of water to enter the lungs. This is referred to as "dry drowning".

(5) Aspiration is more likely in submersion victims who enter very cold water (10°C). An involuntary gasp reflex occurs, with a marked increase in respiratory rate and depth. This individual is at increased risk of aspiration, and subsequent drowning.

b. Submersion in very cold water will quickly results in hypothermia.

(1) Rapid induction of total body hypothermia (core temperatures of 89.6°F or less) may be protective in submersion incidents. Significant decreases in oxygen requirement result, especially in the case of the brain itself. Marked decreases in cerebral blood flow and oxygen demand have been demonstrated (see Hypothermia lesson). This protective effect of hypothermia has been used to explain the many cases of complete recovery after extended periods of submersion, which have only occurred in cold water.

(2) The "Mammalian Diving Reflex" slows the heart rate, shunts blood to the brain and closes the airway when more primitive marine mammals are exposed to extremely cold water. General consensus holds that this reflex is not a significant response in human beings. However, it has been found to be more prominent in new-born human infants. It has also been speculated that this "Reflex" may be more significant in those individuals who have survived extended submersion times.

c. Fatalities in submersion victims associated with hypothermia are primarily due to three known mechanisms:

(1) Immersion syndrome. Sudden death due to cardiac arrest secondary to massive vagal stimulation. Usually associated with sudden total body exposure to frigid water.

(2) Excessive fatigue and confusion leading to fatal errors of judgment.

(3) Direct hypothermic effects. Thermal conductivity in water is approximately 25 times that of air. Despite the evidence showing hypothermia's protective effects, the complications associated with hypothermia (cardiac arrhythmias, CNS depression, loss of motor control) can lead to submersion and drowning.

d. Some authorities emphasize distinguishing between "wet" and "dry" drowning. In wet drowning, aspiration of water into the lung occurs. (85-90% of casualties). In dry drowning, laryngospasm prevents aspiration of water (10-15% of cases). Ultimately, the end result is the same: obstruction to respiration, oxygen depletion, and asphyxiation. The morbidity and mortality associated with the two types of drowning are not significantly different.

e. Fresh-water and salt-water drowning are somewhat different.

(1) In fresh-water aspiration, the osmotic difference between hypotonic fresh-water and the relatively hypertonic blood leads to net water movement into the vasculature.

This flow across the alveolar membrane also leads to "washout" of surfactant, which leads to collapse of the alveolar spaces. Hypoxemia follows.

(2) In sea-water aspiration, the osmotic shift is reversed, leading to fluid outpouring from the blood stream flooding the alveoli and the interstitium. This leads to decreased lung compliance as well as direct damage to pulmonary capillaries. Again, hypoxemia results.

(3) Hypoxemia is the end result in both types of drowning. No significant difference in morbidity in survivors has been shown between fresh-water and salt-water drowning. The only bodies of water shown to have a worsening effect on prognosis are the hyper-saline bodies, like the Dead Sea and the Great Salt Lake. Otherwise, prognosis is similar.

f. Hypoxemia is the primary insult after submersion. Efforts to correct this problem should be vigorous and sustained. All clinically dead submersion casualties should receive CPR, or ACLS if available, immediately upon removal from the water, regardless of water temperature or duration of submersion. Since the overwhelming majority of naturally occurring water bodies in the world have temperatures below 92°F, all submersion victims should be treated aggressively as hypothermic casualties. Survivals after submersion times up to 66 minutes have been documented. Remember: No one is dead until they are warm and dead.

g. In the case of prolonged submersion or significant aspiration, pulmonary derangements occur rapidly. In lesser insults, the onset of symptoms may be delayed for as long as 24 hours. Physiological changes include:

(1) Cardiovascular. Casualties develop respiratory acidosis and hypoxemia, which may trigger dysrhythmias. Electrolyte changes are not usually significant in survivors of submersion incidents and should correct once spontaneous respirations are re-established.

(2) CNS. C-Spine injuries are common in submersion incidents, most commonly a result of shallow-water diving. Injury patterns include wedge, tear-drop and burst fractures of the upper C-Spine, frequently resulting in tetraplegia. Concussive head injuries also occur. Drug and alcohol use should always be considered.

(3) Metabolic. Respiratory acidosis is the initial problem. Acid/base derangement should correct after resuscitation of the casualty without specific intervention.

3. **SURVIVAL TECHNIQUES IN COLD WATER**.

a. Swimming motion increases heat loss more than it increases heat production. The direct hypothermic effects of cold-water immersion make it impossible to swim for long periods of time. The average swimmer is unlikely to swim more than a kilometer in 50°F water. Swimming is advised only when a rescue vehicle is nearby, or the distance to shore is less than ½ mile.

b. Treading Water and Drown Proofing.

(1) Treading water results in reduced survival time due to increased heat loss secondary to increased activity.

(2) "Drown Proofing" is a technique whereby the swimmer relaxes in the water, floating facedown, arms spread, allowing his face to submerge between breaths. "Drown proofing" is the least desirable survival technique in cold water since the head and neck (high heat loss areas) are maximally exposed.

c. Personal Floatation Device (PFD). Without the aid of the PFD, the swimmer caught in cold water is forced to use less desirable methods to stay afloat.

(1) If wearing a PFD, the individual can utilize the "Heat Escape Lessening Posture" (HELP). This position minimizes the exposure of the swimmer's high heat loss areas because the arms are folded across the chest and pressed to the sides, the knees are drawn up to the chest and the legs are crossed at the ankles. Unfortunately, this position is fatiguing and becomes difficult to hold.

(2) If there is more than one person in the water, the "Huddle" position should be used. Similar to a football huddle, the chest, abdomen and groin area should be pressed together. Children or injured individuals should be placed in the middle of the huddle, since they become hypothermic more quickly.

d. Increasing Survival Time: (FMST.07.12b)

(1) Try to enter the water in a lifeboat or raft.

(2) Locate and don a PFD as quickly as possible.

(3) Wear several layers of clothes, especially to protect high heat loss areas.

(4) If you must enter the water, enter gradually rather than jumping or diving.

(5) If afloat in the water, avoid unnecessary movement. Assume the HELP or Huddle position.

(6) Don't try to swim.

(7) Remember the situation is not hopeless. Have the will to SURVIVE!

4. **RESCUE TECHNIQUES**.

a. Reaching the Casualty. The basic order for water rescue is: Reach and Pull, Throw, Tow and, as a last resort, Go.

(1) Reach and pull. When the casualty is responsive and close to the shore, begin by holding out an object for him to grab then pull him from the water. Floating line is considered the best choice, but if it is not available, a tree branch, fishing pole, oar or other object will do as well.

(2) Throw. Should the casualty be alert, but too far away for you to reach and pull, throw an object that will float to them. A PFD or ring buoy is best, if available. The primary course of action is to throw anything that will float, and to do it as quickly and as accurately as possible.

(3) Tow. Once the conscious casualty has an object to hold onto, try to find a way to tow him to shore. From a secure position, tow the casualty ashore.

(4) Go. Swim to the casualty as a last resort. You should be a strong swimmer, trained in water rescue, in order to avoid the dangers associated with this form of rescue. If the casualty is unconscious or unresponsive, you must go to the casualty.

5. **SIGNS AND SYMPTOMS**.

a. History of Incident. Several questions need to be answered, as they will help determine the approach to the casualty.

(1) When a person is found floating face down on the water surface or is brought to the surface by a rescuer, there usually is little question about the cause of unconsciousness.

(2) When near-drowning is unwitnessed or occurs after a dive or other water activity (surfing), a head injury and a C-spine injury should be presumed, and C-Spine precautions should be instituted.

(3) Determine how long the casualty was submerged. This helps determine prognosis.

(4) How cold was the water? Very cold water (less than 70°F/21°C) may be protective.

(5) What was the casualty's health prior to submersion? Drowning could have been secondary to alcohol ingestion, cardiac arrest or other illness/injury.

b. Casualties may be symptomatic or asymptomatic at the scene:

(1) Respiratory signs and symptoms may include dyspnea, tachypnea (SOB), sore throat, substernal burning, pleuritic chest pain, persistent cough, frothy sputum, cyanosis, rales, rhonchi and wheezes.

(2) Cardiovascular signs may include tachycardia and hypotension.

(3) Neurological findings may include anxiety, restlessness, lethargy, confusion, convulsions, incontinence, hyporeflexia and coma.

(4) Abdominal distention is common as is fever and hypothermia.

c. Favorable prognostic signs: (FMST.07.12c)

(1) Casualty is alert.

(2) Basic or advanced life support is received at the scene.

(3) Good response to initial resuscitation.

(4) Hypothermic.

(5) Older child or adult.

(6) Brief submersion time.

d. Unfavorable prognostic signs: (FMST.07.12d)

(1) Age under three years.

(2) Fixed dilated pupils.

(3) Submersed five minutes or longer.

(4) No resuscitation attempts for the first ten minutes.

(5) PH less than 7.10.

(6) Cardio-pulmonary arrest as initial presentation.

(7) Resuscitation time greater than 30 minutes.

6. **FIELD MANAGEMENT**.

a. The scene of a submersion incident is frequently chaotic and first consideration should be the safety of the rescuers, then retrieval of the casualty from the water.

b. Priorities: (FMST.07.12e)

(1) Establish an airway and check breathing and circulation (ABC's). Rescue breathing, can be performed by a trained rescuer, while the casualty is still in the water.

(2) Observe C-Spine precautions, since cervical spine and head injuries are common, especially in diving accidents.

(3) Begin CPR as soon as possible, if necessary, when the casualty reaches the shore, or suitable rescue platform.

(4) The symptomatic patient should be given high-flow 02 and IV access should be established, if available.

(5) Medevac ASAP.

(6) It is recommended that all victims of submersion be observed in a medical facility for a minimum of 6-8 hours. It is not uncommon for asymptomatic near-drowning victims to refuse further care. All that can be done is attempt to get a contact number in order to follow-up with the victim via telephone 8 hours later.

c. Vomiting is common and should be anticipated by turning the patient's head and having suction ready, if available. Frequently, the cause of vomiting is gastric distention.

(1) The American Heart Association recommends that gastric distention be relieved only when it interferes with CPR.

(2) Gastric decompression can be achieved by "breaking" the patient. This maneuver is performed by placing the patient in the prone position and lifting upwards at the abdomen.

(3) Another method is to turn the patient to a lateral recumbent position and compress the abdomen with the palm of one hand.

(4) The preferred method of gastric drainage is to pass an NG tube and relieve distention with gentle suction.

(5) Gastric drainage should not be attempted without suction equipment available, due to the risk of aspirating vomitus.

7. **PROGNOSTIC PREDICTORS**.

a. 0 - 2 "Unfavorable Prognostic Indicators" yield a greater than 90% survival rate (see above).

b. 3 or more "Unfavorable Prognostic Indicators" yields a less than 5% survival rate.

c. Corm's Criteria: Based on the status of the patient 1 hour after initial resuscitation.

(1) A - Awake	All survive (99+%)
(2) B - Blunted	Most survive (95+%)
(3) C – Comatose	
C1 - Flexor Response	Good survival with ICU care (80+%)
C2 - Extensor Response	Good survival with ICU care (80+%)
03 - Flaccid	Poor survival
C4 - Arrest	Dismal survival

UNITED STATES MARINE CORPS
Mountain Warfare Training Center
Bridgeport, California 93517-5001

FMST.07.14

04/02/02

STUDENT HANDOUT

WILDERNESS ORTHOPAEDIC/ TRAUMA INJURIES

TERMINAL LEARNING OBJECTIVE: Given a casualty in a mountainous enviornment and, necessary equipment and supplies, manage common orthopedic injuries in a mountainous enviornment to prevent death or further injury per the reference. (FMST.07.14)

EABLING LEARNING OBJECTIVES:

(1) Without the aid of references, select from a given list the phases of clearing a cervical spine in the field in accordance with the references. (FMST.07.14a)

(2) Without the aid of references, select from a given list, the correct treatment for an open book pelvic fracture in accordance with the references. (FMST.07.14b)

(3) Without the aid of references, select from a given list, six considerations for orthopedic patient evacuation in accordance with the references. (FMST.07.14c)

OUTLINE

1. INTRODUCTION.

Orthopedic injuries occuring in the wilderness are not significantly different from those suffered in garrison. Bones break, ligaments are torn and strained, and joints are dislocated in both settings. The key to preparing for and managing wilderness orthopedic injuries is understanding that the provider will have significant limitations with regard to diagnostic tools and initial management resources at his/her disposal. The personal risks each expedition member is exposed to can rise dramatically if the team is needed to transport a victim out of the back-country. Determination of a casualty's mobility becomes a major consideration for the wilderness medical provider. Assessment of the terrain, creative use of available natural resources, and effective leadership are all necessary components of caring for orthopedic injuries in the wilderness.

10

←

2. GENERAL: Making the diagnosis.

The wilderness "doc", whatever his/her training background, needs to have good physical exam skills. NO X-RAY and NO LABS will be available in the back-country. These basic tools are not available in the field. The provider will need to rely on his/her observation, palpation, and auscultation skills to make a diagnosis. This person will also need to use the patient, if possible, to uncover the likely injury by taking a thorough history which includes careful review of the mechanism of injury.

A. HISTORY: Critical information can be gained from talking to your patient.

- Mechanism of Injury - Determine:
    1. High vs. low-velocity accident
    2. Direction of force
    3. Nature of activity
    4. Loss of consciousness

- Ask the casualty:
    1. Did they hear a pop, snap, crack, or breaking sound?
    2. How long ago did the injury occur?
    3. Can they move the injured body part?
    4. Was there swelling within 15 minutes of injury or did it take longer?
    5. Was there immediate disability or was there a gradual loss of function?
    6. Was there immediate discoloration of a joint?
    7. Can they bear weight on the injured extremity?

- Other important information:
    1. Evaluate the environment. (i.e. Hot/Cold, Wet/Dry)
    2. Allergies.
    3. Medications currently being taken.
    4. Previous injuries/surgeries to the site in question.

B. PHYSICAL EXAMINATION

Your powers of observation become crucial.

- Look for:
    1. Swelling.
    2. Discoloration and bruising.
    3. Obvious deformity/angulation.
    4. Open wounds. (with or without protruding bone fragments)
    5. Differences right vs. left.

6. Muscle spasm surrounding injured site.
7. Point tenderness.
8. Crepitus.

- Beware:
  Consider the environment while assessing your patient
  - Don't undress the casualty in the cold!
  - Protect your patient from the elements. (Sun, Rain, Cold)
  - Talor your physical exam to meet the constraints of the environment.
    * palpate under clothing.
    * visualize one region at a time, then re-dress.
    * set up temporary shelter from the elements.

3. Head Injuries:
   a. A blow to the head may lead to increased intracranial pressure (ICP) or intracranial bleeding neither of which are managable in the wilderness. The job of the clinician is to diferentiate a serious life threatening injury from a minor one.

      1. Minor injury: No loss of consiousness (LOC), or LOC of less than 15 seconds with immediate return to full alertness. The casualty can not be on medications which increase risk of bleeding or have a history of bleeding disorders. Patient may be monitored every two hours for mental status changes, lethargy, irritibility, persistent nausea and vomiting, changes in speech or visual changes.

      2. Serious Injury: LOC greater than 15 seconds, and/or persisitent confusion or memory loss; signs or symptoms of increased ICP: Debilitating headache, mental status changes, persistent nausea and vomiting, appearance of clear fluid in external auditory canal, Battle sign, raccoon eyes, or seizures.

      3. Field treatment of serious head injury: Suspect injury of C-spine, manage airway-be able to clear vomit, elevate head 30 degrees. Evacuate to treatment facility ASAP.

4. Spinal Injuries.

   A. Cervical Spine

      1. High Risk Activities:

C-spine injuries in the wilderness usually occur after either a fall from a significant height or high-velocity accident from skiing. Common winter activities predisposing participants to C-spine injuries are skiing, snow-boarding, and snow-mobiling.

      2. Anatomy:

The cervical spine consists of seven cervical vertebrae interposed between the base of the skull and the thoracic spine. The cervical spine has a great deal of mobility to allow maximal range of motion for the skull. This increased mobility comes at a high price; this portion of the spine is less stable than the rest of the vertebral column, and hence more subject to injury. The C-Spine is stabilized by three longitudinal ligaments: The anterior longitudinal ligament (ALL) runs longitudinally along the anterior surface of the vertebral bodies. It is broad and very strong and helps to prevent hyper-extension of the head and neck. The posterior longitudinal ligament (PLL) also runs longitudinally, this time along the posterior surface of the vertebral bodies, within the vertebral canal itself. This ligament is relatively narrow and somewhat weaker than the ALL, and helps to prevent hyper-flexion at the neck. Third is the supraspinous ligament, connecting the spinous processes. This, too, is a strong ligament, acting to inhibit hyper-flexion.

The peripheral nerves innervating the muscles and the sensory nerves of the upper extremities originate from the cervical portion of the spinal cord and these are some of the nerves likely to be affected in the event of a C-Spine injury. Hence, it is important to assess the neurologic status of the upper extremities if a cervical injury is suspected.

3. C-Spine Injury Statistics:

-Most common injury: Flexion injury at C5/6
(so look for a deficit in the C6 distribution)

-28% of C-Spine injuries have another spinal fracture associated with it.
(so examine entire spine)

-10% of Head injuries/Facial fractures also have a C-Spine injury, especially if there was LOC.

4. Remember:

a. Head injury -> Assume C-Spine injury.
b. Perform complete Neurovascular exam
   -Motor, Sensory, Reflexes, Pulses, Babinski reflex.

B. Clear a C-Spine. (FMST.07.14a)

When faced with a possible C-Spine injury in the Wilderness, clearing it without X-ray should be considered only in extreme situations. However, there are situations which demand that every effort be made to make a casualty ambulatory, because the movement of the individual would demand herculean effort from the rest of the team, potentially placing more individuals at risk of injury or prolonging exposure to extreme environmental conditions.

Clearing a C-Spine in the field should follow a three-phased assessment. If the patient fails the assessment at any one of the phases, then the team is obligated to maintain spinal precautions and transport the patient. The phases are outlined below, and expanded upon in the lecture.

PHASE ONE:

-ensure no alteration of mental status.
-maintain in-line traction.
-perform complete neurovascular exam (focusing on the upper extremities)
-palpate spinous processes one-by-one while asking casualty whether palpation causes pain (supraspinous ligament)
-distinguish between pain over spinous process and muscle soreness associated with paraspinal musculature.
-only if palpation was completely pain-free can you move on to ...

PHASE TWO:

-continue to maintain in-line traction
-palpate anteriorly to the left and right of trachea. along the anterior vertebral bodies (anterior longitudinal ligament).
-Assess for bony tenderness under the angle of the jaw to the clavicle.
-Distinguish between tenderness of the sternocleidomastoid muscle and actual vertebral pain.
-only if palpation was pain-free can you progress to...

PHASE THREE:

-while loosely maintaining in-line traction, have patient move head through active range of motion

*first flexion/extension
*followed by rotational movement

-stop test at the first indication of pain with movement

Note: If you are able to move the patient's head through a full range of motion without pain, you have effectively cleared his C-Spine. Have the patient resume activity slowly and cautiously; evaluate any other complaints noted.

B. Thoracic and Lumbar Spine:

-Look for a T-L Spine injury with calcaneal fractures. Approximtely 10% of patients with a calcaneal fracture will also have an associated lumbar fracture.

C. Pelvis:

-Think HEMORRHAGE/SHOCK if faced with a Pelvic Fracture
Up to 6 liters of blood loss possible – (internal iliac areteries)

Place gentle constricting wrap around pelvis if open book fracture is suspected. (FMST.07.14b)

4. FRACTURES.

Some common wilderness fractures:

a. Metacarpal/Phalangeal	-the hands are your means of interfacing with the environment, sometimes it's a forceful meeting.
b. Distal Radius	-Very common wilderness fracture.
c. Scaphoid	-check for snuffbox tenderness, pain with axial load to thumb.
d. Lunate	-also common, pain dorsally at base of 3rd MC.
f. Clavicle	-need good NV exam, check for pneumothorax.
g. Long bone	-Radius/Ulna: if proximal, check elbow -Humerus: radial nerve runs in spiral groove -Tib/Fib: think compartment syndrome- Fib fx may ambulate with cane -Femur: think Hemorrhage, think traction.
f. Ankle	-commonly a fracture/dislocation, check NV
g. Hip	- Leg will typically be externally rotated

REMEMBER:

- Always think HEMORRHAGE with Long Bone fractures
- Always perform NEUROVASCULAR Exam before and after treatment
- Always SPLINT AND PAD for stability, comfort, function and transport.
- Indications for REDUCTION are: NV deficit , severe angulation, severe pain, if angulation prediposes to open fx or makes transport difficult.
- OPEN FRACTURES: Have 6-8 hours to get to surgery. Field Tx- Gently wash off with betadine/ iodine solution to get off obvious dirt- don't scrub. Wrap in sterile gauze. Don't place exposed bone back under skin. Give antibiotics: Augmentin, 2nd or 3rd generation cephalosporin, a quinolone or tetracycline.

THE REASONS WE PLACE FEMUR FRACTURES UNDER TRACTION.

1. Re-establishing length tightens fascia and tamponades bleeding.
2. Dramatic pain relief.
3. Helps prevent open fractures.

4. Helps reduce secondary soft tissue damage.
5. Increases ease and safety of transport.

SOME BASIC RULES OF EXTREMITY SPLINTING:

- Splint all fractures before moving the casualty, unless in imminent danger.
- Splint all fractures as they are found, unless severe angulation

complicates transport or is causing neurovascular deficits.
- Splint to include the joint above and below.
- Construct splints using uninjured extremity, then splint in the position of function, comfort or stability.

5. DISLOCATIONS.

a. Rapid Diagnosis and Reduction is imperative, if evacuation time is > 1-2 hours.

- Easier to reduce immediately after injury. (muscle spasm)
- Makes transport much easier. (increased patient comfort)
- Dramatic pain relief.
- Early reduction reduces risk of Neurovascular injury.
- Reduction could make difference between an ambulatory vs. litter patient

the safety of the entire expedition could be placed at risk

b. Signs and Symptoms: Remember, NO X-RAYS!

- Decreased Range of Motion.
- Obvious Deformity. (compare right vs. left)
- Usually NO Crepitus.
- Typical Posture.

c. Reduction Techniques:

- Phalanges	Linear traction, buddy tape. Can not reduce volar displacement of 1st phalanyx.
- Shoulder	Bedsheet method- counter traction. Prone with weights method. External rotation. Snowbird technique.
- Patella	Extension with medially directed pressure.
- Ankle	Linear traction with knee flexion.

- Hip	Anchor pelvis, hip flexed at 30 degrees, upward traction.
- Sternoclavicular	Roll between scapulae, and apply forceful downward pressure.
-Knee	Should only be reduced as an absolute last resort. Apply anterior traction gently.

REMEMBER: Analgesia and Sedation should be utilized when available before attempting joint or fracture reduction, unless multiple traumatic injuries are present.

6. EVACUATION CRITERIA:

Conditions demanding expeditious evacuation: (FMST.07.14c)

1. Suspected cervical, thoracic, and/or lumbar spine injuries.
2. Pelvic injury with instability and/or significant blood loss.
3. Any open fractures. (6-8 hour window, antibiotics early)
4. Compartment syndromes. (pain with passive motion, pallor, pulseless, paresthesia)
5. Hip or Knee dislocations.
6. Any other injury the medical provider feels he/she is not prepared to manage.

7. OTHER TRAUMA.

Heavy bleeding:	Usually can be treated by direct pressure, if a tourniquet is required,release pressure every five minutes to reassess.
Contusion:	Apply ice 1st 48 hours which will provide pain relief and limit expansion of hematoma.
Lacerations and avulsions:	High pressure irrigation with clean water, may remove debris with flame treated forceps. Use steri-strips for minor clean lacs. Keep dirty or puncture wounds open and give antibiotics.
Animal Bites:	Copious irrigation, and antibiotic, do not close. Consider rabies if bite was unprovoked and animal was acting strange. (bats, raccoons, skunks, canines, and felines) Has never been seen in livestock, rabbits, squirrels, rats or mice. Most common zoonotic source in North America is the skunk.

Severed part:	Can be reattached up to 24 hours after injury if the cut is clean. Gently clean part, wrap in gauze and keep cool without freezing.

UNITED STATES MARINE CORPS
Mountain Warfare Training Center
Bridgeport, California 93517-5001

FMST.07.38
04/02/02

STUDENT HANDOUT

REPTILE AND ARTHROPOD ENVENOMATION BITES AND STINGS

TERMINAL LEARNING OBJECTIVE: Given a simulated casualty, in a wilderness environment, treat reptile and arthropod envenomation, in accordance with the references. (FMST.07.38)

ENABLING LEARNING OBJECTIVES:

1) Without the aid of references, select from a given list the first aid measures used for pit viper envenomation, in accordance with the references. (FMST.07.38a)

2) Without the aid of references, select from a given list the first aid measures used for coral snake envenomation, in accordance with the references. (FMST.07.38b)

3) Without the aid of references, select from a given list the first aid measures used for widow spiders (lactordectus species) bites, in accordance with the references. (FMST.07.38c)

4) Without the aid of references, select from a given list the first aid measures used for brown recluse spider (loxosceles species) bites, in accordance with the references. (FMST.07.38d)

5) Without the aid of references, select from a given list choose the first aid measures used for bee, wasp, and hornet stings, in accordance with the references. (FMST.07.38e)

12

OUTLINE.

1. **PIT VIPERS**. Crotalidae have a triangular head, cat-like vertical pupils, hinged fangs and a heat-sensitive "pit" on each side of the head between the tip of the nose and the eye. Rattlesnakes have a variable number of rattles depending upon age and number of molts. They sometimes strike without rattling. About 60% of this country's venomous bites are attributed to rattlesnakes. Cottonmouths (water moccasins) and copperheads are the other two commonly encountered North American pit vipers. Copperhead and cottonmouth venom are quite similar but weaker than most rattlesnake venom.

a. Anatomy of Pit Vipers.

(1) Venom apparatus: Venom gland; Compressor glandular muscle; Primary duct; Accessory gland; Secondary duct; Fang sheath; and fangs.

(2) Facial pit organs: Highly sensitive paired receptors of infrared radiation that can detect temperature changes of less than 0.2°C; used to detect warm-blooded prey/predators and aim strike; also pit organs may have a role in determining volume of venom injected.

(3) Rattles or buttons: Interlocking keratin rings at tip of tail in rattlesnakes; a new rattle is added with each molting cycle (every 50-400 days); some are lost due to trauma.

(4) The strike: Aimed primarily by facial pits; rarely strike farther than 1/2 their body length; speed - approximately 8 feet/second.

(a) Rattlesnakes generally inject 25-75% of their venom when they bite humans.

(b) Can bite without injecting any venom ("dry bite").

1. 20% - no envenomation.
2. 20-30% - mild envenomation.
3. 30-50% - moderate to severe envenomation.

b. Signs and Symptoms of Snake Venom Poisoning:

(1) Pit Vipers:

(a) Local:

1. Puncture wounds/scratches:

A. Pattern can be misleading (venomous vs. nonvenomous).

B. Must differentiate from other animal/insect bites or plant puncture wounds- usually venomous snakebite wounds are larger, with more bleeding.

C. May be one or many fang marks present; range from a few mm to 4cm apart.

2. Pain, edema, erythema:

A. Usually presents within 30 minutes if envenomation.

B. More severe envenomation generally yields more rapid progression and more severe pain.

C. Pain is usually characterized as burning and immediate in onset.

D. Pain usually confined to bite site.

E. Edema limited to subcutaneous tissues (with no increase in intra-compartmental pressures).

F. Usually starts within 5 minutes (if none in 10 minutes, probably no significant envenomation unless dealing with Mojave rattlesnake).

G. Gradually spreads up extremity for 36 hours.

H. Danger of airway obstruction in bites to head and/or neck.

3. Ecchymosis:

A. Ecchymosis starts within several hours.

B. May involve entire extremity.

4. Lymphangitis:

A. Venom absorbed rapidly through lymphatic system results in lymphatic inflammation.

5. Petechiae, vesicles, hemorrhagic bullae:

A. Will occur in untreated rattlesnake envenomation.

B. Rarely seen when treated early with adequate antivenin.

C. Usually occurs 6-36 hr after bite.

6. Necrosis, tissue destruction:

A. Due to direct venom effects.

B, Can be prevented if adequate antivenin is given within 2 hours of bite.

(b) Systemic:

1. Nausea, vomiting:

A. Common (early onset may indicate severe envenomation).

2. Weakness, diaphoresis, chills, dizziness/vertigo.

A. Syncope is common with all pit viper bites.

B. Frequency is proportional to severity of bite.

3. Change in taste:

A. May complain within minutes of rubbery or metallic taste.

4. Increased salivation.

5. Fever.

6. Tingling, numbness in scalp/face/fingers/toes:

A. Can occur within 10 minutes and indicates moderate to severe envenomation.

7. Fasciculations in face/neck/back/ or other involved extremity:

A. Can occur early and indicates severe envenomation.

8. Visual disturbances:

A. Blurred vision, yellowing of vision and blindness.

9. Tachycardia, bradycardia.

10. Hemorrhage, hemolysis, thrombosis and DIC:

A. Bleeding from wounds, gingival bleeding, epistaxis, Hematuria, hematemesis, Melena, lower GI bleeding, hemoptysis, peritoneal hemorrhage and cerebral hemorrhage.

B. Can occur as early as 6 hours; more common 12-72 hours after bite.

C. Systemic bleeding only occurs in moderate to severe envenomation.

D. Most coagulopathic effects secondary to proteases acting at various sites in the coagulation cascade.

11. Hypotension, shock:

A. Can occur rapidly in severe envenomation.

B. Early, due to pooling of blood.

C. Late may be due to loss of volume.

D. Generally little effect from decreased cardiac contractility.

12. Pulmonary edema:

A. Common in severe envenomation.

B. Secondary to toxin and inflammatory injury of pulmonary capillary membranes and pooling of blood in major vessels and capillary beds of the lung.

C. Compounded by direct cardiodepressant factor in some venoms.

13. Oliguria, anuria:

A. Hypotension with resultant decreased GFR. (#1).

B. Hemoglobinuria and myoglobinuria with renal tubular obstruction.

C. Direct toxic effect of venom on kidneys.

D. Acute tubular necrosis from rhabdomyolysis.

14. Paresis, Paralysis:

A. Seen with Eastern Diamondback and Mojave rattlesnakes.

B. Uncommon with other rattlesnakes.

15. Respiratory failure:

A. Uncommon with pit viper bites (except Mojave).

B. Complicated by cardiovascular failure.

16. Seizures:

A. Probably due to hypertension and hypoxia.

17. Coma:

A. Usually sensorium is clear.

B. When coma occurs, it is secondary to cerebral anoxia.

18. Death:

A. Generally occurs in 6 to 48 hours in untreated severe envenomation.

B. Can occur sooner with IV envenomation.

c. First Aid Measures for Pit Viper Envenomation. (FMST.07.38a)

(1) Do not try to kill the snake and bring it in for identification.

(a) Only vital to identify the snake when Mojave rattlesnake or coral snake bites are a possibility, as management is altered.

(b) Risk of a second bite to the victim or rescuer.

(c) Should never delay transportation of the victim.

(2) Remove any jewelry, which could become a tourniquet with progression of edema.

(3) Calm and reassure patient to decrease heart rate and circulation of venom.

(4) Constriction band.

(a) Most widely accepted first aid measure in the literature.

(b) Should be more than 1/2 inch wide.

(c) Apply to proximal end of extremity affected and wrap to approximately 5 cm above swelling (site of envenomation). Then wrap back towards proximal end of extremity from where the wrap started.

(d) Apply only tight enough to occlude lymphatic and superficial venous return.

(e) Probably no benefit when applied 30 min or more after bite.

(f) Maintain until antivenin is started.

(5) Incision and suction.

(a) Very controversial; should only be used by experienced medical personnel.

(b) Do not incise wound.

(c) Suction best applied by mechanical means (Sawyer extractor).

(d) Do not use mouth (unknown oral lesions may expose rescuer to risk).

(e) Should be started within 30 minutes of envenomation.

(f) Continue suction for 30 to 60 min.

(6) Rest and immobilization.

(a) Put victim at rest and splint extremity in position of function, at or just below heart level.

(b) Allow room for swelling in splint.

(7) Ice.

(a) Avoid any method of cooling.

(8) Watch for adverse reactions.

(a) Rare documented cases of anaphylaxis caused by snake venom in patients previously bitten or otherwise exposed to snake venom.

(b) Some evidence that venom can activate the alternate-complement system directly and cause a similar reaction.

(9) Rapid transport to a medical facility.

(a) The most important measure in first aid.

(b) No measures should delay this; antivenin is the only proven definitive therapy.

2. **CORAL SNAKES**. The elapidae species of coral snakes are brightly colored, with black noses and alternating red-yellow-red-black bands around their bodies (remember "red on yellow, kill a fellow"). They have relatively small mouths with fixed fangs. From southern Mexico through tropical South America, the rules for distinguishing coral snakes are highly unreliable. Unless you are a knowledgeable herpetologist, it is best not to pick up colorful snakes in the tropics.

a. Signs and Symptoms of Coral Snake Envenomation:

(1) Fang marks may be hard to see.

(2) Frequent delay in onset of symptoms, followed by extremely rapid progression.

(3) Little or no pain at bite site.

(4) No local necrosis.

(5) Earliest evidence may be drowsiness or euphoria.

(6) N/V, increased salivation, paraesthesias at bite site.

(7) Bulbar paralysis progressing to peripheral paralysis.

(8) Paraesthesias, and fasciculation.

(9) Occasional seizures.

(10) Mild Hypotension.

(11) Death is due to respiratory and cardiovascular failure.

b. First Aid Measures For Coral Snake Envenomation. (FMST.07.38b)

(1) None of proven benefit except rest, reassurance and rapid transportation to medical facility for antivenom administration.

3. **GILA MONSTERS**. No direct injection mechanism for their venom, but have powerful jaws that allow them to chew and tear at their victims. Drooling venom produces a substantial amount of pain and possible complications resembling pit viper envenomation. Follow the recommendations for pit viper envenomation. Fatal Gila monster encounters are extremely rare.

4. **VENOMOUS ARTHROPODS**. Arthropods represent the largest phylum in the animal kingdom. It contains spiders, scorpions, insects, ticks, kissing bugs, water

bugs, caterpillars, moths, butterflies, grasshoppers, centipedes, and millipedes, among others. Some arthropods sting (bees, ants, scorpions, etc.), others bite (spiders, centipedes, kissing bugs, etc.), while still others discharge a secretion that is toxic (millipedes, caterpillars, etc.). Although the number of arthropods that are sufficiently venomous to endanger humans is not known, these animals are implicated in far more bites and stings than all other phyla combined. Almost all of the 20,000 species of spiders are venomous, but luckily for, man, only relatively small numbers have fangs long and strong enough to penetrate the human skin. There are some 500 species of scorpions and all are venomous, although only a small number pose a significant danger to humans.

a. The number of deaths from arthropod stings and bites is not known, nor do most countries keep records of the incidence of such injuries. In Mexico, parts of Central and South America, and in North Africa, deaths from scorpion stings may exceed several thousand per year. Spider bites probably account for no more than 200 deaths per year, worldwide. The number of deaths from arthropod bites or stings in the temperate countries is far greater than the number of deaths from snakebite. Almost all of these deaths, however, are from anaphylactic reactions.

5. **SPIDERS**. -There are at least 200 species of spiders that have been implicated in significant bites to humans (Russell, 1988a). Spiders got the misplaced blame for the bites seen in one large series of case studies. Of some 600 suspected spider bites, 80% were found to be caused by other arthropods. The arthropods most frequently involved in these mis-diagnoses were ticks (including their embedded mouth parts), kissing bugs, mites, bedbugs, fleas, flies, beetles, water bugs and various other stinging arthropods.

a. Lactordectus species (Widow spiders): It is most commonly referred to as the Black Widow, Brown Widow or Frog-legged spider depending on the species. There are many other commons names such as: hourglass, poison lady, deadly spider, red-bottom spider, T-spider, Gray Lady spider, or shoe-button spider. Although both male and female widow spiders are venomous, only the latter have fangs large and strong enough to penetrate the human skin. Mature females range in body length from 10 to 18mm, whereas males are from 3 to 5mm.

(1) Signs and Symptoms of spider envenomation.

(a) Sharp, pinprick-like bite, but in some cases the incident is so minor that it goes unnoticed.

(b) Initial bite-pain is often followed by dull-numbing pain in the affected extremity, and cramps in one of the large muscle masses.

(c) Muscle fasciculations can be seen within 30 minutes of the bite.

(d) Sweating is common.

(e) Complaints of weakness and pain in the regional lymph nodes.

(f) Pain in the lower back, thighs, or abdomen is common and rigidity of the abdominal muscles is seen in severe envenomations.

(g) Severe paroxysmal muscle cramps may occur accompanied by arthralgias.

(h) Tissue reaction around the bite may vary, depending on the species involved. In most cases in the United States, there is no reaction at the bite site and the puncture wounds may be impossible to find. There may be an immediate 2 to 4 mm blanched area around the puncture wounds, surrounded by a slightly erythematous area, which in time becomes pallid. This may persist for 20 to 30 minutes.

(i) Hypertension is a characteristic finding in moderate to severe poisonings. It usually appears 2 to 3 hours post envenomation.

(j) Elevation of body temperature.

(k) Increased salivation.

(1) Anorexia.

(2) Treatment for Widow Spider envenomation.
(FMST.07.38c)

(a) First aid measures are supportive in nature. If the pain is intense, ice can be placed over the wound until the patient arrives at a medical facility.

(b) Antivenin is given in severe cases at medical facilities.

b. Loxosceles species (Violin spiders): Are known in the United States as fiddle-back, or brown recluse spiders. There are over 100 species of Loxosceles. The abdomen of these spiders varies in color from gray to orange or reddish-brown to dark brown. The violin on the cephalothorax is brown to blackish and distinct from the pale yellow to reddish-brown background of the cephalothorax. This spider has three pairs of eyes, forming a curved row. Females average 8 to 15 mm in body length whereas males average 6 to 10 mm. Both are venomous.

(1) Signs and Symptoms of Violin spider envenomation.

(a) Pain; about the same degree of a bee sting but patients may be completely unaware of the bite.

(b) Localized burning sensation around the site of injury.

(c) Pruritus is often present and the area begins to appear red, with a small-blanched area around the immediate bite site.

(d) The reddened area enlarges during the subsequent 1- 8 hours. It often becomes irregular in shape and, as time passes, hemorrhages may develop throughout the area.

(e) A small bleb or vesicle forms at the bite site and increases in size.

(f) Vesicle subsequently ruptures and a pustule forms.

(g) The whole area becomes swollen and painful.

(h) Lymphadenopathy may develop.

(i) Necrosis can develop from superficial to deep layers.

(j) Rare systemic signs and symptoms include:

1. Malaise.
2. Stomach cramps.
3. Nausea and vomiting.
4. Jaundice.
5. Spleen enlargement.
6. Hemolysis.
7. Hematuria.
8. Thrombocytopenia.

(2) Treatment for Loxosceles spider envenomation: (FMST.07.38d)

(a) There are no first-aid measures of value. In fact, all first-aid measures should be avoided, because the natural appearance of the lesion is most important in determining the diagnosis.

(b) An ice pack may be placed over the wound if the pain is severe.

(c) Excision of the bite site is no longer recommended as this has been demonstrated to cause worsening of necrosis and tissue damage. Definitive therapy includes use of dapsone, antibiotics, tetanus prophylaxis and possible steroid use.

c. There are many other species of spiders that include Steatoda species (cobweb spiders), Phidippus species (jumping spiders), Chiracanthium species (running spiders), Lycosa species (wolf spiders), and the newest member to the United States is the Hobo spider. There have been several reported deaths in the Northwestern United States. Originally a European species of brown spider, the Hobo spider was introduced into the U.S. in 1936, and little research has been done on it to date.

6. **SCORPIONS**. Approximately 75 of the 650 species of scorpions can be considered of sufficient importance to warrant medical attention. In the United States, members of the genera Hadrurus and Vejovis are capable of inflicting a painful and erythematous, but non-lethal lesion. However, species of the genus Centruroides are sufficiently dangerous to warrant definitive medical care. Most stings are inflicted by the Vejovis species.

a. Centruroides species. There are approximately 30 species of this genus confined to the New World. Of these, eight are of considerable medical importance, and most of these species are found in Mexico. In the United States, they are commonly referred to as "bark scorpions", because of the preference for hiding under the loose bark of trees, in dead trees, or in logs. Their general color is straw to yellowish-brown or reddish-brown, and they are often easily distinguishable from other scorpions in the same habitat by their long thin tail with its "telson" on the end, the pedipalps, or pincer-like claws. They can reach lengths up to 55 mm.

(1) Signs and symptoms of envenomation.

(a) Initial pain, rarely severe.

(b) Area becomes sensitive to touch (tap test).

(c) Tachycardia may ensue after one hour.

(d) Respiratory rate increases.

(e) Fasciculations may be seen over the face or large muscle masses.

(f) Complaints of generalized muscle weakness.

(g) Respiratory distress may progress to paralysis.

(h) Slurring of speech may be present.

(i) Convulsions have been reported.

(2) Treatment for scorpion envenomation.

(a) There is no first-aid measure of value for scorpion stings.

(b) Ice may be placed over the wound to reduce pain.

(c) In severe cases, antivenin may be recommended.

b. Vejovis species is closely related to the Urectonus. Both genera are ground scorpions. Whereas Urectonus usually inhabits mountain habitats from southern California to Oregon, Bejovis has a wide distribution from the southern portions of Canada, south through Wyoming and Colorado to Texas and west to California. Some species measure up to 85mm, but most are from 30 to 35mm. Vejovis spinigerus, a common western U.S. species, is often called the striped-tail scorpion.

c. Hadrurus species is native to North America. These are the longest and most stout-bodied of our scorpions, generally referred to as the giant hairy scorpions because of their size and the conspicuous bristles on their legs, pedipalps and caudal segments. Adults may can measure up to 135 mm. These are burrowing species and may be found as deep as 2 feet in sandy soil. They are native to Arizona, California and parts of Utah, Nevada and Idaho, as well as Mexico.

7. **BEES, WASPS, HORNETS AND ANTS**. While it may take over 100 bees to inflict a lethal dose of venom in most adults, one sting can cause a fatal anaphylactic reaction in a hypersensitive person. There are 3 to 4 times more deaths in the United States from bees than from snakebites. In the few fatalities that have resulted from multiple bee stings, death has been attributed to acute cardiovascular collapse.

a. Treatment for bee stings: (FMST.07.38e)

(1) The stingers of many bees may remain in the skin and should be removed by teasing or scraping rather than pulling, to avoid even more venom being injected into the patient.

(2) Ice may be applied to the sting site.

(3) Persons with known hypersensitivity to stings should carry a Bee-Sting Kit, containing an antihistamine and epinephrine when in endemic areas.

(4) Desensitization can be carried out using whole-body antigens or, preferably, whole-venom antigens.

8. **TICKS AND MITES**. Carry many diseases. In North America, some species of Dermacentor and Amblyomma cause tick paralysis.

a. Signs and symptoms of Tick Paralysis.

(1) Restlessness and irritability.

(2) Lethargy and anorexia.

(3) Generalized muscle weakness.

(4) Incoordination and ataxia.

(5) Nystagmus.

(6) Ascending flaccid paralysis with loss of deep tendon reflexes.

(7) Bulbar or respiratory paralysis may develop.

b. Treatment for Tick Paralysis:

(1) Remove the tick.

(2) Treat symptomatically (ventilator support may be necessary).

(3) Lesions should be cleaned.

(4) Antivenin and/or corticosteroids are used in severe cases.

9. **LEPIDOPTERA (CATERPILLARS)**. Caterpillars commonly cause envenomation in humans. There are at least ten families of caterpillars that are venomous.

a. MEGALOPYGE OPERCULARIS (Puss Caterpillar or Woolly Slug) The most commonly seen venomous caterpillar in the United States.

(1) Description: Hairy, flat, ovoid shaped: 30-35 millimeters in length.

(2) Location: Texas north to Maryland and Missouri.

(3) Stinging apparatus: Consists of spines intermingled with the hairs of the body:

b. SIGNS AND SYMPTOMS.

(1) Needling pain (Can be intense).

(2) Redness.

(3) Swelling.

(4) Nausea.

(5) Headache.

(6) Fever.

(7) Vomiting.

(8) Lymphadenopathy.

(9) Hypotension.

(10) Shock.

(11) Bleeding of the mucus membranes.

a. Treatment

(1) Apply contact tape to affected area, and remove tape. This will effectively remove spines from affected area.

(2) Group I corticosteroid cream and ointment.

(3) Codeine or Demerol may be required for pain.

(4) Antiemitics may be needed to control vomiting.

Note: Anaphylactic reactions have been reported, but are rare.

UNITED STATES MARINE CORPS
Mountain Warfare Training Center
Bridgeport, California 93517-5001

FMST.07.10

04/02/02

STUDENT HANDOUT

FREEZING AND NEAR FREEZING TISSUE INJURIES

TERMINAL LEARNING OBJECTIVE: Given a casualty in a cold weather environment, and necessary equipment and supplies manage common cold weather injuries in a cold weather environment to prevent death or further injury per the reference. (FMST.07.10)

ENABLING LEARNING OBJECTIVES.

1) Without the aid of references, from a given list select the correct definition of frostbite, in accordance with the references. (FMST.07.10a)

2) Without the aid of references, from a given list select the correct two mechanisms of injury for frostbite, in accordance with the references. (FMST.07.10b)

3) Without the aid of references, from a given list select the correct signs and symptoms of frostbite, in accordance with the references. (FMST.07.10c)

4) Without the aid of references, from a given list select the correct favorable post thaw signs of a frostbite injury, in accordance with the references. (FMST.07.10d)

5) Without the aid of references, from a given list select the correct unfavorable post thaw signs of a frostbite injury, in accordance with the references. (FMST.07.10e)

6) Without the aid of references, from a given list select the correct field treatment for frostbite, in accordance with the references. (FMST.07.10f)

12

7) Without the aid of references, from a given list select the correct primary field treatment consideration when treating frostbite in the field, in accordance with the references. (FMST.07.10g)

8) Without the aid of references, from a given list select the correct four methods to reduce the chance of becoming a frostbite casualty, in accordance with the references. (FMST.07.10h)

9) Without the aid of references, from a given list select the correct definition of immersion foot, in accordance with the references. (FMST.07.10i)

10) Without the aid of references, from a given list select the correct mechanism of injury for immersion foot, in accordance with the references. (FMST.07.10j)

11) Without the aid of references, given a list of categories and characteristics match the category of immersion foot to its characteristics, in accordance with the references. (FMST.07.10k)

12) Without the aid of references, from a given list select the correct field treatment of immersion foot, in accordance with the references. (FMST.07.10l)

OUTLINE.

1. **FROSTBITE**

a. Definition: Frostbite is the actual freezing of tissue. (FMST.07.10a)

(1) Frostbite usually occurs as a result of exposure to subfreezing temperatures for a long period of time. However, brief exposure to high wind-chill or very cold temperatures can also cause frostbite. Frostbite usually occurs after more than twelve hours of exposure to subfreezing temperatures. Skin temperature of the extremities is dependent on heat production and is more a function of underlying muscle mass than of circulation. Thus, the areas of the body most prone to frostbite (hands, feet, ears, nose, chin and molar regions) have small amounts of underlying muscle tissue.

(2) Incidence. Frostbite, unlike hypothermia tends to spare the extremes of age. Thus, it is most prevalent in the middle years of life and like hypothermia, the greatest incidence occurs in urban areas. Other factors predisposing a person to frostbite include: poor nutritional status, dehydration, prior cold injury, individuals from the southeastern part of the United States, diabetics, individuals with atherosclerotic disease, and use of tobacco products.

(3) Although the actual numbers are not known, persons born in the south are 3.7 times more susceptible to frostbite. Persons with a history of prior cold injury are 1.7 times more prone to have a second injury. This includes parts of the body other than the originally frostbitten region.

b. Mechanisms of Tissue Injury. Tissue injury resulting from frostbite is produced in two ways: (FMST.07.10b)

(1) *Actual freezing of the tissues.* Ice crystal formation in the extra-cellular space leads to an osmotic gradient across the cell membrane drawing water out of the cell. This process, known as crenation, causes a marked change in intra-cellular electrolyte and enzyme concentrations, which can lead to cell death or recovery depending on the duration of exposure to freezing conditions and post thaw care of the tissue. Intra-cellular ice crystal formation is much more destructive, however, this requires much faster freezing rates.

(2) *Obstruction of blood supply to the tissues.* Cell injury and death caused by ice crystal formation leads to the release of inflammatory mediators, which damage vascular endothelial cells. The net result is increased vascular permeability. Fluid leaks out of the vascular space and into the intercellular space causing tissue edema and increased blood viscosity. Elevated viscosity combined with platelet aggregation, to the walls of damaged blood vessels, lead to vascular obstruction with resultant tissue damage.

c. Signs/Symptoms. (FMST.07.10c) Initially there is discomfort and pain in the affected tissues, which can progress to loss of sensation. The skin appears white and bloodless. There is decreased or no capillary refill. The skin may feel waxy and will not glide freely over a bony prominence. With minor injury, blisters form, usually in the first 24 hours, and edema may also be present. Deep injury is not associated with substantial formation of blisters. There is also little edema seen in deep injury.

(1) Favorable prognostic signs of frostbitten tissue. These signs indicate probably little or no tissue loss, post thaw. (FMST.07.10d)

(A) Warmth of tissue.

(B) Normal tissue color.

(C) Preservation of sensation.

(D) Blister formation. *A* blister extending the full distal length of the injured digit is a good prognostic sign. However, a blister limited to the line of demarcation between normal and frostbitten tissue is a less favorable sign.

(E) Edema persisting more than 24 hours.

(F) Rapid capillary refill.

(2) Unfavorable prognostic signs. These indicate a poor prognosis with probable tissue loss, post thaw: (FMST.07.10e)

(A) Complete absence of edema.

(B) Cyanotic tissue.

(C) Continued loss of sensation.

(D) Affected areas continue to stay cold.

(3) Since it is extremely difficult to predict ultimate tissue loss from initial examination, any classification must be done retrospectively. Classically, frostbite is divided into four degrees of severity:

(A) First degree - Erythema, edema, transient tingling and/or burning.

(B) Second degree - Blisters, edema, anesthesia and/or paresthesias.

(C) Third degree - Involves the entire thickness of skin and extends into subcutaneous tissue.

(D) Fourth degree - Involves the entire thickness of the skin along with muscle and bone.

(F) Frostnip is a completely reversible injury that is technically not considered frostbite. Frostnip classically presents as a white patch and numbness and is readily warmed in the field by placing the affected extremity in one's axilla or groin. No resultant skin changes occur.

d. Field Management (FMST.07.10f)

(1) The correct treatment for frostbite is as follows:

(A) Once an extremity has suffered frostbite, the primary consideration of treatment is not to rewarm the region if there is a chance of refreezing. (FMST.07.10g) This occurs because the amount of tissue damage is greatly increased if the part is thawed then allowed to re-freeze. Generally, the sooner a frostbitten region is thawed the lower the probability and extent of damage. Rapid rewarming is the preferred method of thawing a frostbitten region. This consists of immersion of an affected part in circulating warm water of 104°-108°F until the tissues are soft and pliable, which usually takes about 40-60 minutes.

(B) If slow rewarming methods (such as allowing a part to sit at room temperature) is used, some refreezing of the melted fluid occurs and the new ice crystals are actually larger than those formed during the original freezing causing further tissue destruction.

(C) Out dated methods such as rubbing with snow or rewarming next to an open fire, etc., only enhance tissue damage.

(D) Once thawing has taken place no benefit of continued rapid rewarming can be demonstrated.

(E) Cold injuries that are old and have thawed spontaneously should not be rapidly re-warmed again.

(F) Nicotine use should be prohibited since it is a vasoconstrictor and may enhance tissue loss.

(G) There is no need for ointments, salves, etc.

(H) Blisters should be allowed to remain intact.

(I) Medevac all cases of frostbite, as soon as the tactical situation permits.

(J) One final note; in military tactical situations such as ours, the victim often must walk down below the snow line to reach a BAS or base camp where rapid rewarming can be carried out. In this situation let the victim walk if he has frostbitten feet. Once his feet have thawed, he will be unable to walk and will have to be carried out.

(K) Hospital and long-term care: After re-warming the frostbitten tissue, attention must turn to prevention of infection and functional restoration. The patient's tetanus status should be determined and meticulous care should be taken to keep the injury clean. Daily whirlpool therapy for 30-45 minutes has been shown to be beneficial. The use of post-thaw medications is controversial but there is evidence that topical aloe vera and NSAIDS to control inflammation are probably beneficial. Some studies have demonstrated benefit from the use of allopurinol as a free radical scavenger. Also, animal studies have shown a definite benefit from the use of 90% DMSO gel, however, this use is not FDA approved on humans and it may be difficult or impossible to obtain this medication. The wide scale use of heparin and dextran 40 has been suggested but a definitive recommendation that supports their use still cannot be made. Studies suggesting the use of thrombolytics are very small and not well controlled therefore there is no current recommendation for use of thrombolytics. Lastly, it has been shown that there is a clear benefit for the use of early and aggressive physical therapy to preserve function and prevent contracture.

e. Prevention (FMST.07.10h)

(1) Some general principles for the prevention of frostbite are listed below

(a) Dress to keep comfortably cool. The head radiates 30-50% or more of all body heat and so goes the old adage, "If your feet are cold put on your hat".

(b) Nutrition. Eat the right type and quantity of foods.

(c) Avoid tight fitting and constrictive clothing. This decreases blood flow and increases the chance of frostbite. Rings, jewelry, etc., conduct heat away from the skin and increase the chance of frostbite.

(d) Wear mittens instead of gloves. They are much warmer. For work involving a high level of dexterity use contact gloves.

(e) Do not touch metals with bare hands at extremely low temperatures. Your flesh will stick to the surface and this will result in subsequent frostbite and tissue loss. If you find yourself in such a predicament pour a warm fluid over the metal to thaw your flesh off the metal.

(f) Avoid nicotine.

2. IMMERSION FOOT.

a. Definition. (FMST.07.10i) Immersion foot (AKA trenchfoot) is a non-freezing injury of the extremities in which the tissues are damaged.

b. Usually a disease of the feet (but it can occur in the hands), it occurs when the extremity is exposed to a cold-wet environment (external temperatures above freezing and less than

$50^{o}F$) for long periods of time (usually greater than 12 hours). Prolonged standing enhances the pathologic potential for injury due to wet and cold conditions.

c. Mechanism of Injury. (FMST.07.10j) Even though this injury is called immersion foot, actual immersion of the feet in water is not necessary. This injury can also occur in the hands.

(1) The circulation of blood to the affected extremity is reduced because the victim's extremity is cold, wet and the peripheral blood vessels have constricted to reduce overall heat loss.

(2) Because prolonged cooling damages sensitive tissues, nerve injuries are also involved. This nerve damage is responsible for the severe pain and paresthesias. Damaged skin and subcutaneous tissues result in erythema, edema and possible blister formation.

d. Signs/Symptoms Early signs include numbness and paresthesias followed by complete numbness and sometimes leg/arm cramps. The skin at first appears erythematous but becomes progressively pale and mottled then grayish-blue as the condition progresses. Often there is a large amount of edema and blister formation. In severe cases the arterial pulses are absent. Immersion foot is classified by severity: (FMST.07.10k)

CATEGORY	CHARACTERISTICS
Minimal	Reddening of skin; slight sensory changes
Mild	Edema; sensory changes (reversible)
Moderate	Edema, redness, blisters, intra-cutaneous hemorrhage; irreversible nerve damage
Severe	Severe edema, massive intra-cutaneous hemorrhage; necrosis, gangrene

e. Field Management (FMST.07.10l) Treatment consists of:

(1) Pat drying of extremity.

(2) Gentle rewarming.

(3) Elevation of affected extremity.

(4) Bed rest.

(5) Treatment after this is supportive. For more severe injuries, debridement or resection of a portion of the foot may be necessary. However, this is best accomplished at a higher echelon of care. Do not be fooled by this injury. These patients can get

extremely ill from wet-gangrene which may necessitate amputation of tissue. However, the initial treatment should be as conservative as possible. DO NOT TREAT THESE PATIENT'S WITH WATER IMMERSION THERAPY. If infection is present obtain surgical consult early, and plan on small sequential amputations.

3. **CHILBLAINS (PERNIO)**.

a. Definition. Chilblains are a non-freezing injury that leads to red inflamed, pruritic skin.

b. Mechanism of Injury. Pernio is caused by intermittent exposure to above freezing temperatures and high humidity. (More extreme cold is usually associated with lower humidity). Classically, chilblains occur in young individuals and most commonly on the anterior surface of the legs and the dorsum of the hands. The cause of pernio is not clearly known, but it is thought to reflect an abnormal vascular response to the environment.

c. Signs/Symptoms. Skin is red, tender, and pruritic, edematous and blanches with pressure. Itching may be severe and hyperhydrosis can occur.

d. Management. The field treatment is massage and gentle heat. This treatment is usually self-limiting and will not recur without re-exposure. Topical steroids are effective for itching and burning.

4. **CORNEAL FROSTBITE**

a. Definition. Actual freezing of the cornea.

b. Mechanism of injury. This occurs when large amounts of cold air pass over and cool the cornea, resulting in corneal clouding and loss of vision. This can occur as the result of even short periods of exposure to activities associated with high wind chill i.e. snowmobiling.

c. Signs/Symptoms. On examination the cornea will be clouded over.

d. Treatment and Prevention. Antibiotic ointment and DO NOT PATCH THE EYES. Prevention is the key.

UNITED STATES MARINE CORPS
Mountain Warfare Training
Center Bridgeport, California
93517-5001

FMST.07.22
04/03/02

STUDENT HANDOUT

LAND NAVIGATION REVIEW

TERMINAL LEARNING OBJECTIVE In a mountainous environment, navigate in mountainous terrain, in accordance with the references. (FMST.07.22)

ENABLING LEARNING OBJECTIVES

1) Given a map and compass, orientate the map, in accordance with the references. (FMST.07.22a)

2) Given a map and a protractor locate a six-digit grid coordinate, in accordance with the references. (FMST.07.22b)

3) Given a map, compass, and protractor perform a resection to locate an unknown point from three known points, in accordance with the references. (FMST.07.22c)

4) Without the aid of references, select from a given list the correct definition for true north, grid north, and magnetic north, in accordance with the references. (FMST.07.22d)

OUTLINE.

1. **THE MAP**. Before you can properly use a map, there is some basic information you will need to understand.

a. Definition. A map is a reduced or scale drawing of the ground and important things on the ground as seen from the air. Essentially it is a picture of the surface of the earth, as it would appear looking at it from an aircraft.

b. Marginal Information. As you take a quick look at the borders, you will notice that there is a large amount of information. When using a map, the instructions are placed around the outer edges of the map and are known as marginal information. All maps are not the same, some having more marginal information than others, and the information being located in different places around the border. It is important that you know how to use this information if you plan to get the most out of your map. The different information on the margins is as follows:

(1) Sheet name. The sheet name is found in two places: the center of the upper margin and lower right margin of the map. The map is usually named after it's outstanding cultural or geographical features. Whenever possible, the name of the largest city is used.

(2) The sheet number. The sheet number is found in the upper right margin of your map and is used as a reference number for the map sheet.

(3) Series name and scale.

(a) The series name is found in the upper left margin of your map. A map series usually comprises a group of similar maps on the same scale and on the same sheet lines or format designed to cover a particular geographical area. It may also be a group of maps designed to serve a common purpose. The name given a series is of the most prominent area.

(b) The scale notation represents the ratio of the maps distance to the corresponding (actual) distance on the earth's surface. For example, the scale notation of 1:50,000 indicates that one unit of measure on the map equals 50,000 of the same units of measure on the actual surface of the ground (1 map inch = 50,000 earth inches).

(4) Series number. The series number appears in the upper right margin and lower left margin. It is a comprehensive reference expressed either as a four digit number or as a letter followed by three or four numbers.

(5) Index to adjoining sheets. The index to adjoining sheets is found in the lower right margin of your map.

(6) Bar scales. The bar scale is located in the center lower margin of your map.

(7) Contour interval. The contour interval is located in the center lower margin of your map, just below the bar scales. It states the vertical distance between adjacent contour lines on the map. For example, every contour line represents an increase in elevation as indicated by the distance of the contour interval.

(8) Northern References. There are 3 different northern references located on the declination diagram. It is located in the right portion of the lower margin of your map.

(a) True North. A line from any position on the earth's surface to the North Pole, which is the exact top of the earth.

(b) Grid North. The north direction established by the vertical grid lines on the map.

(c) Magnetic North. The direction to the North Magnetic Pole.

(9) Legend. Is located in the lower left margin. It illustrates and identifies some of the symbols on the map. There is not enough room on a map sheet to show the true outlines of objects and land features so the mapmaker uses a set of standard symbols to represent them. Sometimes the symbols are not the same, or standard types of symbols typically used. This is because of the type of map, the scale, or the origin. Because of this possible difference, to prevent errors in symbol identification, the legend should be referred to every time a map is used. (The example on the legend may state, "unimproved road, hard surface road, jeep trail," etc. The right side of the symbol will have three symbols; the first being "unimproved road", the second "hard surface road", and the third being "jeep trail".)

(10) Map colors. The colors you see on your map are not there to make the map pretty or enjoyable to the user. They have important meaning. The colors vary with different types of maps, but on a standard, large scale map, there are five basic colors:

(a) Black. Represents most of the man-made features.

(b) Red. Represents major roads.

(c) Blue. Represents water.

(d) Green. Represents vegetation.

(e) Brown. Represents all terrain features, such as hills, draws, fingers, depressions, and saddles by the use of contour lines. These contour lines are shown in brown.

(f) Other Colors. Occasionally other colors are used to show special information. These, as a rule, will be indicated in the marginal information. Currently, there are changes being made to the color system, which may result in changes of the marginal information in regards to map colors.

2. **HORIZONTAL SCALES**.

a. Bar Scale. This special ruler, located at the bottom of the map sheet, enables you to measure a distance between 2 points on your map and change that map distance into the actual ground distance.

(1) The primary scale. The primary scale is the part of the scale that starts at the zero (0) and extends to the right.

(2) The extension scale. The second part of the bar scale is the extension scale. It starts at the zero (0) and extends left.

b. Measuring Straight-Line Distance. If you are given a mission that requires you to move from your present position to a different position some distance away and you plan to travel in a straight line between these two points, the following method is used to determine the distance you will have to travel:

(1) First locate your present position and your destination on your map. Then you will lay a straight edged piece of paper on the map so that the edge of the paper touches the location of your present position and the location of your destination. Make a tick mark on the edge of the paper at each point. Now move the paper down to the bar scale. Place the left tick mark on the zero (0) of the scale so that the other tick mark will fall somewhere on the primary scale.

c. Measuring a Curved Line Distance. If you are given a mission that requires you to move from your present position to a different position and this time you choose to move along a road to reach your destination, the following method is used to work out the ground distance you will have to travel:

(1) Find your present position, your destination, and the road you will follow, on your map. Position a straight edged piece of paper at your location and make a tick mark,. Now, rotate the paper so that it runs along the road until the road turns, and make a tick mark on both the paper and the map. Keeping these 2 tick marks together, rotate the paper again until the road turns, and again mark the paper and the map. Continue this procedure until you have reached your destination. Now using your bar scale as previously stated, find the actual ground distance your squad will be traveling.

3. **ELEVATION AND RELIEF**. In order to begin learning the subject of elevation and relief, let's first look at some definitions.

a. Contour lines. Of the several ways to indicate elevation and relief on maps, the most common is through the use of contour lines. A contour line represents an imaginary line on the ground along which all points on this line are the same elevation above sea level.

b. Characteristics of Contour Lines. There are a few characteristics of a contour line that you should know that will help you understand what they represent on your map. These characteristics are as follows:

(1) Contour lines never cross.

(2) Contour lines connect on themselves.

c. Types of Contour Lines. In addition to knowing the definition, purpose, and characteristics of a contour line, you need to realize that there are different types of contour lines.

(1) Index contour lines. Starting at zero elevation, every 5th contour line is drawn with a heavier line.

(2) Intermediate contour lines. These are the contour lines falling between the Index contour lines.

(3) Supplementary contour lines. These lines are shown as dashed brown lines.

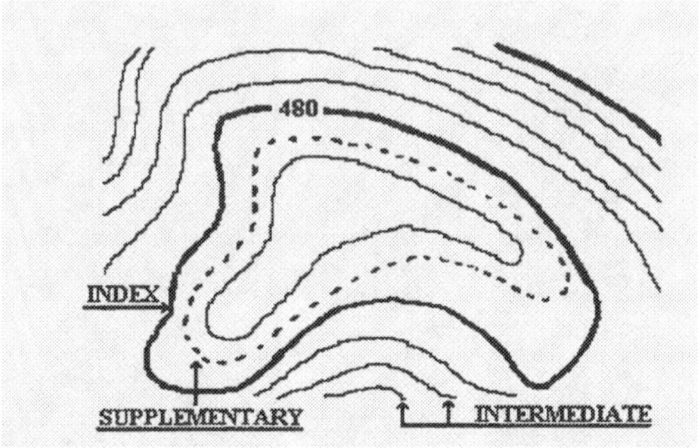

d. Determining Elevation of a Point. Now, you are ready to learn how to determine the elevation of any point of the ground represented on your map.

(1) Given elevation nearest the point, examine the area around your point until you find the nearest point that has a marked elevation. There are a few different ways that contour lines are marked, and they are:

(a) Labeled index contour line. Index contour lines are broken with the elevation of the line printed in the break.

(b) Spot elevations. At various places on your map, usually on prominent landforms such as road junctions or hilltops, a spot elevation will be given. It will be printed in either brown or black.

(c) Bench marks. Similar to Spot elevations, but more accurate. They are marked on your map by a black "x", with the letters "BM", and the elevation.

(d) A point on a contour line. If a point is located on a contour line, its elevation will be the same as that contour line.

(e) A point between contour lines. If a desired point is less than 1/4 the distance between 2 lines, its elevation is considered to be the same as the closest contour line. If the point is from 1/4 to 3/4 the distance, its elevation is considered to be in the middle of the 2 lines.

(f) A point on top of a hill. To estimate the elevation on the top of an unmarked hill, add 1/2 of the contour interval to the elevation of the highest contour line around the hill.)

4. **TERRAIN FEATURES**. There are many types of terrain features that you must be able to identify. By interpreting the arrangement of contour lines, you should be able to get a mental picture of how these terrain features would appear if you were actually out there on the ground looking at them. Some of the features you need to be able to recognize are as follows:

a. Hills. Defined as a point or small area of high ground. A hill is represented on your map by a number of contour lines each of which circles around and connects onto itself. These rough circles get smaller as they progress toward the top of the hill. You need to carefully compare the arrangement of these contour lines to the actual appearance of the hill.

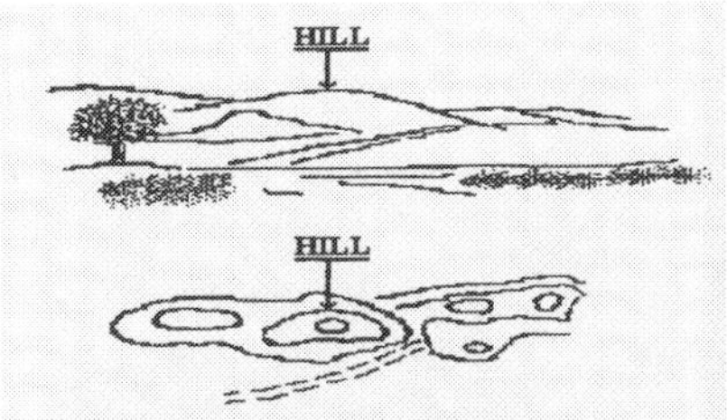

b. Spur. Defined as a line of high ground. The points along the top or crest of the spur are higher than the ground on both sides. A spur is represented on the map by a number of contour lines, each of which form a U-Shaped or V-Shaped design. The end or bottom of the U or V-Shaped lines point down to a lower elevation.

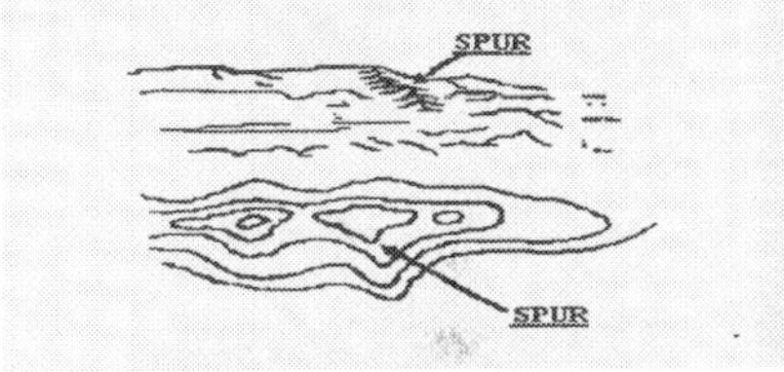

c. Draw. Defined as a line of low ground. The points located in the bottom of a draw are at a lower elevation than the ground to either side. A draw usually runs between 2 spurs. It is sometimes difficult to tell the difference between the contour lines that represent a draw from those that represent a spur because they are both U or V-Shaped lines. Remember that the curved portion representing a spur points down to ground of lower elevation, and the curved portion representing a draw points to ground of higher elevation. Now by looking closely at the area on your map, you will be able to tell which of the 2 you are dealing with. In many cases, draws will contain a stream, while spurs usually never have a stream running down their centers.

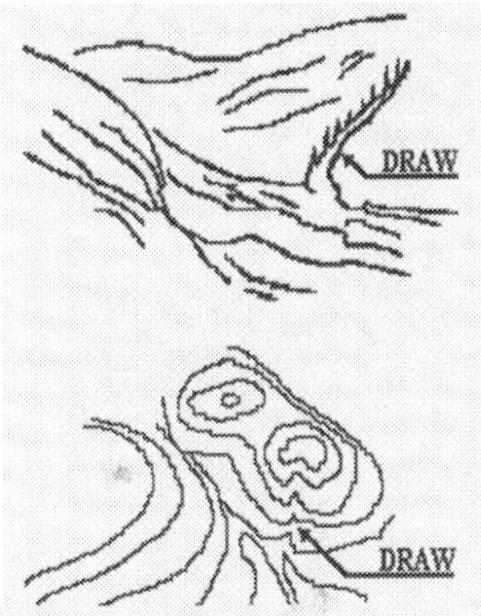

d. Saddle. Defined as the low ground between 2 hilltops, or a dip along the crest of a spur. On your map a saddle is represented by contour lines that form 2 hilltops that are located closely to each other. The area between these 2 hilltops being the location of the saddle.

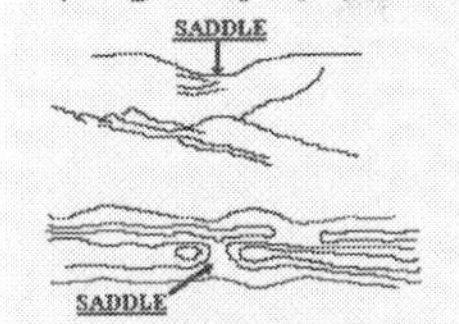

e. Depression. Defined as a point or area of low ground surrounded on all sides by higher ground. A depression is represented by contour lines forming circles within a small area on the map. These rough circles look identical to hilltops except for the presence of hachures. These are short stubby lines that are connected to the contour line at one end and point away from the contour lines at right angles in a downhill direction.

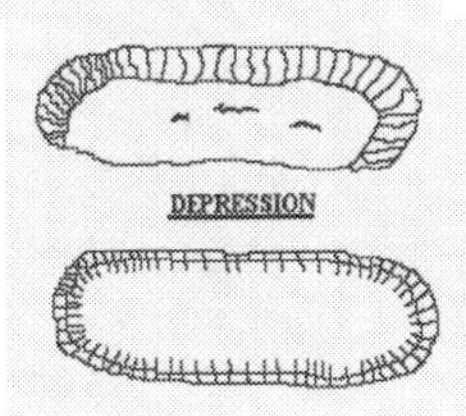

f. Cut. A cut is a man-made feature by which the bed of a road or railroad is leveled by cutting through high ground.

A cut is represented by a sudden ending of contour lines, the ends being connected by a straight line. Hachures are connected to this straight line and point downhill toward the roadbed. This sudden ending of contour lines occurs on both sides of the road bed.

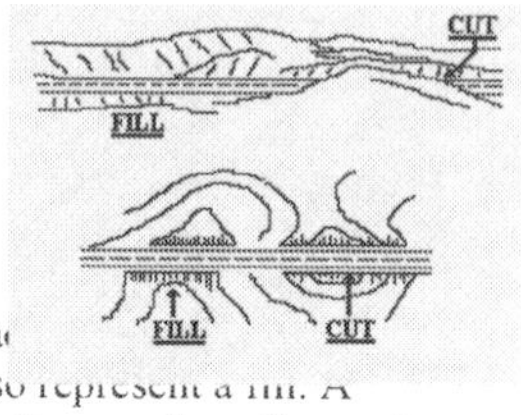

g. Fill. A fill is also a man-made feature by which the bed of a road filling in the low area. A sudden ending of contour lines can also represent a fill. A straight line connects these ends with Hachures pointing downhill away from the road bed. Again, this occurs on both sides of the road bed.

5. **SLOPES.** The rate of rise or fall of a ground form is known as its slope. In combat, this factor of slope must be considered when making plans for an operation. You must have the ability to recognize the ground forms in your area of operation by studying your map and the spacing between the contour lines on your map.

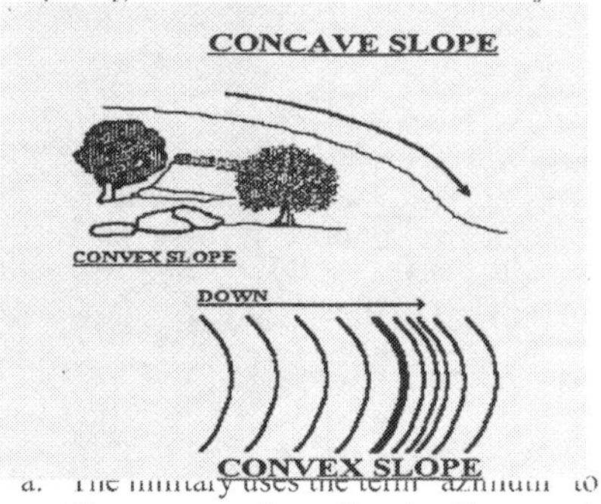

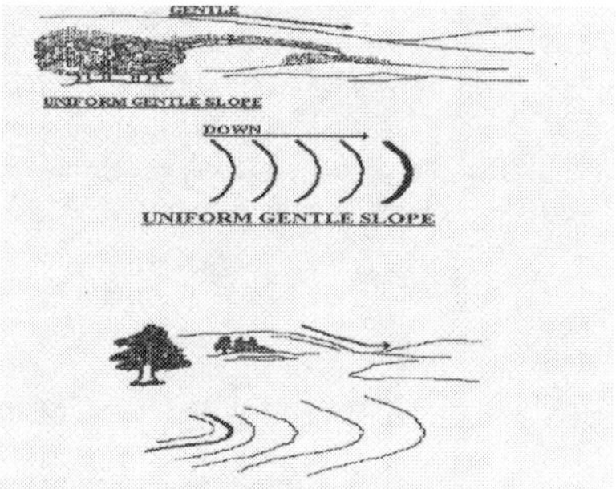

a. The military uses the term azimuth to describe the direction from one point to another. There are two base direction lines used to measure the value of an azimuth. They are as follows:

(1) Grid north. It is used when measuring direction from your map.

(2) Magnetic north. Used when measuring direction from your compass.

b. Units of Measurement. Now that you have learned about the two base direction lines that are used to measure direction, we need to learn the units of measurement that are used to measure the angle from one of these direction lines to the direction line.

(1) Degrees. The degree is the unit of measurement that is used most often. Anytime you need an azimuth to move from one point to another, you will use degrees as your unit of measurement. The base direction line, grid north or magnetic north, will be zero (0) degrees and measuring clockwise from this base direction line, the angle increases until you reach 360 degrees which brings you back around in a full circle to the base direction line.

(2) Mils. The mil is also used as a unit of measurement for direction. It is used when working with indirect fire weapons such as mortars and artillery. The azimuth you send will have to

be in mils, not in degrees. Remember, you are still measuring the angle between the base direction line and the line of sight from you to the distant object only you are using a different unit to measure the angle. Moving clockwise from the base direction line, which would be zero (0) mils, the angle back to the direction line. Now you know there are 360 degrees and also 6400 mils in the circle. This means that each degree equals approximately 17.5 mils.

c. Converting Map and Compass Directions. (WMC.5.2d) Earlier you learned that an azimuth was always measured from a base direction line. You also learned that we can use two different base direction lines. Grid north is the base direction line used when you are working with the map and getting a map direction or a grid azimuth. Magnetic north is the base direction line used when you are using the compass to get a compass direction or a magnetic azimuth. Because these two different base direction lines are used and because inmost cases there is a difference of one or more degrees between them, it is necessary for you to learn how to convert from a grid azimuth to a magnetic azimuth or from a magnetic azimuth to a grid azimuth. In order to learn this procedure we must take another look at the declination diagram.

(1) Declination diagram. A diagram that shows the relation between three different norths true north, grid north and magnetic north. It is the difference between grid north and magnetic north that we need to look at. The difference between these two north's is known as the grid-magnetic angle or G-M angle.

(a) Grid-Magnetic angle. The grid-magnetic angle is the angular difference between grid north and magnetic north. As you can see on this diagram a line that connects grid north and magnetic north represents the G-M angle. The size of the angle is stated off to the left side of the diagram in both degrees and mils. The G-M angle on your maps is 18 degrees or 320 mils. It is important to know that the G-M angle will change depending on where you are located on the earth's surface.

(2) Converting. When you receive a mission as a squad leader to move your squad from your present position to some distant point, you take out our map, a protractor and figure out the map direction or the grid azimuth to that distant point. However, before you can use your compass to direct you to your destination, you will have to convert the grid azimuth you have from your map, to a magnetic azimuth that you can set on your compass.

(a) The first step in making this conversion is to examine the declination diagram on your map. The G-M angle on your map is 18 degrees. You know that you will have to add the G-M angle to, or subtract it from the azimuth you measured on your map, or shot on your compass.

(b) The instructions as to whether you will add or subtract this angle are written on the map depending on whether you are going from grid to magnetic or vise-versa.

d. Orientation of a Map. (WMC.6.14a) In order to gain the best use of your map and to avoid confusion while navigating across terrain, you will need to know how to orient your map to the ground. *A* map is oriented when it is lying flat and level with it's north and south pointing in the same directions as north and south on the ground. There are two methods you can use to orient a map.

(1) Inspection. The first method used to orient the map is by inspection. By carefully examining your map and the ground around the area in which you are located to find linear features such as roads, railroads and powerless, you can align the feature on the map with the same feature on the ground. If there is only one linear terrain feature in the area in which you are located, you will have to be careful not to reverse your directions. Even though you have a road running north-south on your map aligned with the same road on the ground, the north end of your map may be pointed in a southerly direction down the road. To avoid this from occurring, find another terrain feature in the area that you can also identify on the map. Once you have aligned your map along the linear feature, check to insure that the second feature is positioned on the same side of the linear feature on your map as it is on the ground.

(2) Compass. This is the second method of orienting a map. You begin by placing your map on the ground so it lies flat and level. Open your compass so that the cover and eyepiece are completely open. Align the graduated straight edge of the compass along one of the north-south grid lines on your map with the cover end of the compass pointing toward the top of the map. This will place the stationary index line parallel to the grid north. Since the north seeking arrow on the dial of your compass will always point to magnetic north, you set your compass up just like a declination diagram. Grid north being the stationary index line and magnetic north being your north seeking arrow. Now rotate your map, insuring that the graduated straight edge of your compass remains aligned with the north-south grid line, until the face of the compass appears exactly as the declination diagram in the lower margin of the map. Your map is now oriented. For the Sonora Pass map sheet you would rotate the map until the north seeking arrow on the compass lies 16.5 degrees to the right of the stationary index line.

7. **GRID SYSTEM.** (WMC.6.14b) In most combat situations it is very important to be able to pinpoint your location, the location of another unit, or the location of an objective. In order to do this, it is necessary to have a location system that will allow you to accurately determine a specific location and state that location in such a way that will allow others to find that location quickly. This system is known as a grid system.

a. Grid Box. This box gives basic instructions on reading grids in determination of specific points on the map.

b. Grid Lines. The straight black lines that run across, and up and down the surface of your maps form the 1000 meter grid squares. These lines are called grid lines. If you look around the border of your map you will see that both the horizontal and vertical grid lines are identified by a two digit number.

c. Grid Square Identification. Each 1000 meter grid square on your map has it's own unique number. This number is called a grid coordinate. A four digit number is the result of combining the two digit number identification of the vertical grid line and the two digit number identification of the horizontal grid line that cross to form the lower left corner of the grid square. To find the grid coordinate of any grid square on your map you will always read your map right and up. You find the two digit number of that vertical grid line and you will have the first 2 digits of your coordinate. Next, find the two digit number of the horizontal grid line and place it after your vertical number and you now have the 4 digit grid coordinate that identifies your grid square.

d. Locating Terrain Features Within a Grid Square. If you are given a 4 digit grid coordinate to find a certain point on the earth's surface, that grid coordinate would put you within 1000 meters of that point. If you are told to find a certain feature within that grid square and you are good at reading your map, a 4-digit coordinate may get you close enough to your point to enable you to find it. However, there are many situations in which a 4 digit number will not get you close enough to a given position to allow you to accomplish your task. It is necessary to be able to locate points within a grid square and using a six-digit grid coordinate does this.

(1) Six Digit Grid Coordinate. The 1000-meter grid square will have to be broken down further into 100-meter grid squares. Because the surface of your map would become too cluttered by all these additional lines, a special instrument has been made to help you divide up a 1000-meter grid square. This instrument is called a coordinate scale or protractor.

8. **RESECTION**. (WMC.6.14c)

a. You will face a situation where you need to pinpoint the location of your position. This may not be possible by associating the terrain in your area with the map because you may have a poor map or the terrain in your immediate area is flat and has no distinguishing features. When you find yourself in this situation you will need to know how to perform a resection.

(1) Examine the terrain around your position and find two terrain features or objects that you can also identify on your map.

(2) Use your compass to find the magnetic azimuth to one of the distant terrain features you have identified.

(3) Convert the magnetic azimuth you found into a grid azimuth.

(4) Change the grid azimuth you found into a back azimuth.

(5) Drawing a line on your map from a distant point you identified in step one back toward your present position. Place a small mark on your map at this point. Using a straight edge, draw a line from where you plotted the distant object to the mark you made by the edge of your protractor. This line should be extended to insure it continues through the area where you are located.

(6) Repeat, finding the magnetic azimuth to the second distant object. That is at least 60 degrees from the first.

(7) When you have drawn your second line on the map, the point where the second line crosses the first line is the location of your position. The process that has just been covered is called a two-point resection.

(8) If your position is located on a linear line feature such as a road, power line, railroad or river, then it is only necessary to perform a one-point resection in order to pinpoint your location. Instead of identifying two distant points, you will only need to select one. Where the single line that you drew on your map crosses the linear feature on which you are located will be the location of your position.

9. **DETOURING AROUND AN OBSTACLE**.

a. There will be occasions when it will become necessary to move off your plotted route in order to avoid an obstacle such as a small lake or swamp. You must be able to move around these obstacles and return to the route you were following before you took your detour. This can be accomplished by using one of three detour methods: Far side landmark method, nearside landmark method, and the 90-degree offset method.

(1) Far side landmark. Used during hours of daylight when you are able to identify some feature on the far side of the obstacle that lies on your line of march and is obvious enough so that you can find it once you have moved around to the far side.

(2) Nearside landmark. Used during hours of daylight when you cannot find a landmark on the far side. However, you can identify a feature on the nearside that is on your line of march and you will still be able to see once you have moved around to the far side. Again, let us use the example of being on an azimuth of 198 degrees, and it brings you to a small lake. From the nearside you cannot locate any features on the far side. There is, however, a small pier on the nearside that is located on your line of march. You can now move around the lake using the easiest and safest route. Move to an area on the far side that is as close to being opposite of the nearside landmark as possible. You then figure out the back azimuth of your route azimuth, which in this case would be 18 degrees. Shoot an azimuth back across the lake to the pier. Move to the right or left until you are in a position where the pier lies at an azimuth of 18 degrees. This puts you back on your line of march and you can continue on following your original azimuth of 198 degrees.

(3) 90 degree offset. Used during hours of reduced visibility or during daylight hours when there are no visible landmarks along the line of march on either side of the obstacle. This time you are moving along a 360 degree azimuth and you encounter a large, impassable area of thick brush. The brush is too high to see any landmarks from either side of the thicket. The following procedure should be taken to detour around the thicket.

(a) Stop the pace count you have been making for movement along your assigned route.

(b) Decide which way you will detour around the thicket. Let's say in this case the right flank appears to be the easiest and safest. Add 90 degrees to your original azimuth of 360 degrees in order to move your patrol to the right.

(c) Move on this 90 degrees azimuth until you have reached a point where you can clear the right flank of the thicket. You will keep a new pace count starting from the point where you made your 90 degrees turn to the right to the point where you can clear the right flank of the thicket.

(d) Make another change of direction. This time you want to return to a 360 degrees azimuth so you will parallel your original course and move by the right flank of the thicket. In order to return to the 360 degrees azimuth you will now subtract 90 degrees.

(e) Move on the 360-degree azimuth until you have by passed the thicket. At the point of your second 90 degree turn, you will pick up the pace count you were making before you began making your detour around the obstacle. Even though you are not on your original line of march, you are moving parallel to your original line of march and therefore must include the distance you cover during this leg of the offset in your pace count for your route.

(f) Once you have bypassed the thicket, you will make another change of direction. This time, you want to move back toward your original line of march. Since that means you will be turning toward the left, you will subtract 90 degrees from the 360 degrees azimuth you have just completed. Your new direction will be 270 degrees.

(g) Pace back toward the original line of march following the 270-degree azimuth, the same number of paces as you made off the original line of march. In doing this you will have returned to your original line of march.

(h) Once you have moved back to your original line of march you will make your last change of direction. By adding 90 degrees to the original azimuth of the last leg of the offset, which in this case was 270 degrees, you will be back on the original azimuth of 360 degrees.

(i) Return to your original line of march pace count.

UNITED STATES MARINE CORPS
Mountain Warfare Training Center
Bridgeport, California 93517-5001

FMST.07.39
04/03/02

STUDENT HANDOUT

SEARCH AND RESCUE MANAGEMENT

TERMINAL LEARNING OBJECTIVE: In a mountainous environment, execute a search and rescue, in accordance with the references. (FMST.07.39)

ENABLING LEARNING OBJECTIVES:

(1) Without the aid of references and from a given list, choose the definition of a search, in accordance with the references. (FMST.07.39a)

(2) Without the aid of references and from a given list, choose the definition of a rescue, in accordance with the references. (FMST.07.39b)

(3) Without the aid of references and from a given list, choose one the four methods used to determine a search area, in accordance with the references. (FMST.07.39c)

(4) Without the aid of references and from a given list, choose one of the five methods for confining a search area, in accordance with the references. (FMST.07.39d)

(5) Given a lost individual scenario in a mountainous environment, plan a search and rescue in accordance with the references. (FMST.07.39e)

OUTLINE.

1. **GENERAL**. When discussing a search and rescue situation we must first be able to differentiate between the two facets of the operation. A "search" means that you are looking for an overdue missing person (FMST.07.39a). A "rescue" 'is the extraction of a known subject from a known location (FMST.07.39b). In either of these situations time is critical for several reasons:

a. Subject may need emergency medical care.

b. Size of the search area is limited.

c. Time and weather destroys clues.

d. Subject may need protection from the environment or from himself.

2. **SEARCH METHODS**. When conducting a search you will be searching for clues, instead of searching for the victim. There are two reasons for this:

a. There are more clues than victims. Anyone moving through an area will leave signs.

b. Clue detection greatly reduces search difficulty by reducing the search area.

c. This determination will assist the search manager in the use of assets, the justification for requesting outside agency assistance, and other management decisions. There are eight points that must be considered in determining the search urgency:

1. Subject profile.
2. Weather profile.
3. Equipment available to subject.
4. Subject's training, experience, and mobility.
5. Terrain hazards.
6. History of incidents in the area.
7. Time.
8. Political sensitivity.

3. **SUBJECT PROFILE**. The subject profile is based on the subject's capabilities and mobility in the terrain and conditions of the search area. The three points are age, medical condition, and number of subjects. Discussing this will assist us is rating the relative urgency:

a. Age: Very young/old, other.

b. Medical condition: Injured, ill, mental problems, healthy, or dead.

c. Number of subjects: Single or group.

d. Subject experience profile: No experience/no knowledge of area, no experience/knows area, experienced/no knowledge of area, or experienced/knows area.

e. Weather profile: Past or current bad weather, bad weather in next 8 hours, bad weather in next 24 hours, or no bad weather predicted. Temperature and visibility.

f. Equipment profile: Inadequate, questionable, or adequate for environment or weather.

g. Terrain/Hazard profile: Known hazardous terrain or, hazards or few or no hazards.

h. History of incidents in the area: Outcome of other searches for individuals with similar profiles in similar situations.

i. Time: Time since the subject went missing and the effects of time on clues.

j. Political sensitivity: External sources; VIPs, politicians, relatives, media, or higher authority in your chain of command.

4. **DETERMINING THE SEARCH AREA**. There are four methods used to determine the search area by rescue managers. The best application is to use a combination of these methods by personnel with search experience. (FMST.07.39c)

a. Theoretical: This is the distance that the victim could have traveled in the elapsed time since he/she was reported missing. (PI x R Squared)

b. Statistical: Historical data on searches for similar individuals in similar conditions.

c. Subjective: Evaluation of limiting factors for this specific incident and location.

d. Deductive reasoning: Sherlock Holmes version, based on analysis of circumstances involving the specific incident.

5. **BEHAVIOR ASPECTS OF A LOST PERSON**. General aspects of lost person behavior will give you an awareness of the actions of the individuals you are searching for by separating them into different categories. Some factors that could effect the search strategy are the general health of the subject, his/her past experiences, physiological effects of the environment, and the victims' mental state or biological cycle. Also, looking at the subject's ability to assist in his/her own survival with simple skills such as fire, shelter, signaling, and land navigation will determine strategy. The psychological impact on the victim should also be taken into account. Will the subject be overcome by Stress? Fear? Panic? Embarrassment? Or will the subject be calm and develop a positive "I'll be OK attitude? A further point to consider is the circumstances under which the subject became lost. Was the subject at a known location such as a campground? Enroute to a specific location or on a trail? Traveling in a wilderness area with no specific location or destination in mind? Or does the subject fall under one of the following categories:

a. Children age 1 to 3: Unaware of concept of being lost, no navigation skills or sense of direction, wander aimlessly, might seek out a convenient place to take a nap. i.e. hollow log, under bush/rock/picnic table.

b. Children age 3 to 6: More mobile, more inquisitive and easily distracted, will attempt to "go home" once they decide that they are lost, taught to avoid strangers and may hide from searchers.

c. Children 6 to 12: Good navigation skills and sense of direction in familiar terrain but become disoriented in unknown surroundings, may have intentionally run away and do not want to be found, may not answer when called, same fears as adults but no experiences to fall back on, once night sets in more likely to respond to searchers and except assistance.

d. Mentally retarded (all ages): Act similar to children 6 to 12, generally will not respond to name, will hide from searchers or seek shelter, once shelter found might remain in place, will do little or nothing to assist in their own survival or rescue.

e. Elderly (above 65): Possibly suffering from senility or Alzheimer's, easily confused or attracted to something that strikes their fancy, may not be oriented to the present, likely to over-extend themselves physically, and often have hearing impairments.

f. Despondent: Often are seeking solitude and will not respond to searchers, usually found near prominent terrain/scenic views, within sight and sound of civilization, rarely found in the underbrush.

g. Hikers: Rely on trails and have a set destination in mind, have poor land navigation once they loose the trail, often get confused at trail intersections or in springtime when trails are covered with snow, hiking parties of different ability levels get separated, will miss trails and head down wrong drainages. Hunters: Concentrate on game more than navigation, often pursue game into drainages/deadfalls/underbrush/deep snow, tend to over-extend themselves, typically not prepared for foul weather.

h. Cone pickers/Berry pickers/Photographers/ rockhounds/etc: Intention to stay or go to a specific location, carry little provisions or survival gear, go in good weather so they don't carry foul weather gear, often are not aware of surrounding as they are intent on pictures or the ground if pickers, these people are an extremely high risk group.

i. Fishermen: Well oriented once they arrive at lake or river, most often late due to accident, these are often recovery operations.

j. Climbers: Generally well equipped and self sufficient, tend to remain on marked routes and leave good plans with friends, primary factors are weather and hazardous conditions such as falling debris or avalanches, once located these searches often become technical rescues. 1. Marines: Subject profile will depend on individual's MOS, mission, training, and abilities learned as a civilian.

6. **BEGINNING THE SEARCH**. Once we have determined that a subject is missing we must determine whether or not he/she is in our search area. The sheriffs department will initiate a "Bastard Search", this is to ensure that the subject is not simply at the local tavern, shacked up with an old friend, or was not in an automobile accident enroute to the area. Next we will begin our search at one of the four following area:

a. Last Known Point (LKP): This is the last place that you have physical evidence to support that the subject was there. i.e. Subject's vehicle at the trailhead, name in a logbook, campsite with gear, etc.

b. Place Last Seen (PLS): The last place he was physically seen. i.e. On the trail by other hikers, at the hunting cabin, at a restaurant or gas station, or back home in San Diego.

c. Departure Point: The place where the subject was supposed to enter the search area. Usually a trailhead or specific point relating to the individual's activity.

d. Clues: As clues are discovered, and you confirm that they belong to the subject, you will update or adjust the search area.

7. **MATTSON**. Once we have determined that we do have an overdue subject and can assume that he/she is in our search area we need to break the search area into manageable sized segments. Depending upon the area to be searched, special training required to search high hazard areas, and assets available, we will divide the area by hazard or by terrain features. The lines of division should follow terrain features so that the searchers on the ground can easily navigate and stay within their zones. We will then determine the POA using the Mattson approach. This is done with each of the five or six members of the management team with the most search experience assigning a percentage to each of the segments using his individual judgment on the likelihood of the subject being in that area. Always include an extra column to represent the "rest of the world". Once all personnel have made their determinations, divide the totals and this will give you your areas to search in order of precedence:

NAME	A	B	C	D	E*	TOTAL
DUNSTON	40%	20%	10%	20%	10%	100%
KEOGH	50%	30%	5%	10%	5%	100%
KOFFEL	40%	25%	15%	10%	10%	100%
AYRES	60%	20%	10%	5%	5%	100%
TOTAL	190%	95%	40%	45%	30%	400%
AVERAGE	47.50%	23.75%	10.00%	11.25%	7.5%	
PRIORITY	1 ST	2ND	4TH	3RD	5TH	

8. **COMMAND SYSTEM**. The Office of Emergency Service (OES), has developed a command system that they use for all emergency situations. This system is called the

Incident Command System (ICS). The breakdown is very similar to the military chain of command and the duties correspond to those of the various "S" shops in the battalion. The personnel in each duty area will complete their assigned duties as you would in a military unit with the Incident Commander (IC) being overall responsible for the entire action. Normally, in California, the IC is the County Sheriff.

a. Incident Commander. CO

b. Operations-Tactics, coordination, insert resources. S-3

c. Plans-Strategy, intelligence, briefing and debriefing teams. S-2

d. Logistics-Beans, Band-Aids, beds, heads, motor-T, Special equipment. S-4

e. Communication-COC and field, landlines, frequency coordinator. S-6

f. Finance-Documentation of all man-hours, damages, legal issues.
S-1

g. Liaisons-inter agency, media, safety, family. PAO/Chaplain

9. **CONFINING A SEARCH AREA**. There are several reasons to confine the search area. The most important is that the longer the subject has free movement, the farther he/she can travel. (Remember the Theoretical Method) Additionally, this will ensure that the subject does not walk out without you being aware of it. It's real dumb to be conducting a search for a subject that is on his/her way home. The use of confinement will not interfere with the other searchers, decreases the chances of a more massive search and if promptly initiated has been shown to be highly effective. There are several methods that we can use. (FMST.07.39d)

a. Sign-cutters - Searchers that hike or ski the perimeter of the entire search area, then the perimeter of the segments. This will let you know if he/she moves from one area to another and allows you to concentrate limited assets in a specific area.

b. Trail-watchers/Road Blocks/Camp-ins - Searchers that camp at trailheads. They can alert you if the subject walks out and interview other individuals concerning possible sightings of the subject.

c. Look-outs - These personnel are already in place in fire towers. Simply notify them of the situation.

d. Track Traps - Clearing areas of the trail that are checked periodically for fresh tracks from the subject.

e. String-Lines - Spools of line with paper arrows attached that point to a safe area of a trail-watcher camp.

10. **USING CLUES**. The use of clues cannot be underestimated. They can tell you if a subject was or was not in the area, by freshness you can tell how long ago the subject was there, by the type of clue you can determine physical and mental condition, and by following clues you can determine a direction and speed of travel. Once you have located a CONCLUSIVE clue, you will re-mattson the area, using the clue as the LKP. The clues we will be looking for fall under four categories:

a. Physical - Footprints, equipment, candy wrappers, cigarette butts, etc.

b. Recorded - Summit logs or trail registers.

c. People - Witnesses who have seen the subject.

d. Event - Mirror flashes, camp fires, whistle signals, etc.

11. **ADDITIONAL ASSETS**. There are a variety of assets available to you in the search and rescue business. Some are volunteers, some charge money and some (Like the military) require higher authority to grant permission for their use. Remember that in addition to active searchers you can also use passive measures such as lights, noise, balloons, fires, smoke generators, sirens, or P.A. systems. Some examples of active searchers are:

a. Scent dogs.

b. Tracking dogs.

c. Human trackers.

d. Hasty teams.

e. Grid teams.

f. Helicopters.

g. Civil Air Patrol.

h. Snowmobile clubs.

i. Horse teams.

UNITED STATES MARINE CORPS
Mountain Warfare Training Center
Bridgeport, California 93517-5001

FMST.07.16
04/03/02

STUDENT HANDOUT

TRIAGE

TERMINAL LEARNING OBJECTIVE: Given simulated casualties, in a wilderness environment, triage casualties, in accordance with the references. (FMST.07.16)

ENABLING LEARNING OBJECTIVES:

1) Without the aid of references and from a given list, choose when actual triage most often begins, in accordance with the references. (FMST.07.16a)

2) Without the aid of references and from a given list, identify one of the four-triage categories used by NATO, in accordance with the references. (FMST.07.16b)

3) Without the aid of references and from a given list, choose the primary cause of death in a mass casualty situation, in accordance with the references. (FMST.07.16c)

OUTLINE.

1. **DEFINITION**: Triage or sorting (often used interchangeably) is the evaluation and classification of casualties for the purpose of establishing the priority for treatment and evacuation.

NOTE: Casualty sorting is one of the most important tasks for the entire medical service. No other job requires more informed judgment, hard work and courage. The Officer and the Corpsman responsible for triaging wounded Sailors and Marines must exercise sound judgment so that casualties receive proper, expedient care. Triage is a continuing process and any individual assigned should be the most capable and experienced available.

a. Decision Factors. The factors concerning casualty triage are based upon the following:

(1) The need for immediate resuscitation.

(2) The need for emergency surgery.

(3) The futility of surgery due to obviously lethal wounds.

(4) Time needed to perform surgery compared with:

(a) Probability of success.

(b) Number of other casualties needing treatment.

b. The actual triage begins most often with the casualty himself. (FMST.07.16a). This usually occurs when the injured individual determines that "Buddy-Aid" or "Self-Aid" will not be adequate.

(1) The corpsman, once presented with the casualty, must make several determinations:

(a) Is the casualty a walking-wounded or a litter case?

(b) If severely wounded, are resuscitative efforts required immediately?

(c) Does the casualty need any other immediate treatment (bleeding, asphyxia or pain control)?

(2) Once the casualty has been transported to a Battalion Aid Station (BAS), triage continues. The dental officer will commonly perform the duties of the Triage Officer at the echelon I or echelon II level, if present.

2. **NATO TRIAGE CATEGORY CODES**. (FMST.07.16b)

a. At each medical treatment facility (MTF) in the area of operations (AO), incoming casualties are classified by level of treatment required. Four triage category groups have been universally adopted for use by both United States and NATO Forces. These categories are listed and defined as they appear in standardization agreement (STANAG) No 2879:

(1) Immediate Treatment (Group T1) (RED Tag). To include those requiring emergency life-saving surgery. These procedures should not be time consuming and should concern only those patients with high chances for survival. Some examples are:

(a) Respiratory obstruction.

(b) Accessible hemorrhage.

(c) Emergency amputation.

(2) Delayed Treatment (Group T2) (YELLOW Tag). To include those badly in need of time-consuming major surgery, but whose general condition permits delay in surgical treatment without unduly endangering life. To mitigate the often, critical effects of delay in surgery, sustaining treatment, such as stabilizing I.V. fluids, splinting, administration of antibiotics, catheterization, gastric decompression, and relief of pain will be required. Examples are:

(a) Large muscle wounds.

(b) Fractures of major bones.

(c) Intra-abdominal and/or thoracic, head or spinal injuries.

(d) Uncomplicated major burns.

(3) Minimal Treatment (Group T3) (GREEN Tag). To include those with relatively minor injuries who can effectively care for themselves or who can be helped by untrained personnel. Some examples are:

(a) Minor lacerations.

(b) Abrasions.

(c) Fractures of small bones.

(d) Minor burns.

(4) Expectant Treatment (Group T4) (BLUE Tag). This group comprises patients who have received serious and often multiple injuries, and whose treatment would be time-consuming and complicated with a low chance of survival. If fully treated, they may make heavy demands on medical manpower and supplies. Until the mass casualty situation is under control, they will receive appropriate supportive treatment. The extent of treatment depends on available supplies and manpower, and may involve the use of large doses of narcotic analgesics. These patients should not be abandoned, and every effort should be devoted to their comfort. The possibility of their survival, despite alarming injuries, must always be kept in mind. Examples are:

(a) Severe multiple injuries.

(b) Severe head or spinal injuries.

(c) Large doses of radiation.

(d) Widespread severe burns.

3. **TREATMENT CONCEPTS**.

a. The triage officer must be familiar with all aspects of combat casualty care (airway management, hemorrhage control, use of anesthesia, surgical procedures, fluid resuscitation and post-operative care, to name a few), but ever-present in his mind should be the following concepts:

(1) Asphyxia and hemorrhage are initially the primary causes of death in any mass casualty situation. (FMST.07.16c)

(2) Once at a definitive care facility, uncontrolled hemorrhage is the leading cause of death.

(3) Life has priority over limb, and preserving function overrides the correction of superficial anatomic defects.

(4) A casualty is in a constant state of change until the wound or injury has been repaired.

(5) Systemic disturbances caused by the wound continue until healing is complete.

(6) Combat conditions will dictate the modifications necessary in triage and care of mass casualties.

b. Priorities of Treatment. With the six concepts of treatment in mind, we can now address three basic priorities for treatment.

(1) First Priority:

(a) Asphyxia:

1. Mechanical obstruction of the airway.

2. Sucking chest wound.

3. Tension pneumothorax.

4. Maxillofacial trauma.

5. Massive hemo-pneumothorax.

(b) Shock:

1. Exsanguinating hemorrhage.

2. Major internal hemorrhage (thoracic, abdominal, and pelvic).

3. Visceral injuries/evisceration.

4. Cardio-pericardial injuries.

5. Massive muscle injury leading to Fat Embolization to the lung.

6. Full-thickness or deep-partial burns of 10 % or greater.

(2) Second Priority:

(a) Visceral injuries:

1. Perforation of G.I. Tract.

2. Pancreatic and Biliary system.

3. Genitourinary tract.

4. Thoracic wounds w/o asphyxia.

(b) Vascular injuries requiring repair:

1. Tourniquet cases.

(c) Closed head injuries:

1. Increasing loss of consciousness.

(d) Major burns:

1. Partial-thickness burns of greater than 25% TBSA in the low risk group, 20% TBSA in the high risk group.

2. Burns involving the poor risk group.

3. Burns involving the face, hands, feet, genitalia/perineum.

4. Circumferential limb burns.

(3) Third Priority:

(a) Brain and spinal injuries - requiring burr hole decompression.

(b) Soft tissue injuries.

(c) Debridement necessary with less than major muscle damage.

(d) Lesser fractures and dislocations.

(e) Eye injuries.

(f) Maxillofacial injuries w/o asphyxia.

(g) Burns under 20% in other locations other than priority 2.

4. **TRIAGE TOOLS**.

a. Trauma Score/Glasgow Coma Scale. In recent years, many methods have been proposed to speed up the classification and sorting of casualties. One method that has gained wide acceptance is the Trauma Score/Glasgow Coma Scale. Several studies have been performed using various parameters, i.e., respiratory rate, chest wall expansion, BP, capillary refill, etc. By assigning a number scale to each specific response, the likelihood of survival as a percentage can be calculated.

1. Glasgow Coma Scale.

a. Eye opening:

1) Spontaneous = 4.
2) To voice = 3.
3) To pain = 2.
4) None = 1.

b. Verbal Response.

1) Oriented = 5.
2) Confused = 4.
3) Inappropriate words = 3.
4) Incomprehensive words = 2.
5) None = 1.

c. Motor Response.

1) Obeys Command = 6.

2) Localized Pain = 5.
3) Withdraw Pain = 4.
4) Flexion (pain) = 3.Extension (pain) = 2.
5) None = 1.

d. Total the points up and they equal:

1) 13-15= 4.
2) 9-12 = 3.
3) 6-8 = 2.
4) 4-5 = 1.
5) <3 = 0

2. Trauma Scale.

a. Respiratory Rate:

1) 12-29 min = 4.
2) 10-12 min = 3.
3) 6-9 min = 2.
4) 1-5 min = 1.
5) None = 0.

b. Systolic BP:

1) >89 mmHg = 4.
2) 76 mmHg = 3.
3) 50-75 mmHg = 2.
4) 1-49 mmHg = 1.
5) None = 0.

c. Total the points.

3. Total trauma Scale + GCS = % survival (1-12)

12 = 99
10 = 88
8 = 67
6 = 63
4 = 33
2 = 29

1 = 25
0 = 4

UNITED STATES MARINE CORPS
Mountain Warfare Training Center
Bridgeport, California 93517-5001

FMST.07.08
04/03/02

STUDENT HANDOUT

PREVENTIVE MEDICINE AND WATER PURIFICATION

TERMINAL LEARNING OBJECTIVE. Given a unit in a mountainous environment and the necessary equipment and supplies, perform preventive medicine, in accordance with the references. (FMST.07.08)

ENABLING LEARNING OBJECTIVES.

1) Without the aid of references, from a given list select the five important areas of personal hygiene, in accordance with the references. (FMST.07.08a)

2) Without the aid of references, from a given list select the correct steps taken to ensure proper immunization prior to deployment, in accordance with the references. (FMST.07.08b)

3) Without the aid of references, from a given list select the three methods of water purification, in accordance with the references. (FMST.07.08c)

4) Without the aid of references, from a given list select the four forms of halogens used for water purification, in accordance with the references. (FMST.07.08d)

5) Without the aid of references, from a given list select the definition of giardia, in accordance with the references. (FMST.07.08e)

6) Without the aid of references, from a given list select the most common drug used to treat giardia infection in the United States, in accordance with the references. (FMST.07.08f)

7) Without the aid of references, from a given list select the three important reasons for proper waste disposal, in accordance with the references. (FMST.07.08g)

8) Without the aid of references, from a given list select the correct methods of waste disposal, in accordance with the references. (FMST.07.08h)

16

1. **PERSONAL HYGIENE** Your body's defense mechanisms break down due to the physical demands of the arduous terrain, environment, stress and metabolic changes. Personal

hygiene becomes the key to prevent the spread of communicable diseases. Five important areas are: (WMC.6.17a)

a. Body.

(1) The body should be washed frequently in order to minimize the chances of small cuts and scratches developing into full blown infections and as a defense against parasitic infections.

(2) A daily bath or shower consisting of soap and hot water is ideal. However, when this is not possible, you should:

(a) Give yourself a "sponge bath" using soap and water, making sure particular attention is given to body creases, i.e., armpits, groin area, face ears, hands and feet.

(b) If water is in extremely short supply, you should take an "air bath" by:

(1) Remove all clothing and hang it up to air.

(2) Expose the body for two hours to the sunlight, which is ideal, but the effects will basically be the same if done indoors or during an overcast day. BE CAREFUL NOT TO SUNBURN!

(3) Shaving. Shaving at night before bedtime is preferable because shaving will remove facial oils, which aid in protecting the face from a harsh cold environment during the day.

b. Hair.

(1) Clean frequently.

(2) Inspect at least once a week for parasites. Have your mountain buddy inspect the areas not within your view. This is particularly important during the summer season. Any time the area you are operating in an area that contains wild life there is the possibility of parasites like fleas and ticks.

c. Hands. Hand washing with soap and water will cut down in the transmission of viral and parasitic infections via the person-person and fecal-oral route.

(1) Keep fingernails trimmed and clean to prevent accidentally scratching yourself and to prevent harborage for bacteria.

d. Feet. Are your prime sources of transportation, so take care of them!

(1) Inspect frequently for:

(a) Blisters.

(b) Infections, bacterial and fungal.

(2) Preventive measures:

(a) Keep your feet dry, change socks frequently.

(b) Use antifungal foot powder.

(c) Antiperspirants. Pretreatment for 2 weeks prior to field with an antiperspirant containing Aluminum Chlorhydrate. This will aid in the prevention of athlete's foot.

e. Oral Hygiene. The mouth and teeth should be cleaned at least daily to prevent tooth decay and gum disease.

(1) Ideally with:

(a) Tooth brush.

(b) Tooth paste.

(c) Dental floss.

(2) If these items are not available:

(a) A "chew stick" made from a clean twig about 8" long and finger width. Chew one end until it becomes frayed and brush-like, then use as a toothbrush.

(b) Field expedient dental floss can be pulled from the center lines of paracord.

(c) The gums should be stimulated at least once daily by rubbing them vigorously with a clean index finger.

2. **IMMUNIZATIONS (NAVMEDCOMINST 6230.3)**. In addition to being required, immunizations are your best front line defense against communicable disease.

a. Eight weeks prior to deployment: (WMC.6.17b)

(1) Ensure Basic Series Completed (BSC) on required immunizations.

(2) Determine if additional immunizations/chemophylaxis will be required in that geographical area of operations.

(3) Identify and flag health risk personnel: i.e. TB converters (NAVMEDCOMINST 6224.1), hepatitis (NAVMEDCOMINST 6230.1A), mononucleosis, parasitic infections and anemic personnel.

(4) Review medical intelligence reports:

(a) EPMU, Navy Environmental Preventive Medicine Unit.

(b) G-2, USMC.

(c) NAVDISVECTOECOLCONCEN, Navy Disease Vector Ecology and Control Centers.

(d) CDC - Center For Disease Control.

(e) WHO - World Health Organization.

3. **WATER.** A satisfactory source is one where quantity and quality is enough to supply the needs of all the troops. Water or ice taken from the environment should be considered contaminated and must be purified with available means. Be aware that cold weather slows the chemical reaction time of purification measures.

a. Sources.

(1) Rivers

(2) Streams

(3) Ponds

(4) Lakes

(5) Ice

(6) Snow

b. Water Source Selection. The choice of a water source is influenced by:

(1) Not contaminated by: (Check all to one mile upstream).

(a) Sewage

(b) Enemy pollution

(c) Chemicals

(d) Biologicals

(e) Radiologicals

NOTE: Check condition of vegetation or for dead animals.

(2) Quantity.

(3) Ease of procurement.

(4) Ease of purification.

(5) Freedom from turbidity.

(6) Excessive organic or non-organic contamination.

(7) Should be easily protectable.

4. **PURIFICATION.** Simply consists of removing or destroying enough impurities to make water safe and pleasant to drink.

a. The first step in purifying is to select a treatment method. (All figures are based on one quart of water.) The three different purifying methods are: (WMC.6.17c)

(1) Boiling. The standard recommendation is to bring the water to a rolling boil for 1 minute at sea level for complete sterilization of the water. Recent research showed that additional boiling time for sterilization was not necessary and that boiling for two minutes at any altitude rendered water safe for consumption.

(2) Halogens. The number of tablets and contact time must be considered. All charts based on one-quart canteen. (WMC.6.17d)

(a) Iodine tablets.

TYPE WATER	TABLETS	CONTACT TIME
CLEAR	1	25 MIN
CLOUDY	2	25 MIN
COLD	2	25 MIN

(b) Chlorine bleach. The strength of the solution, number of drops and contact time must be considered.

PERCENT	CLEAR WATER	CLOUDY WATER	COLD WATER
1%	10GTTS/30 MIN	20GTS/30 MIN	10GTTS/60MIN
4-6%	2GTTS/30 MIN	4GTTS/30 MIN	2GTTS/60 MIN
7-10%	1 GTTS/30 MIN	2GTTS/30 MIN	1 GTTS/60 MIN

(c) Iodine solution. The strength, number of drops and contact time must be considered.

PERCENT	CLEAR WATER	CLOUDY WATER	COLD WATER
2%	5GTTS/30 MIN	10GTTS/30 MIN	8GTTS/60 MIN

(d) Betadine solution.

CLEAR WATER	CLOUDY WATER	COLD WATER
8GTTS/30 MIN	16GTTS/30 MIN	8GTTS/60 MIN

(1) Filtration Most filters are not small enough to completely disinfect water. They remove particulate debris, which allows lower halogen dose and improved appearance and taste of "dirty" water. Filters clog quickly if the water is dirty or has a lot of suspended particles. NEVER USE FILTERS AS YOUR SOLE METHOD OF WATER PURIFICATION.

(a) Maximum Filter Pore Size:

1) Parasitic eggs and larvae: 20 micrometers (um).

2) Giardia, E. histolytica: 5 um.

3) Cryptosporidium: 1 um

4) Enteric bacteria: 0.2 um.

5) Viruses: 0.01 um (too small for field water filtration).

5. **GIARDIA**

a. Definition. (WMC.6.17e) An intestinal flagellate protozoa that can cause clinical illness in humans.

(1) Found worldwide in tropical and temperate areas, the infectious cysts are passed by humans and animals and can be transmitted by direct fecal-oral route or waterborne contamination (cysts maintain viability for months in cold water). Due to these factors, Giardia is both a common cause of diarrhea in travelers to foreign countries, and campers in the United States where it has become endemic and is now considered the most common human parasite.

(2) It is found with increasing frequency in formerly "pristine" mountain streams, rivers and lakes and outbreaks have occurred in remote recreational areas and resorts throughout the western mountain states.

(3) Giardiasis is currently a risk for anyone drinking untreated surface water in the western mountains and for communities with water treatment techniques inadequate for eliminating the cysts.

b. Pathophysiology. The organism has two forms: Trophozoite and cyst.

(1) These may be passed in the stool of an infected person. The trophozoites die within one hour outside the body, however the cysts are very hardy, retaining infectivity more than two months in cold water. If the cysts are ingested, excystation is encouraged by stomach acid, and the motile trophozoite attaches by suction disc to the wall of the duodenum and proximal jejunum, where they may cause symptoms of illness.

(2) Water, ice, or food prepared with contaminated water, are the major sources of infection, but direct person-to-person transmission can occur from sexual contact or hand-to-mouth contamination.

(3) Incubation time: 5 - 25 days (7 - 10 medium). Most infections are thought to be asymptomatic carrier states, however, a wide array of clinical syndromes may occur.

c. Symptoms.

(1) A number of people have abrupt onset of explosive diarrhea accompanied by abdominal cramps, foul gas, vomiting, fever and malaise. This commonly lasts three to four days before developing the more common subacute syndrome. In most individuals, the onset is more insidious, and symptoms are persistent or recurrent.

(2) Stools become mushy, greasy and malodorous. Over a period of weeks, watery diarrhea may alternate with soft stools and even constipation. Upper GI symptoms accompany stool changes, mid and upper abdominal cramping, substernal burning, acid indigestion, sulfurous belching, nausea and foul flatus, distention with an early "full" feeling are all typically increased after a meal.

(3) Symptoms of anorexia, fatigue and weight loss are common, but fever and vomiting are infrequent except during an acute onset of illness.

d. Laboratory Parameters.

(1) The "typical" victim should suggest a diagnosis of giardiasis. This individual may have a history of camping, vacationing at a mountain resort, or foreign travel. Symptoms of upper abdominal discomfort, foul gas and soft stools arose one or two weeks after a short trip or during a lengthy one and have fluctuated for weeks. Sometimes improving, the individual held off visiting the doctor, only to have the symptoms recur. Work or school is not interrupted, however, physical discomfort and embarrassment from the gas are concerns.

(2) Laboratory confirmation of giardiasis can be difficult. Stool samples are the first means of diagnosis. Trophozoites may be found in fresh, watery stools, but disintegrate rapidly. Cysts remain in fresh stools for at least 24 hours. Cysts passage, however, is extremely variable and not related to clinical symptoms. Three samples, taken every other day, should be examined for a definite diagnosis.

(3) Average duration of symptoms has varied from three to ten weeks. Acquired resistance has been shown. Immunity is suspected, but not proven and human reinfection definitely occurs.

e. Treatment. (WMC.6.17f) Complete cure of most Giardia infections can be achieved 90% of the time with *Metronidazole* (flagyl) 250 mg three time s a day for 5 days.

f. Prevention There is no safe or effective chemoprophylactic drug and prevention should be practiced by travelers and campers. The purity of American wild waters can no longer be assumed and in developing countries, even tap water should be suspected. In addition to choosing a technique for individual field water disinfection, individuals must become more responsible for prevention of water contamination. This requires attention to personal latrines and camper trailer waste disposal, a subject that most people do not want to touch.

6. **WASTE**. The importance of proper waste disposal cannot be over emphasized.

a. It serves to: (WMC.6.17g)

(1) Eliminate harborage for rodents and vermin.

(2) Prevents attraction of rodents and vermin.

(3) Prevents a pathogenic contamination source.

b. Two basic types of waste are organic and non-organic. The basic ways to dispose of them are burning, burying, or hauling it away. (WMC.6.17h)

(1) Organic wastes:

(a) Human waste - Burn or haul away to a designated waste pit area.

(b) Urine - Use only assigned marked areas.

(c) Edible garbage - Burn. It should not be left exposed for animals, vermin or the enemy.

(d) Paper -Burn.

(e) Contaminated bandages - Sterilize/burn.

(2) Non-organic waste:

(a) Metals - Haul away or bury.

(b) Liquids - Burn or bury.

UNITED STATES MARINE CORPS
Mountain Warfare Training Center
Bridgeport, California 93517-5001

FMST.07.19
04/03/02

STUDENT OUTLINE

REQUIREMENTS FOR SURVIVAL

TERMINAL LEARNING OBJECTIVE Given a survival situation in any type of environmental condition and minimal equipment and resources, apply the requirements for survival to facilitate individual or group survival and recovery per the references. (FMST.07.19)

ENABLING LEARNING OBJECTIVES

1. Without the aid of references and from a given list, choose one of the survival stressors, in accordance with the references. (WMC.A1.a)

2. Without the aid of references and from a given list, choose one of the priorities of work in a survival situation, in accordance with the references. (WMC.A1.b)

1. REQUIREMENTS FOR SURVIVAL

a. This mental "mind-set" is important in many ways. We usually call it the "will to survive" although you might call it "attitude" just as well. This basically means that, if you do not have the right attitude, you may still not survive.

A1

b. A guideline that can assist you is the acronym " SURVIVAL".

(1) **S**ize up.

(a) Size up the situation.

1. Conceal yourself from the enemy.

2. Use your senses to hear, smell, and see to determine and consider what is developing on the battlefield before you make your survival plan.

(b) Size up your surroundings.

1. Determine the rhythm or pattern of the area.

2. Note animal and bird noises and their movement.

3. Note enemy traffic and civilian movement.

(c) Size up your physical condition.

1. Check your wounds and give yourself first aid.

2. Take care to prevent further bodily harm.

3. Evaluate condition of self and unit prior to developing survival plan.

(d) Size up your equipment.

1. Consider how available equipment may affect survival senses; tailor accordingly.

(2) **U**ndue haste makes waste.

(a) Plan your moves so that you can move out quickly without endangering yourself if the enemy is near.

(3) **R**emember where you are.

(a) If you have a map, spot your location and relate it to the surrounding terrain.

(b) Pay close attention to where you are and where you are going. Constantly orient yourself.

(c) Try to determine, at a minimum, how your location relates to the following:

1. The location of enemy units and controlled areas.

2. The location of friendly units and controlled areas.

3. The location of local water sources.

4. Areas that will provide good cover and concealment.

(4) <u>V</u>anquish fear and panic.

(a) Realistic and challenging training builds self-confidence and confidence for a unit's leadership.

(b) The feeling of fear and panic will be present. The survivor must control these feelings.

(5) <u>I</u>mprovise and improve.

(a) Use tools designed for one purpose for other applications.

(b) Use objects around you for different needs. (i.e. use a rock for a hammer)

(6) <u>V</u>alue living.

(a) Place a high value on living.

(b) Refuse to give into the problem and obstacles that face you.

(c) Draw strength from individuals that rise to the occasion.

(7) <u>A</u>ct like the natives.

(a) Observe the people in the area to determine their daily eating, sleeping, and drinking routines.

(b) Observe animal life in the area to help you find sources of food and water.

NOTE: Remember that animal reactions can reveal your presence to the enemy. Animals cannot serve as an absolute guide to what you can eat and drink.

(8) <u>L</u>ive by your wits, **but for now**, learn basic skills.

(a) Practice basic survival skills during all training programs and exercises.

2. **STRESS**. Stress has many positive benefits. Stress provides us with challenges: it gives us chances to learn about our values and strengths. Too much stress leads to distress. While many of these signs may not be self-identified, it remains critical that all survivors remain attentive to each other's signs of distress. Listed are a few common signs of distress found when faced with too much stress:

a. Difficulty in making decisions (do not confuse this sign for a symptom of hypothermia).

b. Angry outbursts.

c. Forgetfulness.

d. Low energy level.

e. Constant worrying.

f. Propensity for mistakes.

g. Thoughts about death or suicide.

h. Trouble getting along with others.

i. Withdrawing from others.

j. Hiding from responsibilities.

k. Carelessness.

3. **SURVIVAL STRESSORS**. (FMST.07.19a) **Any** event can lead to stress. Often, stressful events occur simultaneously. These events are not stress, but they produce it and are called "stressors". In response to a stressor, the body prepares either to "fight or flee". Stressors add up. Anticipating stressors and developing strategies to cope with them are the two ingredients in the effective management of stress. It is essential that the survivor be aware of the types of stressors they will encounter.

a. Injury, Illness, or Death: Injury, illness, and death are real possibilities a survivor has to face. Perhaps nothing is more stressful than being alone in an unfamiliar environment where you could die from hostile action, an accident, or from eating something lethal. Illness and injury can also add to stress by limiting your ability to maneuver, get food and drink, find shelter, and defend yourself.

b. Uncertainty and Lack of Control: Some people have trouble operating in settings where everything is not clear-cut. The only guarantee in a survival situation is that nothing is guaranteed. This uncertainty and lack of control also add to the stress of being ill, injured, or killed.

c. Environment: A survivor will have to contend with the stressors of weather, terrain, and the variety of creatures inhabiting an area. Heat, cold, rain, winds, snow, mountains, insects, and animals are just a few of the challenges awaiting the Marine working to survive. Depending on how a survivor handles the stress of environment, his surroundings can be either a source of food and protection or a cause of extreme discomfort leading to injury, illness, or death.

d. Hunger and Thirst: Without food and water a person will weaken and eventually die. Getting and preserving food and water takes on increasing importance as the length of time in a survival setting increases. With the increased likelihood of diarrhea, replenishing electrolytes becomes critical. For a Marine used to having his provisions issued, foraging can be a significant source of stress.

e. Fatigue: It is essential that survivors employ all available means to preserve mental and physical strength. While food, water, and other energy builders may be in short supply, maximizing sleep to avoid deprivation is a very controllable factor. Training data collected at MCCDC demonstrates that individuals who lack sleep for 24 hours or more suffer a 25% decrease in effective performance. Further, sleep deprivation directly correlates with increased fear. Effective survival is highly unlikely when fatigue builds to a level where staying awake becomes a stressful evolution.

f. Isolation: Although Marines complain about higher headquarters, we become used to the information and guidance it provides, especially during times of confusion. Being in contact with others provides a greater sense of security and a feeling someone is available to help if problems occur. A significant stress in survival situations is that often a person or team has to rely solely on its own resources.

4. **NATURAL REACTIONS**. Man has been able to survive many shifts in his environment throughout the centuries. His ability to adapt physically and mentally to a changing world kept him alive. The average person will have some psychological reactions in a survival situation. These are some of the major internal reactions you might experience with the survival stresses.

   a. Fear: Fear is our emotional response to dangerous circumstances that we believe have the potential to cause death, injury, or illness. Fear can have a positive function if it encourages us to be cautious in situations where recklessness could result in injury. Unfortunately, fear can also immobilize a person. It can cause us to become so frightened that we fail to perform activities essential for survival.

   b. Anxiety: Anxiety can be an uneasy, apprehensive feeling we get when faced with dangerous situations. When used in a healthy way, anxiety urges us to act to end, or at least master, the dangers that threaten our existence. A survivor reduces his anxiety by performing those tasks that will ensure his coming through the ordeal alive. Anxiety can overwhelm a Marine to the point where he becomes easily confused and has difficulty thinking.

c. Anger and Frustration: Frustration arises when a person is continually thwarted in his attempts to reach a goal. The goal of survival is to stay alive until you can reach help or until help can reach you. To achieve this goal, Marines must complete some tasks with minimal resources. One outgrowth of frustration is anger. Getting lost, damaged or forgotten equipment, the weather, inhospitable terrain, enemy patrols, and physical limitations are just a few sources of frustration and anger. Frustration and anger encourage impulsive reactions, irrational behavior, poorly thought-out decisions, and, in some instances, an "I quit" attitude. If the Marine does not properly focus his angry feelings, he can waste much energy in activities that do little to further either his chances of survival or the chances of those around him.

d. Depression: Depression is closely linked with frustration and anger when faced with the privations of survival. A destructive cycle between anger and frustration continues until the person becomes worn down-physically, emotionally, and mentally. When a person reaches this point, he starts to give up, and his focus shifts from "What can I do" to "There is nothing I can do." If you allow yourself to sink into a depressed state, then it can sap all your energy and, more important, your will to survive.

e. Loneliness and Boredom: Man is a social animal and enjoys the company of others. Loneliness and boredom can be another source of depression. Marines must find ways to keep their minds productively occupied.

f. Guilt: The circumstances leading to your survival setting are sometimes dramatic and tragic. It may be the result of an accident or military mission where there was a loss of life. Perhaps you were the only, or one of a few, survivors. While naturally relieved to be alive, you simultaneously may be mourning the deaths of others that were less fortunate. Do not let guilt feelings prevent you from living.

5. **PRIORITIES OF WORK IN A SURVIVAL SITUATION.** (FMST.07.19b) Each survival situation will have unique aspects that alter the order in which tasks need to be accomplished. A general guideline is to think in blocks of time.

a. First 24 hours: The first 24 hours are critical in a survival situation. You must make an initial estimate of the situation. Enemy, weather, terrain, time of day and available resources will determine which tasks need to be accomplished first. They should be the following:

(1) Shelter.

(2) Fire.

(3) Water.

(4) Signaling.

b. Second 24 hours: After the first 24 hours have passed, you will now know you can survive. This time period needs to be spent on expanding your knowledge of the area. By completing the following tasks, you will be able to gain valuable knowledge.

(1) Tools and weapons: By traveling a short distance from your shelter to locate the necessary resources, you will notice edible food sources and game trails.

(2) Traps and snares: Moving further away from your shelter to employ traps and snares, you will be able to locate your shelter area from various vantagepoints. This will enable you to identify likely avenues of approach into your shelter area.

(3) Pathguards: Knowing the likely avenues of approaches, you can effectively place noise and casualty producing pathguards to ensure the security of your shelter area.

c. Remainder of your survival situation. This time is spent on continuously improving your survival situation until your rescue.

6. **GROUP SURVIVAL.** In-group survival, the group's survival depends largely on its ability to organize activity. An emergency situation does not bring people together for a common goal; rather, the more difficult and disordered the situation, the greater are the disorganized group's problems.

a. Groups Morale: High morale must come from internal cohesiveness and not merely through external pressure. The moods and attitudes can become wildly contagious. Conscious, well-planned organization and leadership on the basis of delegated or shared responsibility often can prevent panic. High group morale has many advantages.

(1) Individual feels strengthened and protected since he realizes that his survival depends on others whom he trusts.

(2) The group can meet failure with greater persistency.

(3) The group can formulate goals to help each other face the future.

b. Factors that Influence Group Survival: There are numerous factors that will influence whether a group can successfully survive.

(1) Organization of Manpower - Organized action is important to keep all members of the group informed; this way the members of the group will know what to do and when to do it, both under ordinary circumstances and in emergencies.

(2) Selective Use of Personnel - In well-organized groups, the person often does the job that most closely fits his personal qualifications.

(3) Acceptance of Suggestion and Criticisms - The senior man must accept responsibility for the final decision, but must be able to take suggestion and criticisms from others.

(4) Consideration of Time - On-the-spot decisions that must be acted upon immediately usually determines survival success.

(5) Check Equipment - Failure to check equipment can result in failure to survive.

(6) Acquiring survival knowledge and skills increases survival Knowledge and Skills - Confidence in one's ability to survive.

UNITED STATES MARINE CORPS
Mountain Warfare Training Center
Bridgeport, California 93517-5001

FMST.
07.20
04/03/02

STUDENT OUTLINE

SURVIVAL KIT

TERMINAL LEARNING OBJECTIVE. In a summer mountainous environment, construct a survival kit, in accordance with the references. (FMST.07.20)

ENABLING LEARNING OBJECTIVES

(1) Without the aid of references, choose from a given list in the criteria for a survival kit, in accordance with the references. (FMST.07.20a)

(2) Without the aid of references, choose from a given list one example of each criteria for a survival kit, in accordance with the references. (FMST.07.20b)

1. COMPONENTS FOR A SURVIVAL KIT

a. The environment is the key to the types of items you will need in your survival kit. How much equipment you put in your kit depends on how you will carry the kit. A kit on your body will have to be much smaller than one carried in a vehicle.

b. Always layer your survival kit, keeping the most important items on your body.

c. In preparing your survival kit, select items that can be used for more than one purpose.

A2

d. Your survival kit does not need to be elaborate. You only need functional items that will meet your needs and a case to hold them. The case might be a first aid case, an ammunition pouch, or another suitable case. This case should be:

(1) Water repellent or waterproof.

(2) Easy to carry or attach to your body.

(3) Suitable to accept various sized items.

(4) Durable.

e. When constructing a survival kit, you should have the following components: (FMST.07.20a)

(1) Fire starting items.

(2) Water procurement items.

(3) Food procurement items.

(4) Signaling items.

(5) First aid items.

(6) Shelter items.

2. **<u>ITEMS CONTAINED WITHIN EACH COMPONENT</u>** (FMST.07.20b)

a. <u>Fire Starting Items</u>.

(1) Matches.

(2) Magnifying glass.

(3) Flint and Steel.

(4) Lighter.

(5) Potassium Permanganate, with a container of sugar or anti-freeze.

(6) Prepackaged Tinder.

(a) Commercially Manufactured

(b) Cotton Balls and Petroleum Jelly

b. Water Procurement Items

(1) Water Disinfecting Chemicals.

(a) Iodine Tablets

(b) Betadine Solution

(c) Iodine Solution

(2) Metal Container. (Serves for boiling water)

(a) Canteen Cup

(b) Survival Kit Container

(c) Any suitable can that contained no petroleum products.

(3) Water Carrying Items.

(a) Canteen

(b) Plastic Bag

(c) Plastic/Metal/Glass Container which contained no petroleum products

c. Food Procurement Items

(1) Fish.

(a) Various sized hooks

(b) Various sized sinkers/weights

(c) Metal leaders and swivels

(d) Small weighted jigs

(e) Fishing line

i. Think about the size of fish for that environment when selecting weights and sizes.

(2) Game.

(a) Snares

i. Commercially Manufactured

ii. Aircraft Cable

iii. Tie Wire

(b) Bait

i. MRE Cheese Spread or Peanut Butter Package

(c) 550 Cord for Gill Net and Trap Construction

i. Engineer/Marking Tape

ii. Sling shot rubber and pouch

d. Signaling Items

(1) Day.

(a) Mirror

(b) Whistle

(c) Pyrotechnics (Smoke, Pen Flares)

(d) Air Panels

(2) Night.

(a) Pyrotechnics (Pen Flares, Star Clusters)

(b) Lights (Flashlight, Strobe, Chemlight)

(c) Whistle

e. Shelter Items

(1) Cordage.

(a) 550 Cord

(b) Wire

(c) Communication wire

(d) Tie wire

(2) Finger Saw.

(3) Sewing Kit with Needles for construction/repair of clothing.

(4) Tentage.

(a) Poncho

(b) Tarp

(c) Space blanket

(d) Plastic trash bags

f. First Aid Items

(1) Band-Aids.

(a) Steristrips

(b) Adhesive Tape

(c) Non-stick pads, 4x4's, Gauze, Battle Dressings

(d) Muslin Bandage

(2) Ointments.

(a) Burn

(b) Anti-septic

(3) Miscellaneous.

(a) Salt

(b) Sugar

(c) Eye Wash

(d) Alcohol prep pads

(e) Suture Kit

(f) Scalpel

(g) Vile of Yarrow

g. Miscellaneous items

(1) Fingernail clippers

(2) Compass

(3) Notebook with pen or pencil

(4) Wood eye screws and nails

(5) Surgical tubing

NOTE: It is assumed that the Marine is always carrying a high quality fixed bladed knife, a multi-tool knife, and a sharpening stone.

UNITED STATES MARINE CORPS
Mountain Warfare Training Center
Bridgeport, California 93517-5001

FMST.07.23
04/05/02

STUDENT OUTLINE

SIGNALLING AND RECOVERY

TERMINAL LEARNING OBJECTIVE Given a survival situation in any type of environmental condition and minimal equipment and resources, signal for aid to facilitate individual or group recovery per the references. (FMST 07.23)

ENABLING LEARNING OBJECTIVES

(1) Without the aid of references and from a given list, choose the audio international distress Signal, in accordance with the references. (FMST.07.23a)

(2) Without the aid of references and from a given list, choose the visual international distress signal, in accordance with the references. (FMST.07.23b)

(3) Without the aid of references and given a list of messages, match in writing the international code symbols to each corresponding message, in accordance with the references. (FMST.07.23c)

OUTLINE

1. **SIGNALING DEVICES**. The equipment listed below are items that may be on your body or items inside an aircraft. Generally, these items are used as signaling devices while on the move. They must be accessible for use at any moment's notice. Additionally, in a summer mountainous environment, Marines may experience areas that are snow covered and must be familiar with the effects that snow will have on specific signaling devices.

    a. Pyrotechnics. Pyrotechnics include star clusters and smoke grenades. When using smoke grenades in snow pack, a platform must be built. Without a platform, the smoke grenade will sink into the snow pack and the snow will absorb all smoke. A rocket parachute flare or a hand flare has been sighted as far away as 35 miles, with an average of 10 miles. Pyrotechnic flares are effective at night, but during daylight their detectability ranges are reduced by 90 percent.

A3

b. M-186 Pen Flare: The M-186 Pen Flare is a signaling device carried in the vest of all crew chiefs and pilots. Remember to cock the gun prior to screwing in the flare.

c. Strobe Light: A strobe light is generally carried in the flight vests of all crew chiefs and pilots. It can be used at night for signaling. Care must be taken because a pilot using goggles may not be able to distinguish a flashing strobe from hostile fire.

d. Flashlight: By using flashlights, a Morse code message can be sent. A SOS distress call consists of sending three dots, three dashes, and three dots. Keep repeating this signal.

e. Whistle: The whistle is used in conjunction with the audio international distress signal. It is used to communicate with forces on the ground.

f. AN/PRC-90 & AN/PRC-112: The AN/ PRC 90 survival radio is a part of the aviator's survival vest. The AN/PRC-112 will eventually replace the AN/PRC-90. Both radios can transmit either tone (beacon) or voice. Frequency for both is **282.8 for voice**, and **243.0 for beacon**. Both of these frequencies are on the UHF Band.

g. Day/Night Flare: The day/night flare is a good peacetime survival signal. The flare is for night signaling while the smoke is for day. A red cap identifies the older version flare with three nubbins while the new generation has three rings around the body for identification during darkness. The flare burns for approximately 20 second while the smoke burns for approximately 60 seconds.

NOTE: Once one end is used up, douse in water to cool and save the other end for future use.

h. Signal Mirror: A mirror or any shiny object can be used as a signaling device. It can be used as many times as needed. Mirror signals have been detected as far away as 45 miles and from as high as 16,000', although the average detection distance is 5 miles. It can be concentrated in one area, making it secure from enemy observation. A mirror is the best signaling device for a survivor; however, it is only as effective as its user. Learn how to use one now, before you find yourself in a survival situation.

(1) Military signal mirrors have instructions on the back showing how to use it. It should be kept covered to prevent accidental flashing that may be seen by the enemy.

(2) Any shiny metallic object can be substituted for a signal mirror.

(3) Haze, ground fog, or a mirage may make it hard for a pilot to spot signals from a flashing object. So, if possible, get to the highest point in your area when flashing. If you can't determine the aircraft's location, flash your signal in the direction of the aircraft noise.

AIMING THE SIGNAL MIRROR

2. METHODS OF COMMUNICATION

a. Audio: Signaling by means of sound may be good, but it does have some disadvantages.

(1) It has limited range unless you use a device that will significantly project the sound.

(2) It may be hard to pinpoint one's location due to echoes or wind.

(3) International Distress Signal. (FMST.07.23a) The survivor will make six blasts in one minute, returned by three blasts in one minute by the rescuer.

b. Visual: Visual signals are generally better than audio signals. They will pinpoint your location and can been seen at a greater distances under good weather conditions.

(1) The visual international distress symbol is recognized by a series of three evenly spaced improvised signaling devices. (FMST.07.23b)

3. **IMPROVISED SIGNALING DEVICES**. Improvised signaling devices are generally static in nature. They must be placed in a position to be seen by rescuers. They are made from any resources available, whether natural or man-made.

a. Smoke Generator: The smoke generator is an excellent improvised visual signaling device. It gives the survivor the flexibility to signal in either day or night conditions. This type of signal has been sighted as far away as 12 miles, with an average distance of 8 miles. Smoke signals are most effective in calm wind conditions or open terrain, but effectiveness is reduced with wind speeds above 10 knots. Build them as soon as time and the situation permits, and protect them until needed.

(1) Construct your fire in a natural clearing or along the edge of streams (or make a clearing). Signal fires under dense foliage will not be seen from the air.

(2) Find two logs, 6 - 10 inches in diameter, and approximately five feet long. Place the two logs parallel to each other with 3 - 4 feet spacing.

(3) Gather enough sticks, approximately two inches in diameter and four feet long, to lie across the first two logs. This serves as a platform for the fire.

(4) Gather enough completely dry branches to build a pyramid fire. The pyramid fire should be 4 feet by 4 feet by 2 feet high.

(5) Place your tinder under the platform.

(6) Gather enough pine boughs to lay on top of the pyramid fire. This serves to protect the fire and the tinder.

(7) To light, remove the pine bough and ignite the tinder. If available, construct a torch to speed up the lighting process, especially for multiple fires.

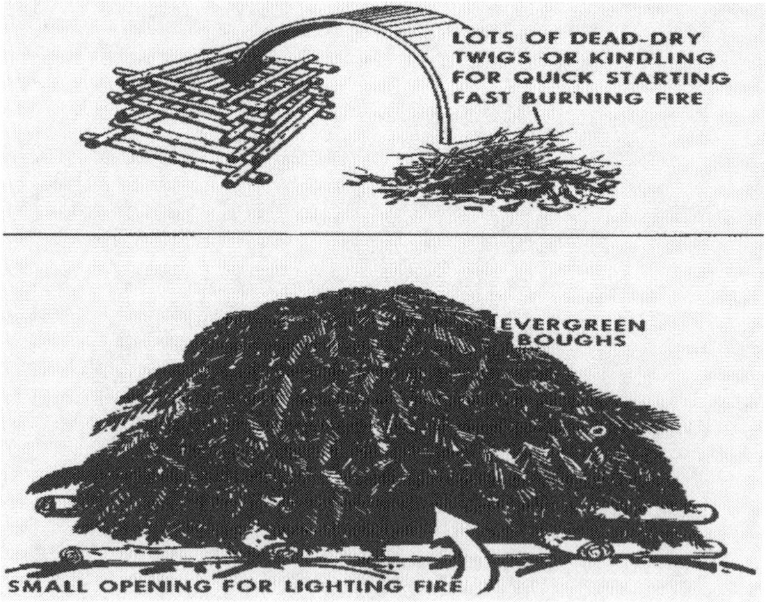

SMOKE GENERATOR

(8) To create a smoke effect during the day light hours, place the pine bough on the ignited fire.

(9) Placing a smoke grenade or colored flare under the platform will change the color of the smoke generated. Remember that you want the fire to draw in the colored smoke, which will create a smoke color that contrasts with the background will increase the chances of success.

b. Arrangement or alteration of natural materials: Such things as twigs or branches, can be tramped into letters or symbols in the snow and filled in with contrasting materials. To attract more attention, ground signals should be arranged in big geometric patterns.

(1) International symbols. The following symbols are internationally known. (FMST.07.23c)

Number	Message	Code symbol
1	REQUIRE ASSISTANCE	V
2	REQUIRE MEDICAL ASSISTANCE	X
3	NO OR NEGATIVE	N
4	YES OR AFFIRMATIVE	Y
5	PROCEED IN THIS DIRECTION	↑

INTERNATIONAL SYMBOLS

(2) Shadows: If no other means are available, you may have to construct mounds that will use the sun to cast shadows. These mounds should be constructed in one of the International Distress Patterns. To be effective, these shadow signals must be oriented to the sun to produce the best shadows. In areas close to the equator, a North—South line gives a shadow anytime except noon. Areas further north or south of the equator require the use of East—West line or some point of the compass in between to give the best result.

(3) Size: The letters should be large as possible for a pilot or crew to spot. Use the diagram below to incorporate the size to ratio for all letter symbols.

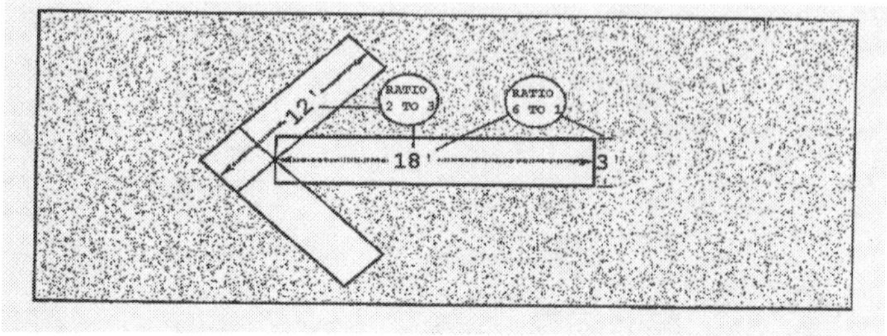

SIZE AND RATIO

(4) Contrast: When constructing letter symbols, contrast the letter from the surrounding vegetation and terrain. Ideally, bring material from another location to build the letter. This could be clothing, air panels, space blanket, etc.

(a) On snow, pile bough or use sea dye from a LPP (Life preserver, personal). Fluorescent sea dye markers have been sighted as far away as 10 miles, although the average detection distance is 3 miles.

4. **AIR TO GROUND COMMUNICATIONS**. Air to ground communications can be accomplished by standard aircraft acknowledgments.

a. Aircraft will indicate that ground signals have been seen and understood by:

(1) Rocking wings from side to side: This can be done during the day or in bright moonlight.

b. Aircraft will indicate that ground signals have been seen but not understood by:

(1) Making a complete, clockwise circle during the day or in bright moonlight.

5. **RECOVERY**. Marines trapped behind enemy lines in future conflicts may not experience quick recovery. Marines may have to move to a place that minimizes risk to the recovery force. No matter what signaling device a Marine uses, he must take responsibility for minimizing the recovery force's safety.

a. Placement Considerations: Improvised signaling devices, in a hostile situation, should not be placed near the following areas due to the possibility of compromise:

(1) Obstacles and barriers.

(2) Roads and trails.

(3) Inhabited areas.

(4) Waterways and bridges.

(5) Natural lines of drift.

(6) Man-made structures.

(7) All civilian and military personnel.

b. Tactical Consideration: The following tactical considerations should be adhered to prior to employing an improvised signaling device.

(1) Use the signals in a manner that will not jeopardize the safety of the recovery force or you.

(2) Locate a position, which affords observation of the signaling device and facilitates concealed avenues of escape (if detected by enemy forces). Position should be located relatively close to extract site in order to minimize "time spent on ground" by the recovery force.

(3) Maintain continuous security through visual scanning and listening while signaling devices is employed. If weapon systems are fire and/or observation should cover available, signaling devices.

(4) If enemy movement is detected in the area, attempt to recover the signaling device, if possible.

(5) Employ improvised signaling devices only during the prescribed times, if briefed in the mission order.

c. Recovery Devices: In mountainous terrain, a helicopter landing may be impossible due to ground slope, snow pack, or vegetation. The survivor must be familiar with recovery devices that may be aboard the aircraft.

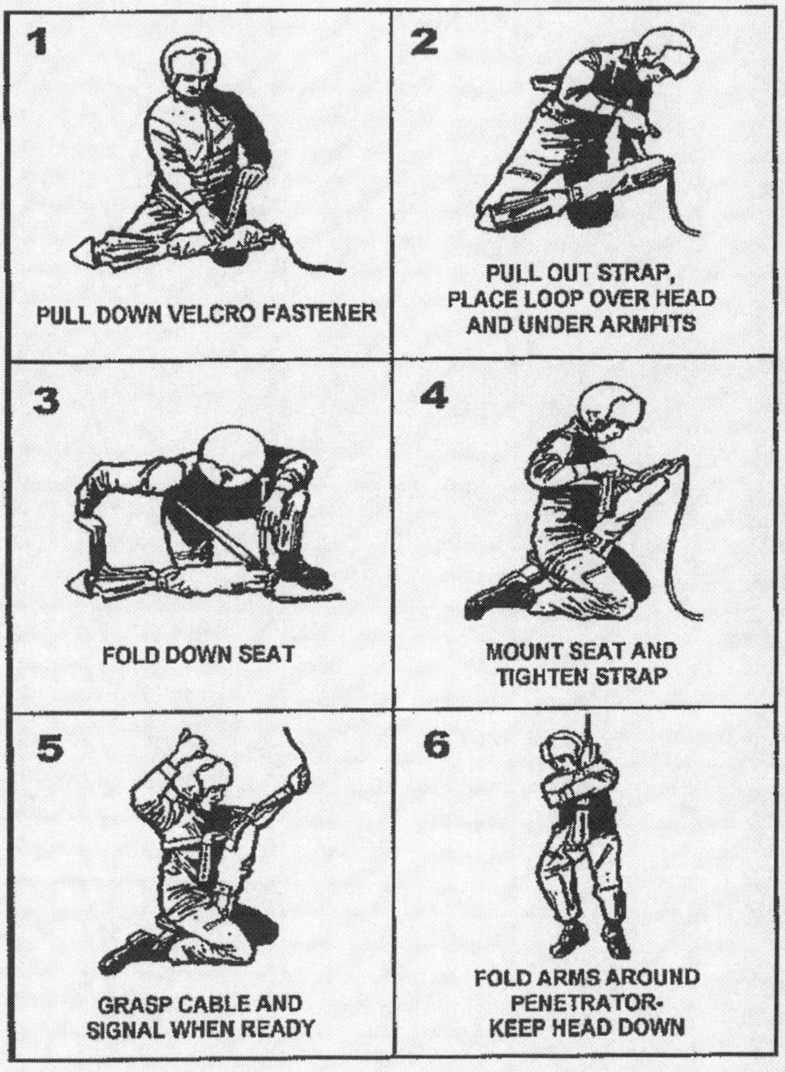

JUNGLE PENETRATOR

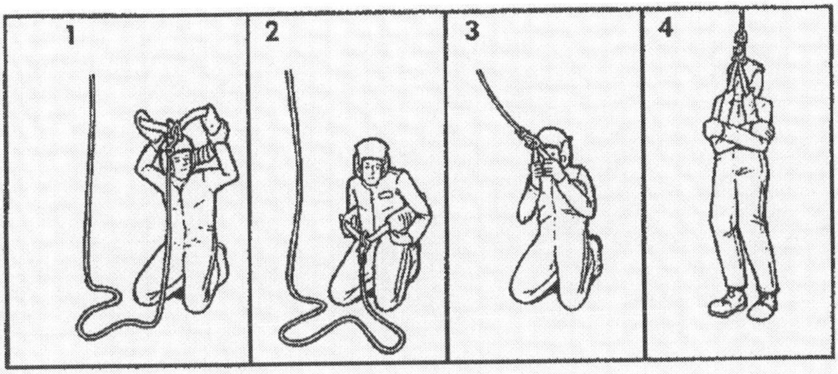

SLING HOIST

d. Recovery by other than aircraft. Recovery by means other than aircraft may occur. Unit Sop's should include signaling and link-up with forces at the following locations:

(1) Border Crossings. The evader who crosses into a neutral country is subject to detention by that country for the duration of the war.

(2) FEBA/FLOT.

(a) Static: Recovery along a static FEBA is always difficult. Under these conditions, enemy and friendly forces can be expected to be densely arrayed and well camouflaged, with good fields of fire. Attempts to penetrate the FEBA should be avoided.

(b) Advancing: Individuals isolated in front of advancing friendly units should immediately take cover and wait for the friendly units to overrun their position.

(c) Retreating: Individuals between opposing forces should immediately take cover and wait for enemy units to pass over their position. After most enemy units have moved on, evaders should try to link up with other isolated friendly elements and return to friendly forces.

(3) Link-up with friendly patrols. Unit authentication numbers and/or locally developed codes may assist the evader to safely make contact in or around the FEBA and when approached by friendly forces.

UNITED STATES MARINE CORPS
Mountain Warfare Training Center
Bridgeport, California 93517-5001

FMST.07.21
FMST.07.04
04/07/02

STUDENT OUTLINE

EXPEDIENT SHELTERS AND FIRES

TERMINAL LEARNING OBJECTIVES

(1) In a summer mountainous environment, construct expedient shelters, in accordance with the references. (FMST.07.21)

(2) In a summer mountainous environment, construct fires, in accordance with the references. (FMST.07.04)

b. **ENABLING LEARNING OBJECTIVES**

(1) Without the aid of references and from a given list, choose the four characteristics of a safe expedient shelter, in accordance with the references. (FMST.07.21a)

(2) Without the aid of references and from a given list, choose the four hazards avoid when using natural shelters, in accordance with the references. (FMST.07.21b)

(3) Without the aid of references and from a given list, choose one of man-made expedient shelters, in accordance with the references. (FMST.07.21c)

(4) Without the aid of references, start a fire using a primitive method, in accordance with the references. (FMST.07.04a)

A4

OUTLINE

1. **BASIC CHARACTERISTICS FOR SHELTER**. (FMST.07.21a) Any type of shelter, whether it is a permanent building, tentage, or an expedient shelter must meet six basic characteristics to be safe and effective. The characteristics are:

    a. Protection from the Elements: The shelter must provide protection from rain, snow, wind, sun, etc.

    b. Heat Retention: It must have some type of insulation to retain heat; thus preventing the waste of fuel.

    c. Ventilation: Ventilation must be constructed, especially if burning fuel for heat. This prevents the accumulation of carbon monoxide. Ventilation is also needed for carbon dioxide given off when breathing.

    d. Drying Facility: A drying facility must be constructed to dry wet clothes.

    e. Free from Natural Hazards: Shelters should not be built in areas of avalanche hazards, under rock fall or "standing dead" trees have the potential to fall on your shelter.

    f. Stable: Shelters must be constructed to withstand the pressures exerted by severe weather.

2. **NATURAL SHELTERS**. Natural shelters are usually the preferred types because they take less time and materials construct. The following may be made into natural shelters with some modification.

    a. Caves or Rock Overhangs: Can be modified by laying walls of rocks, logs or branches across the open sides.

    b. Hollow Logs: Can be cleaned or dug out, then enhanced with ponchos, tarps or parachutes hung across the openings.

    c. Hazards of Natural Shelters: (FMST.04.21b)

        (1) Animals: Natural shelters may already be inhabited (i.e. bears, coyotes, lions, rats, snakes, etc.). Other concerns from animals may be disease from scat or decaying carcasses.

        (2) Lack of Ventilation: Natural shelters may not have adequate ventilation. Fires may be built inside for heating or cooking but may be uncomfortable or even dangerous because of the smoke build up.

        (3) Gas Pockets: Many caves in a mountainous region may have natural gas pockets in them.

(4) Instability: Natural shelters may appear stable, but in reality may be a trap waiting to collapse.

3. **MAN-MADE SHELTERS**. (FMST.04.21c) Many configurations of man-made shelters may be used. Over-looked man-made structures found in urban or rural environments may also provide shelter (i.e. houses, sheds, or barns). Limited by imagination and materials available, the following man-made shelters can be used in any situation.

a. Poncho Shelter.

b. Sapling Shelter.

c. Lean-to.

d. Double lean-to.

e. A-frame Shelter.

f. Fallen Tree Bivouac.

4. **CONSTRUCTION OF MAN-MADE SHELTERS**. To maximize the shelter's effectiveness, Marines should take into consideration the following prior to construction.

a. Considerations.

(1) Group size.

(2) Low silhouette and reduced living area dimensions for improved heat conservation.

(3) Avoid exposed hill tops, valley floors, moist ground, and avalanche paths.

(4) Create a thermal shelter by applying snow, if available, to roof and sides of shelter.

(5) Location of site to fire wood, water, and signaling, if necessary.

(6) How much time and effort needed to build the shelter.

(7) Can the shelter adequately protect you from the elements (sun, wind, rain, and snow). Plan on worst case scenario.

(8) Are the tools available to build it. If not, can you make improvised tools?

(9) Type and amount of materials available to build it.

(10) When in a tactical environment, you must consider the following:

(a) Provide concealment from enemy observation.

(b) Maintain camouflaged escape routes.

(c) Use the acronym BLISS as a guide.

B – Blend in with the surroundings.
L –Low silhouette.
I – Irregular shape.
S – Small.
S – Secluded located.

b. Poncho Shelter: This is one of the easiest shelters to construct. Materials needed for construction are cord and any water-repellent material (i.e. poncho, parachute, and tarp). It should be one of the first types of shelter considered if planning a short stay in any one place.

(1) Find the center of the water-repellent material by folding it in half along its long axis.

(2) Suspend the center points of the two ends using cordage.

(3) Stake the Four Corners down, with sticks or rocks.

c. Sapling Shelter: This type of shelter is constructed in an area where an abundance of saplings are growing. It is an excellent evasion shelter.

(1) Find or clear an area so that you have two parallel rows of saplings at least 4' long and approximately 1 ½' to 2' apart.

(2) Bend the saplings together and tie them to form several hoops, which will form the framework of the shelter.

(3) Cover the hoop with a water-repellent covering.

(4) The shelter then may be insulated with leaves, brush, snow, or boughs.

(5) Close one end permanently. Hang material over the other end to form a door.

d. Lean-to: A lean-to is built in heavily forested areas. It requires a limited amount of cordage to construct. The lean-to is an effective shelter but does not offer a great degree of protection from the elements.

(1) Select a site with two trees (4-12" in diameter) spaced far enough apart that a man can lie down between them. Two sturdy poles can be substituted by inserting them into the ground the proper distance apart.

(2) Cut a pole to support the roof. It should be at least 3-4" in diameter and long enough to extend 4-6" past both trees. Tie the pole horizontally between the two trees, approximately 1 meter off the deck.

(3) Cut several long poles to be used as stringers. They are placed along the horizontal support bar approximately every 1 ½' and lay on the ground. All stringers may be tied to or laid on the horizontal support bar. A short wall or rocks or logs may be constructed on the ground to lift the stringers off the ground, creating additional height and living room dimensions.

(4) Cut several saplings and weave them horizontally between the stringers. Cover the roof with water-repellent and insulating material.

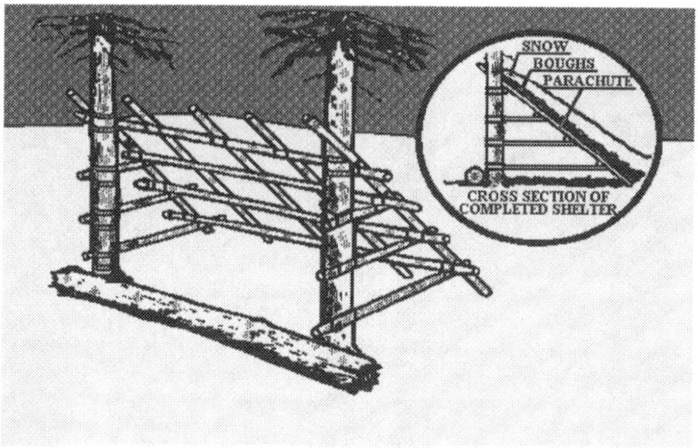

LEAN-TO

e. Double lean-to: The double lean-to shelter is constructed for 2-5 individuals. It is constructed by making two lean-tos and placing them together.

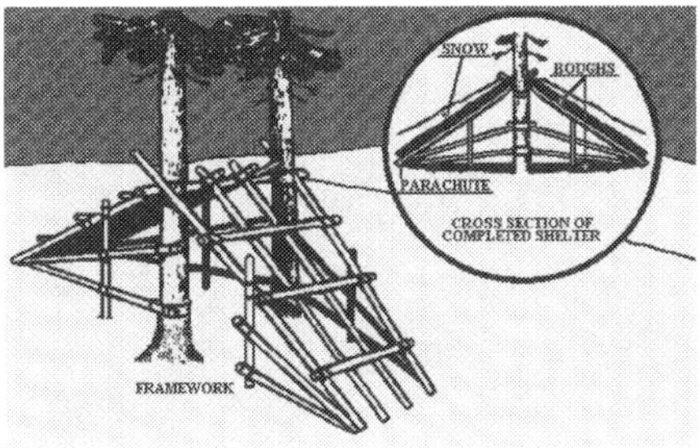

DOUBLE LEAN-TO

f. A-Frame Shelter: An A-Frame shelter is constructed for 1-3 individuals. After the framework is constructed, bough/tentage is interwoven onto the frame and snow, if available, is packed onto the outside for insulation.

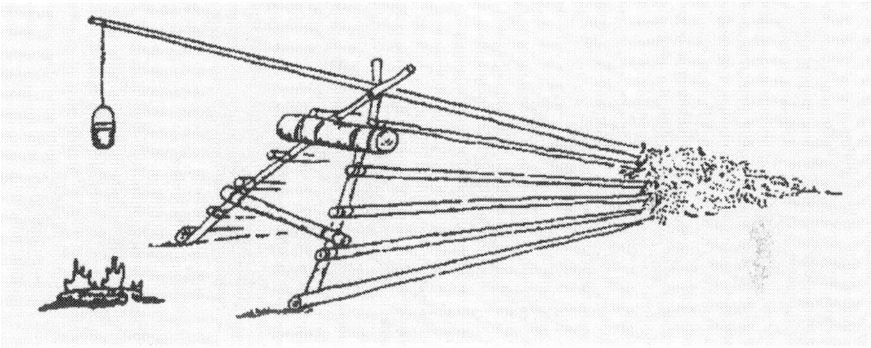

A-FRAME

g. Fallen Tree Bivouac: The fallen tree bivouac is an excellent shelter because most of the work has already been done.

(1) Ensure the tree is stable prior to constructing.

(2) Branches on the underside are cut away making a hollow underneath.

(3) Place additional insulating material to the top and sides of the tree.

(4) A small fire is built outside of the shelter.

FALLEN TREE BIVOUAC

5. **REFLECTOR WALLS.** Heating a shelter requires a slow fire that produces lots of steady heat over a long period of time. A reflector wall should be constructed for all open-ended shelters. A reflector wall is constructed with a flat rock or a stack of green logs propped behind the fire. A surprising amount of heat will bounce back from the fire into the shelter.

6. **FIRES**. Fires fall into two main categories: those built for cooking and those built for warmth and signaling. The basic steps are the same for both: preparing the fire lay, gathering fuel, building the fire, and properly extinguishing the fire.

   a. Preparing the fire lay: There are two types of fire lays: fire pit and Dakota hole. Fire pits are probably the most common.

      (1) Create a windbreak to confine the heat and prevent the wind from scattering sparks. Place rocks or logs used in constructing the fire lay parallel to the wind. The prevailing downwind end should be narrower to create a chimney effect.

      (2) Avoid using wet rocks. Heat acting on the dampness in sandstone, shale, and stones from streams may cause them to explode.

      (3) Dakota Hole: The Dakota Hole is a tactical fire lay. Although no fire is 100% tactical, this fire lay will accomplish certain things:

         (a) Reduces the signature of the fire by placing it below ground.

         (b) Provides more of a concentrated heat source to boil and cook, thus preserving fuel and lessening the amount of burning time.

         (c) By creating a large air draft, the fire will burn with less smoke than the fire pit.

         (d) It is easier to light in high winds.

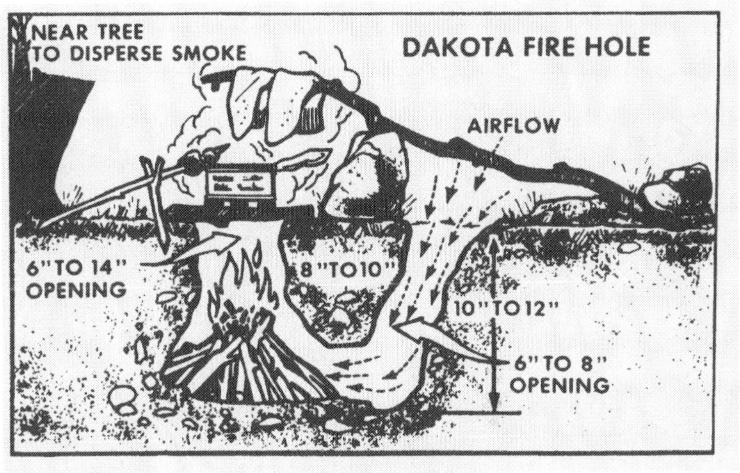

DAKOTA HOLE

b. Gather Fuel: Many Marines take shortcuts when gathering firewood. Taking a few extra minutes can mean the difference between ease and frustration when building a fire.

(1) Tinder: Tinder is the initial fuel. It should be fine and dry. Gather a double handful of tinder for the fire to be built and an extra double handful to be stored in a dry place for the following morning. Dew can moisten tinder enough to make lighting the fire difficult. Some examples are:

(a) Shredded cedar/juniper bark, pine needles.

(b) Dry grass.

(c) Slivers shaved from a dry stick.

(d) Hornet's nest.

(e) Natural fibers from equipment supplemented with pine pitch (i.e., cotton battle dressing).

(f) Cotton balls and petroleum jelly or Char-cloth.

NOTE: Sticks used for tinder should be dry and not larger than the diameter of a toothpick.

(2) Kindling: This is the material that is ignited by the tinder that will burn long enough to ignite the fuel.

(a) Small sticks/twigs pencil-thick up to the thickness of the thumb. Ensure that they are dry.

(b) Due to a typically large resin content, evergreen limbs often make the best kindling. They burn hot and fast, but typically do not last long.

(3) Fuel Wood: Fuel Wood is used to keep the blaze going long enough to fulfill its purpose. Ideally, it should burn slow enough to conserve the woodpile, make plenty of heat, and leave an ample supply of long-lasting coals.

(a) Firewood broken from the dead limbs of standing trees or windfalls held off the ground will have absorbed less moisture and therefore should burn easily.

(b) Refrain from cutting down live, green trees.

(c) Softwoods (evergreens and conifers) will burn hot and fast with lots of smoke and spark, leaving little in the way of coals. Hardwoods (broad leaf trees) will burn slower with less smoke and leave a good bed of coals.

(d) Learn the woods indigenous to the area. Birch, dogwood, and maple are excellent fuels. Osage orange, ironwood, and manzanita, though difficult to break up, make terrific coals. Aspen and cottonwood burn clean but leave little coals.

(e) Stack your wood supply close enough to be handy, but far enough from the flames to be safe. Protect your supply from additional precipitation.

(f) If you happen to go down in an aircraft that has not burned a mixture of gas and oil may be used. Use caution when igniting this mixture.

c. Building the Fire. The type of fire built will be dependent upon its intended use, either cooking or heating and signaling.

(1) Cooking Fires. The following listed fires are best used for cooking:

(a) Teepee Fire: The teepee fire is used to produce a concentrated heat source, primarily for cooking. Once a good supply of coals can be seen, collapse the teepee and push embers into a compact bed.

TEEPEE FIRE

(2) Heating and Signaling Fires.

(a) Pyramid Fire: Pyramid fires are used to produce large amounts of light and heat. They will dry out wet wood or clothing.

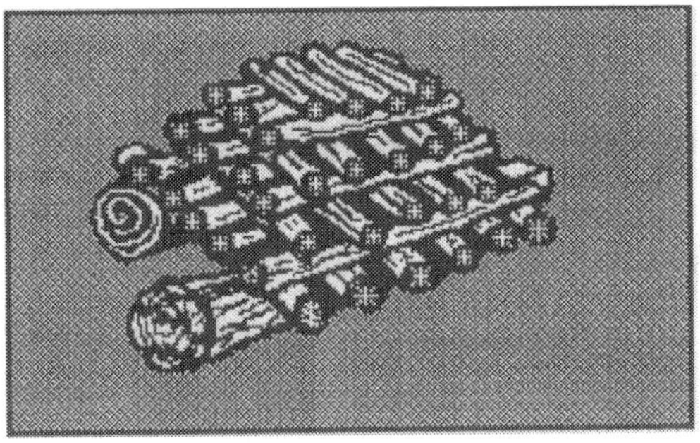

PYRAMID FIRE

d. Starting Fires. Starting a fire is done by a source of ignition and falls into two categories, modern igniters and primitive methods.

(1) Modern Methods: Modern igniters use modern devices we normally think of to start a fire. Reliance upon these methods may result in failure during a survival situation. These items may fail when required to serve their purpose.

(a) Matches and Lighters. Ensure you waterproof these items.

(b) Convex Lens. Binocular, camera, telescopic sights, or magnifying lens are used on bright, sunny days to ignite tender.

(c) Flint and Steel: Sometimes known as metal matches or "Mag Block". Scrape your knife or carbon steel against the flint to produce a spark onto the tinder. Some types of flint & steel designs will have a block of magnesium attached to the device, which can be shaved onto the tinder prior to igniting. Other designs may have magnesium mixed into the flint to produce a higher quality of spark.

(2) Primitive Methods. Primitive fire methods are those developed by early man. There are numerous techniques that fall into this category. The only method that will be taught at MCMWTC is the Bow & Drill.

(3) Bow & Drill. The technique of starting a fire with a bow & drill is a true field expedient fire starting method, which requires a piece of cord and knife from your survival kit to construct. The components of the bow & drill are bow, drill, socket, and fireboard, ember patch, and bird's nest.

(a) Bow: The bow is a resilient, green stick about 3/4 of an inch in diameter and 30-36 inches in length. The bowstring can be any type of cord, however, 550 cord works best. Tie the string from one end of the bow to the other, without any slack.

(b) Drill: The drill should be a straight, seasoned hardwood stick about 1/2 to 3/4 of an inch in diameter and 8 to 12 inches in length. The top end is tapered to a blunt point to reduce friction generated in the socket. The bottom end is slightly rounded to fit snugly into the depression on the fireboard.

(c) Socket: The socket is an easily grasped stone or piece of hardwood or bone with a slight depression on one side. Use it to hold the drill in place and to apply downward pressure.

(d) Fire board: The fire board is a seasoned softwood board which should ideally be 3/4 of an inch thick, 2-4 inches wide, and 8-10 inches long. Cut a depression 3/4 of an inch from the edge on one side of the fireboard. Cut a U-shape notch from the edge of the fireboard into the depression. This notch is designed to collect and form an ember, which will be used to ignite the tinder.

(e) Ember Patch: The ember patch is made from any type of suitable material (i.e., leather, aluminum foil, and bark). It is used to catch and transfer the ember from the fireboard to the bird's nest. Ideally, it should be 4 inches by 4 inches in size.

(f) Bird's Nest: The bird's nest is a double handful of tinder, which will be made into the shape of a nest. Tinder must be dry and finely shredded material (i.e., outer bark from juniper/cedar/sage brush or inner bark from cottonwood/aspen or dry grass/moss). Lay your tinder out in two equal rows about 4 inches wide and 8-12 inches long. Loosely roll the first row into a ball and knead the tinder to further break down the fibers. Place this ball perpendicular onto the second row of tinder and wrap. Knead the tinder until all fibers of the ball are interwoven. Insert the drill half way into the ball to form a partial cylinder. This is where the ember will be placed.

(4) Producing a fire using the bow & drill.

(a) Place the ember patch under the U-shaped notch.

(b) Assume the kneeling position, with the left foot on the fireboard near the depression.

(c) Load the bow with the drill. Ensure the drill is between the wood of the bow and bow string. Place the drill into the depression on the fireboard. Place the socket on the tapered end of the drill.

(d) Use the left hand to hold the socket while applying downward pressure.

(e) Use the right hand to grasp the bow. With a smooth sawing motion, move the bow back and forth to twirl the drill.

(f) Once you have established a smooth motion, smoke will appear. Once smoke appears, apply more downward pressure and saw the bow faster.

(g) When a thick layer of smoke has accumulated around the depression, stop all movement. Remove the bow, drill, and socket from the fire board, without moving the fireboard. Carefully remove your left foot off the fireboard.

(h) Gently tap the fireboard to ensure all of the ember has fallen out of the U-shaped notch and is lying on the ember patch. Remove the fireboard.

(i) Slowly fan the black powder to solidify it into a glowing ember. Grasping the ember patch, carefully drop the ember into the cylinder of the bird's nest.

(j) Grasp the bird's nest with the cylinder facing towards you and parallel to the ground. Gently blow air into the cylinder. As smoke from the nest becomes thicker, continue to blow air into the cylinder until fire appears.

(5) Trouble Shooting the Bow & Drill

(a) Drill will not stay in depression- Apply more downward pressure and/or increase width/depth of depression.

(b) Drill will not twirl- lessens the amount of downward pressure and/or tighten bowstring.

(c) Socket smoking- Lessen the amount of downward pressure. Wood too soft when compared to hardness of drill. Add some lubrication: animal fat, oil, or grease.

(d) No smoke- Drill and fireboard are the same wood. Wood may not be seasoned. Check drill to ensure that it is straight. Keep left hand locked against left shin while sawing.

(e) Smoke but no ember- U-shaped notch not cut into center of the depression.

(f) Bowstring runs up and down drill- use a locked right arm when sawing. Check drill to ensure that it is straight. Ensure bowstring runs over the top of the left boot.

(g) Birds nest will not ignite- Tinder not dry. Nest woven too tight. Tinder not kneaded enough. Blowing too hard (ember will fracture).

e. Extinguishing the Fire. The fire must be properly extinguished. This is accomplished by using the drown, stir, and feel method.

(1) Drown the fire by pouring at water in the fire lay.

(2) Stir the ember bed to ensure that the fire is completely out.

(3) Check the bed of your fire by feeling for any hot spots.

(4) If any hot spots are found, start the process all over again.

UNITED STATES MARINE CORPS
Mountain Warfare Training Center
Bridgeport, California 93517-5001

FMST.07.22
04/07/02

STUDENT OUTLINE

SURVIVAL NAVIGATION

LEARNING OBJECTIVES

TERMINAL LEARNING OBJECTIVE. In a summer mountainous environment, navigate in a survival situation, in accordance with the references. (FMST.07.22)

ENABLING LEARNING OBJECTIVES

(1)Without the aid of references and from a given list, choose one of the considerations for travel, in accordance with the references. (FMST.07.22a)

(2)Without the aid of references and from a given list, choose two methods for locating the North Star, in accordance with the references. (FMST.07.22b)

OUTLINE

1. **CONSIDERATIONS FOR STAYING OR TRAVELLING**. (FMST.07.22a)

    a. Stay with the aircraft or vehicle if possible. More than likely somebody knows where it was going. It is also a ready-made shelter.

    b. Leave only when:

        (1) Certain of present location; have known destination and the ability to get there.

        (2) Water, food, shelter, and/or help can be reached.

        (3) Convinced that rescue is not coming.

A5

c. If the decision is to travel, the following must also be considered:

(1) Which direction to travel and why.

(2) What plan is to be followed.

(3) What equipment should be taken

(4) How to mark the trail.

(5) Predicted weather.

d. If the tactical situation permits leave the following information at the departure point:

(1) Departure time.

(2) Destination.

(3) Route of travel/direction.

(4) Personal condition.

(5) Available supplies.

2. DAYTIME SURVIVAL NAVIGATION

a. Sun Movement. It is generally taken for granted that the sun rises in the east and sets in the west. This rule of thumb, however, is quite misleading. In fact, depending on an observer's latitude and the season, the sun could rise and set up to 50 degrees off of true east and west.

b. The following diagram and terms are essential to understanding how the sun and stars can help to determine direction:

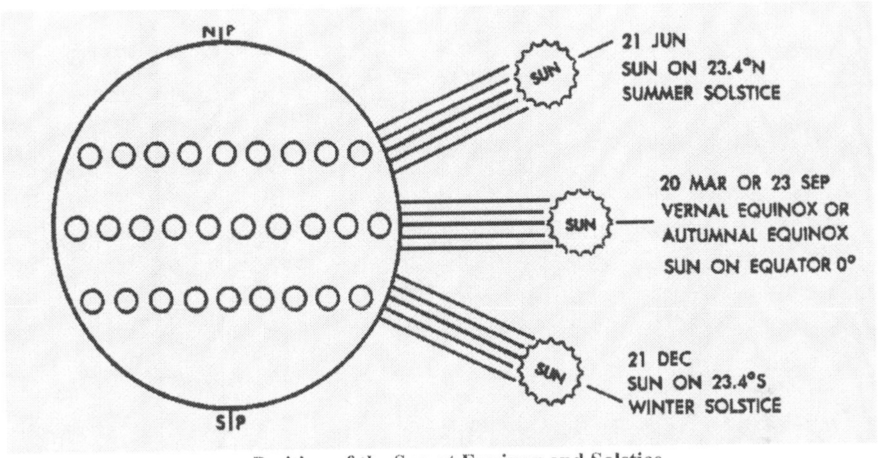

Position of the Sun at Equinox and Solstice

(1) Summer/Winter Solstice: (21 June/21 December) Two times during the year when the sun has no apparent northward or southward motion.

(2) Vernal/Autumnal Equinox: (20 March/23 September) Two times during the year when the sun crosses the celestial equator and the length of day and night are approximately equal.

c. Sun's Movement. As reflected in the diagram above, the sun continuously moves in a cycle from solstice to equinox, throughout each day, however, the sun travels a uniform arc in the sky from sunrise to sunset. Exactly halfway along its daily journey, the sun will be directly south of an observer (or north if the observer is in the Southern Hemisphere). This rule may not apply to observers in the tropics (between 23.5 degrees north and south latitude) or in the Polar Regions (60 degrees latitude). It is at this point that shadows will appear their shortest. The time at which this occurs is referred to as "local apparent noon."

d. Local Apparent Noon. Whenever using any type of shadow casting device to determine direction, "local apparent noon" (or the sun's highest point during the day) must be known. Local apparent noon can be determined by the following methods.

(1) Knowing sunrise and sunset from mission orders, i.e., sunrise 0630 and sunset 1930. Take the total amount of daylight (13 hours), divide by 2 (6 hours 30 minutes), and add to sunrise (0630 plus 6 hours 30 minutes). Based on this example, local apparent noon would be 1300.

(2) Using the string method. The string method is used to find two equidistant marks before and after estimated local apparent noon. The center point between these two marks represents local apparent noon.

e. Sun's Bearing. With an understanding of the sun's daily movement, as well as its seasonal paths, a technique is derived that will determine the true bearing of the sun at sunrise and sunset. With the aid of a circular navigational chart, we can accurately navigate based on the sun's true bearing:

(1) Determine the sun's maximum amplitude at your operating latitude using the top portion of the chart.

(2) Scale the center baseline of the chart where 0 appears as the middle number; write in the maximum amplitude at the extreme north and south ends of the baseline.

(3) Continue to scale the baseline; you should divide the baseline into 6 to 10 tick marks that represent equal divisions of the maximum amplitude.

(4) From today's date along the circumference, draw a straight line down until it intersects the baseline.

(5) The number this line intersects is today's solar amplitude. If the number is left of 0, it is"north" amplitude; if the number is right of 0, it is a "south" amplitude. Use the formula at the bottom of the chart to determine the sun's bearing at sunrise or sunset.

LATITUDE (N OR S)	5	10	15	20	25	30	35	40	45	50	55	60
MAXIMUM AMPLITUDE	24°	24°	24°	25°	26°	27°	29°	31°	34°	38°	44°	53°

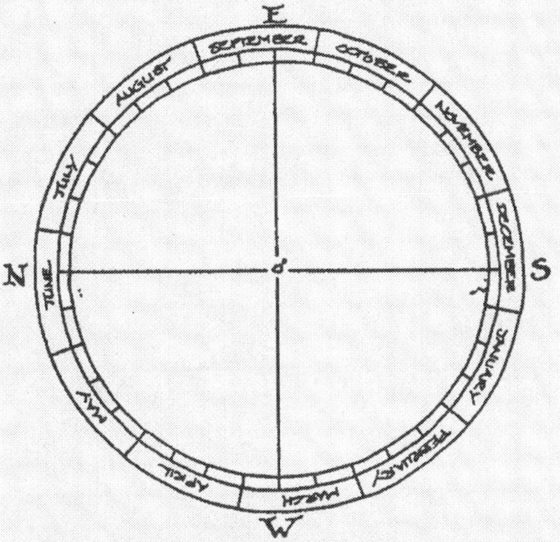

NORTH AMPLITUDES	SOUTH AMPLITUDES
SUNRISE BEARING = 90° − AMPLITUDE	SUNRISE BEARING = 90° + AMPLITUDE
SUNSET BEARING = 270° + AMPLITUDE	SUNSET BEARING = 270° − AMPLITUDE

Circular Navigational Chart

f. Shadow Stick Construction. This technique will achieve a cardinal direction within 10 degrees of accuracy if done within two hours of local apparent noon. Once again, this technique may be impractical near the Polar Regions, as shadows tend to be very long; similarly, in the tropics shadows are generally very small.

(1) Get a straight, 3-6 foot stick free of branches and pointed at the ends and 3-5 small markers: i.e., sticks, rocks, or nails.

(2) Place the stick upright in the ground and mark the shadow tip with a marker.

(3) Wait 10-15 minutes and mark shadow tip again with a marker.

(4) Repeat this until all of the markers are used.

SHADOW STICK METHOD

(5) The markers will form a West—East line.

(6) Put your left foot on the first marker and your right foot on the last marker, you will then be facing north.

3. **POCKET NAVIGATOR** The only material required is a small piece of paper or other flat-surface material upon which to draw the trace of shadow tips and a 1 to 2 inch pin, nail, twig, wooden matchstick, or other such device to serve as a shadow-casting rod.

a. Set this tiny rod upright on your flat piece of material so that the sun will cause it to cast a shadow. Mark the position where the base of the rod sits so it can be returned to the same spot for later readings. Secure the material so that it will not move and mark the position of the material with string, pebbles, or twigs, so that if you have to move the paper you can put it back exactly as it was. Now, mark the tip of the rod shadow.

b. As the sun moves, the shadow-tip moves. Make repeated shadow-tip markings every 15 minutes. As you make the marks of the shadow tip, ensure that you write down the times of the points.

c. At the end of the day, connect the shadow-tip markings. The result will normally be a curved line. The closer to the vernal or autumnal equinoxes (March 21 and September 23) the less pronounced the curvature would be. If it is not convenient or the tactical situation does not permit to take a full day's shadow-tip markings, your observation can be continued on subsequent days by orienting the pocket navigator on the ground so that the shadow-tip is aligned with a previously plotted point.

d. The markings made at the sun's highest point during the day, or solar noon, is the north—south line. The direction of north should be indicated with an arrow on the navigator as soon as it is determined. This north-south line is drawn from the base of the rod to the mark made at solar noon. This line is the shortest line that can be drawn from the base of the pin to the shadow-tip curve.

e. To use your pocket navigator, hold it so that with the shadow-tip is aligned with a plotted point at the specified point. I.e.; if it is now 0900 the shadow-tip must be aligned with that point. This will ensure that your pocket navigator is level. The drawn arrow is now oriented to true north, from which you can orient yourself to any desired direction of travel.

f. The pocket navigator will work all day and will not be out of date for approximately one week.

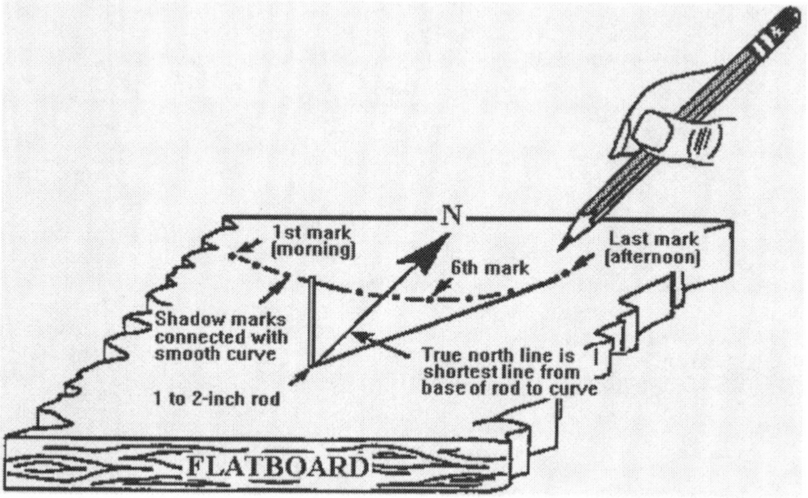

POCKET NAVIGATOR

4. NIGHTTIME SURVIVAL NAVIGATION

a. Mark North. To aid you in navigating at night, it is beneficial to watch where the sun goes down. If you're going to start moving after dark mark the northerly direction.

b. Locating the North Star. There are two methods used in locating the North Star. (FMST.07.22b)

(1) Using the Big Dipper (*Ursa Major*). The best indictors are the two "dippers ". The North Star is the last star in the handle of the Little Dipper, which is not the easiest constellation to find. However, the Big Dipper is one of the most prominent constellations in the Northern Hemisphere. The two lowest stars of the Big Dipper's cup act as pointers to the North Star. If you line up these two stars, they make a straight line that runs directly to the North Star. The distance to the North Star along this line is 5 times that between the two pointer stars.

(2) Using Cassiopeia (*Big M or W*). Draw a line straight out from the center star, approximately half the distance to the Big Dipper. The North Star will be located there.

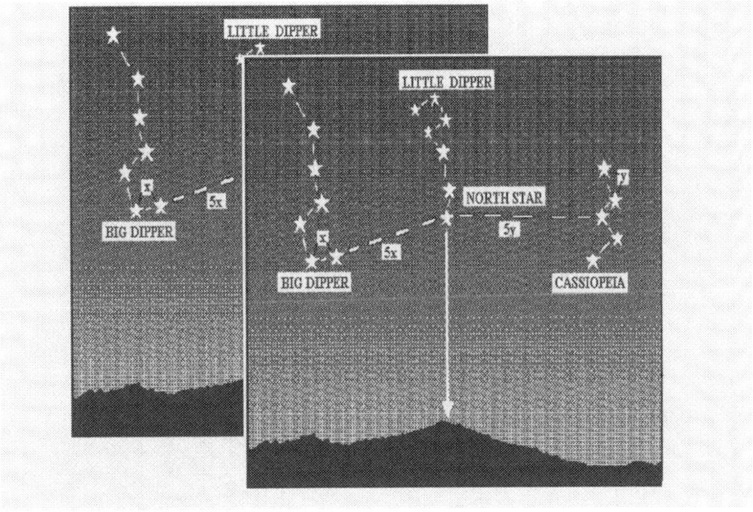

LOCATING THE NORTH STAR

NOTE: Because the Big Dipper and Cassiopeia rotate around the North Star, they will not always appear in the same position in the sky. In the higher latitudes, the North Star is less effective for the purpose of orienting because it appears higher in the sky. At the center of the Arctic Circle, it would be directly overhead and all directions lead south.

c. Southern Cross. In the Southern Hemisphere, Polaris is not visible. There, the Southern Cross is the most distinctive constellation. An imaginary line through the long axis of the Southern Cross, or True Cross, points towards a dark spot devoid of stars approximately three degrees offset from the South Pole. The True Cross should not be confused with the larger cross nearby know as the False Cross, which is less bright, more widely spaced, and has five stars. Two closely spaced, very bright stars that trail behind the crosspiece can confirm the True Cross. These two stars are often easier to pick out than the cross itself. Look for them. Two of the stars in the True Cross are among the brightest stars in the heavens; they are the stars on the southern and eastern arms. The stars on the northern and western arms are not as conspicuous, but are bright.

NOTE: The imaginary point depicted in the picture is the dark spot devoid of stars.

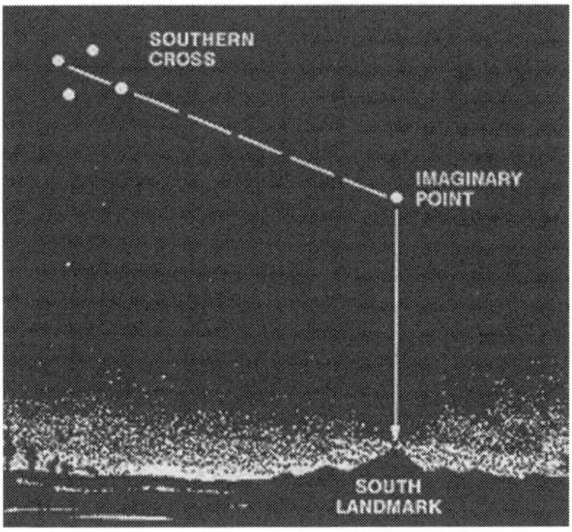

SOUTHERN CROSS

d. Moon Navigator. Like the sun, the moon rises in the east and sets in the west. Use the same method of the shadow stick as you did during the day.

5. **IMPROVISED COMPASSES**. There are three improvised techniques to construct a compass.

a. Synthetic technique. The required items are a piece of synthetic material, (i.e., parachute cloth), and a small piece of iron or steel that is long, thin, and light. Aluminum or yellow metals won't work (only things that rust will do). A pin or needle is perfect, but a straightened paper clip, piece of steel baling wire or barbed wire could also work.

(1) Stroke the needle repeatedly in one direction against the synthetic material. Ensure that you lift the material a few inches up into the air at the end of each stroke, returning to the beginning of the needle before descending for another stroke in the same direction. Do this approximately 30 strokes. This will magnetize the needle.

(2) Float the metal on still water using balled up paper, wood chip, or leaf. Gather some water in a non-magnetic container or a scooped out recess in the

ground, such as a puddle. Do not use a "tin can" which is made of steel. (Aluminum can would be fine.) Place the float on the water, then the metal on it. It will slowly turn to orient itself.

b. Magnet technique. You will achieve the same results by using a magnet. Follow the same steps as you did with the synthetic material. The magnets you are most likely to have available to you are those in a speaker or headphones of a radio.

c. Magnetization through a battery. A power source of 2 volts or more from a battery can be used with a short length of insulated wire to magnetize metal. Coil the wire around a needle. If the wire is non-insulated, wrap the needle with paper or cardboard. Attach the ends to the battery terminals for 5 minutes.

d. Associated problems with improvised compasses. The following are common problems with all improvised compasses.

(1) Soft steel tends to lose its magnetism fairly quickly, so you will have to demagnetize your needle occasionally, though you should not have to do this more than two or three times a day.

(2) Test your compass by disturbing it after it settles. Do this several times. If it returns to the same alignment, you're OK. It will be lined up north and south, though you will have to determine by other means which end is north. Use the sun, stars, or any other natural signs in the area.

(3) Remember that this will give magnetic north. In extreme northern latitudes, the declination angle can be extreme.

6. SURVIVAL NAVIGATION TECHNIQUES

a. Navigator.

(1) Employ a navigation method.

(2) Find the cardinal direction.

(3) Pick a steering mark in the desired direction of travel.

b. Maintain a Log: The possibility may arise when you will not have a map of the area. A log will decrease the chance of walking in circles.

(1) Construction.

(a) Use any material available to you i.e., paper, clothing, MRE box, etc.

(b) Draw a field sketch annotating North, prominent terrain features, and Friendly/enemy position.

(2) Maintenance.

(a) Annotate distance traveled, elevation gained and lost, and cardinal directions.

(b) Maintain and update field sketch as movement progresses.

(c) Ensure readability of your field sketch. (I.e., don't clutter the sketch so much that it can't be read.)

c. <u>During Movement Constantly Refer To</u>.

(1) Log.

(2) Steering marks.

d. <u>Actions If You Become Lost</u>.

(1) Immediate action

(a) Orient your sketch. This will probably make your mistake obvious.

(2) Corrective action

(a) Backtrack using steering marks until you have determined the location of your error.

(b) Re-orient your sketch.

(c) Select direction of travel and continue to march.

UNITED STATES MARINE CORPS
Mountain Warfare Training Center
Bridgeport, California 93517-5001

FMST.07.08
04/07/02

STUDENT OUTLINE

WATER PROCUREMENT

LEARNING OBJECTIVES

TERMINAL LEARNING OBJECTIVE Given a unit in a cold weather environment and the necessary equipment and supplies, perform cold weather preventive medicine, in accordance with the references. (FMST.07.08)

ENABLING LEARNING OBJECTIVES

(1) Without the aid of references from a given list, choose one of the techniques for finding incidental water, in accordance with the references. (FMST.07.08i)

(2) Without the aid of references from a given list, choose one of the four hazardous fluids to avoid substituting for potable water, in accordance with the references. (FMST.07.08j)

(3) Without the aid of references, construct a water-producing device, in accordance with the references. (FMST.07.08k)

OUTLINE

1. WATER INTAKE

a. Thirst is not a strong enough sensation to determine how much water you need.

b. The best plan is to drink, utilizing the OVER DRINK method. Drink plenty of water anytime it is available and particularly when eating.

A6

c. Dehydration is a major threat. A loss of only 5 % of your body fluids causes thirst, irritability, nausea, and weakness; a 10% loss causes dizziness, headache, inability to walk, and a tingling sensation in limbs; a 15% loss causes dim vision, painful urination, swollen tongue, deafness, and a feeling of numbness in the skin; also a loss of more than 15% body fluids could result in death.

d. Your water requirements will be increased if:

(1) You have a fever.

(2) You are experiencing fear or anxiety.

(3) You evaporate more body fluid than necessary. (i.e., not using the proper shelter to your advantage)

(4) You have improper clothing.

(5) You ration water.

(6) You overwork.

2. **INCIDENTAL WATER**. (FMST.07.08.i)

a. During movement, you may have to ration water until you reach a reliable water source. Incidental water may sometimes provide opportunities to acquire water. Although not a reliable or replenished source, it may serve to stretch your water supply or keep you going in an emergency. The following are sources for incidental water:

(1) Dew: In areas with moderate to heavy dew, tying rags or tuffs of fine grass around your ankles can collect dew. While walking through dewy grass before sunrise, the rags or grass will saturate and can be rung out into a container. The rags or grass can be replaced and the process is repeated.

(2) Rainfall: Rainwater collected directly in clean container or in plants that contain no harmful toxins is generally safe to drink without disinfecting. The survivor should always be prepared to collect rainfall at a moment's notice. An inverted poncho works well to collect rainfall.

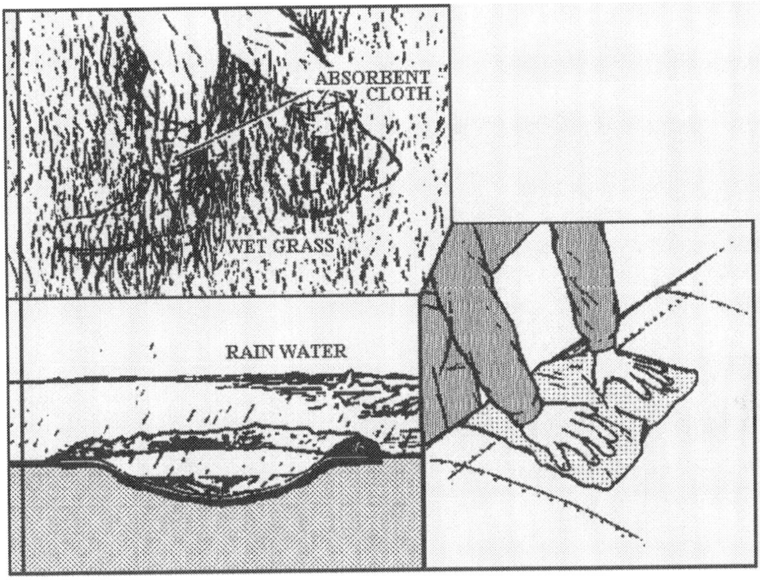

3. **HAZARDOUS FLUIDS** (FMST.07.08j)

a. Survivors have occasionally attempted to augment their water supply with other fluids, such as alcoholic beverages, urine, blood, or seawater. While it is true that each of these fluids has a high water content, the impurities they contain may require the body to expend more fluid to purify them. Some hazardous fluids are:

(1) Seawater: Seawater in more than minimal quantities are actually toxic. The concentration of sodium and magnesium salts is so high that fluid must be drawn from the body to eliminate the salts and eventually the kidneys cease to function.

(2) Alcohol: Alcohol dehydrates the body and clouds judgment. Super-cooled liquid, if ingested, can cause immediate frostbite of the throat, and potential death.

(3) Blood: Blood, besides being salty, is a food. Drinking it will require the body to expend additional fluid to digest it.

(4) Urine: Drinking urine is not only foolish, but also dangerous. Urine is nothing more than the body's waste. Drinking it only places this waste back into the body, which requires more fluid to process it again.

4. **WATER QUALITY**. Water contains minerals, toxins, and pathogens. Some of these, consumed in large enough quantities may be harmful to human health. Pathogens are our primary concern. Pathogens are divided into Virus, Cysts, Bacteria, and Parasites. Certain pathogens are more resistant to chemicals and small enough to move through microscopic holes in equipment (i.e., T-shirt, parachute). Certain pathogens also have the ability to survive in extremely cold water temperatures. Pathogens generally do not live in snow and ice. Water quality is divided into three levels of safety with disinfecting as the most desired level, then purified, followed by potable.

   a. Disinfecting. Water disinfecting removes or destroys harmful microorganisms. Giardia cysts are an ever-present danger in clear appearing mountain water throughout the world. By drinking non-potable water you may contract diseases or swallow organisms that could harm you. Examples of such diseases or organisms are Dysentery, Cholera, Typhoid, Flukes, and Leeches.

   b. Remember impure water, no matter how overpowering the thirst, is one of the worst hazards in a survival situation.

   c. The first step in disinfecting is to select a treatment method. The two methods we will discuss are as follows: (FMST.07.08j)

      (1) Heat. The Manual of Naval Preventive Medicine (P-5010) states that you must bring the water to a rolling boil before it is considered safe for human consumption. This is the most preferred method.

         (a) Bringing water to the boiling point will kill 99.9% of all Giardia cysts. The Giardia cyst dies at 60°C and Cryptosporidium dies at 65C. Water will boil at 14,000' at 86°C and at 10,000' at 90C. With this in mind you should note that altitude does not make a difference unless you are extremely high.

      (2) Chemicals. There are numerous types of chemicals that can disinfect water. Below are a few of the most common. In a survival situation, you will use whatever you have available.

         (a) Iodine Tablets.

         (b) Chlorine Bleach.

         (c) Iodine Solution.

         (d) Betadine Solution.

         (e) Military water purification tablets. These tablets are standard issue for all DOD agencies. These tablets have a shelf life of four years from the date of manufacture, unless opened. Once the seal is broken, they have a shelf life of one year, not to exceed the initial expiration date of four years.

49703 Month / Year / Batch Number

(3) Water Disinfecting Techniques and Halogen Doses.

Iodination techniques Added to 1 liter or quart of water	Amount for 4 PPM	Amount for 8 PPM
Iodine tablets Tetraglcine hydroperiodide EDWGT Potable Aqua Globaline	½ tablet	1 tablet
2% iodine solution (tincture)	0.2 ml 5 gtts	0.4 ml 10 gtts
10% povidone-iodine solution*	0.35 ml 8 gtts	0.70 ml 16 gtts
Chlorination techniques	Amount for 5 PPM	Amount for 10 PPM
Household bleach 5% Sodium hypochlorie	0.1 Ml 2 gtts	0.2 ml 4 gtts
AquaClear Sodium dichloroisocyanurate		1 tablet
AquaCure, AquaPure, Chlor-floc Chlorine plus flocculating agent		8 PPM 1 tablet

*Providone-iodine solutions release free iodine in levels adequate for disinfecting, but scant data is available.

Measure with dropper (1 drop=0.05 ml) or tuberculin syringe

PPM-part per million gtts-drops ml-milliliter

Concentration of Halogen **Contact time in minutes at various water temperatures**

	5 C / 40 F	15 C / 60 F	30 C / 85 F
2 PPM	240	180	60
4 PPM	180	50	45
8 PPM	60	30	15

NOTE: These contact times have been extended from the usual recommendations to account for recent data that prolonged contact time is needed in very cold water to kill *Giardia* cysts.

NOTE: Chemicals may not destroy Cryptosporidium.

d. Purification. Water purification is the removal of organic and inorganic chemicals and particulate matter, including radioactive particles. While purification can eliminate offensive color, taste, and odor, it may not remove or kill microorganisms.

(1) Filtration. Filtration purifying is a process by which commercial manufacturers build water filters. The water filter is a three-tier system. The first layer, or grass layer, removes the larger impurities. The second layer, or sand layer, removes the smaller impurities. The final layer, or charcoal layer (**not the ash but charcoal from a fire**), bonds and holds the toxins. All layers are placed on some type of straining device and the charcoal layer should be at least 5-6 inches thick. Layers should be changed frequently and straining material should be boiled. Remember that this is not a disinfecting method, cysts can possibly move through this system.

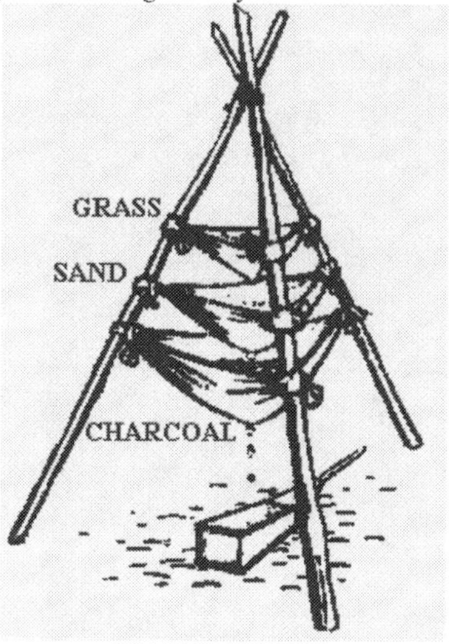

WATER FILTER

(2) Commercial Water Filters. Commercial water filters are generally available in most retail stores and may be with you. Understanding what the filter can do is the first step in safeguarding against future illnesses.

(a) A filter that has a .3 micron opening or larger will not stop Cryptosporidium.

(b) A filter system that does not release a chemical (i.e., iodine) may not kill all pathogens.

(c) A filter that has been overused may be clogged. Usage may result in excessive pumping pressure that can move harmful pathogens through the opening.

e. Potable. Potable indicates only that a water source, on average over a period of time, contains a "minimal microbial hazard," so the statistical likelihood of illness is acceptable.

(1) Sedimentation. Sedimentation is the separation of suspended particles large enough to settle rapidly by gravity. The time required depends on the size of the particle. Generally, 1 hour is adequate if the water is allowed to sit without agitation. After sediment has formed on the bottom of the container, the clear water is decanted or filtered from the top. Microorganisms, especially cysts, eventually settle, but this takes longer and the organisms are easily disturbed during pouring or filtering. Sedimentation should not be considered a means of disinfecting and should be used only as a last resort or in an extreme tactical situation.

5. **SOLAR STILLS.** (FMST.07.08k)

a. Solar stills are designed to supplement water reserves. Contrary to belief, they will not provide enough water to meet the daily requirement for water.

b. AboveGround Solar Still. This device allows the survivor to make water from vegetation. To make the aboveground solar still, locate a sunny slope on which to place the still, a clear plastic bag, green leafy vegetation, and a small rock.

(1) Construction.

(a) Fill the bag with air by turning the opening into the breeze or by "scooping" air into the bag.

(b) Fill the bag half to three-quarters full of green leafy vegetation. Be sure to remove all hard sticks or sharp spines that might puncture the bag.

CAUTION
Do not use poisonous vegetation. It will provide poisonous liquid.

(c) Place a small rock or similar item in the bag.

(d) Close the bag and tie the mouth securely as close to the end of the bag as possible to keep the maximum amount of air space. If you have a small piece of tubing, small straw, or hollow reed, inserts one end in the mouth of the bag before tying it securely. Tie off or plug the tubing so that air will not escape. This tubing will allow you to drain out condensed water without untying the bag.

(e) Place the bag, mouth downhill, on a slope in full sunlight. Position the mouth of the bag slightly higher than the low point in the bag.

(f) Settle the bag in place so that the rock works itself into the low point in the bag.

(g) To get the condensed water from the still, loosen the tie and tip the bag so that the collected water will drain out. Retie the mouth and reposition the still to allow further condensation.

(h) Change vegetation in the bag after extracting most of the water from it.

(i) Using 1-gallon zip-loc. bag instead of trash bags is a more efficient means of construction.

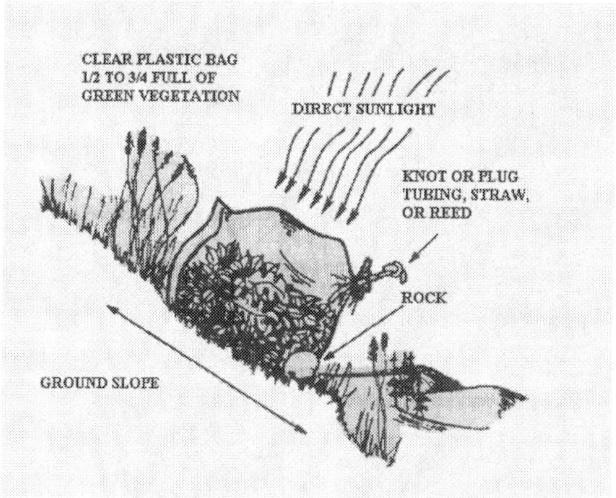

<u>**ABOVE GROUND SOLAR STILL**</u>

c. <u>BelowGround Solar Still</u>. Materials consist of a digging stick, clear plastic sheet, container, rock, and a drinking tube.

(1) Construction.

(a) Select a site where you believe the soil will contain moisture (such as a dry streambed or a low spot where rainwater has collected). The soil should be easy to dig, and will be exposed to sunlight.

(b) Dig a bowl-shaped hole about 1 meter across and 24 inches deep.

(c) Dig a sump in the center of the hole. The sump depth and perimeter will depend on the size of the container you have to place in it. The bottom of the sump should allow the container to stand upright.

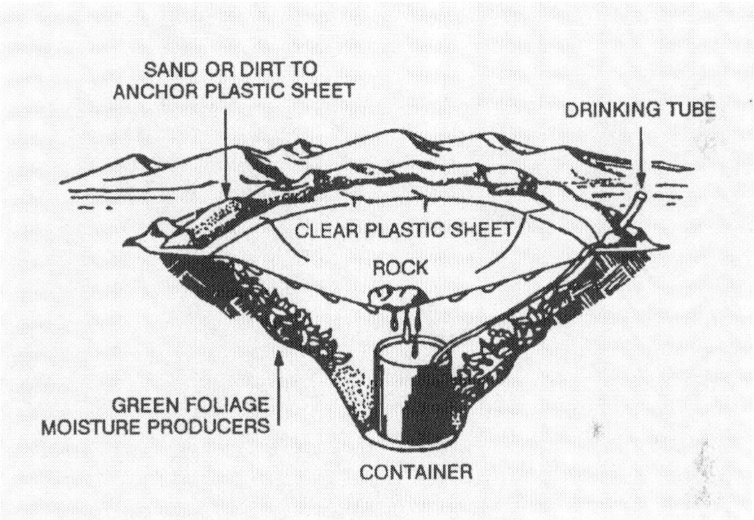

BELOW GROUND SOLAR STILL

(d) Anchor the tubing to the container's bottom by forming a loose overhand knot in the tubing. Extend the unanchored end of the tubing up, over, and beyond the lip of the hole.

(e) Place the plastic sheet over the hole, covering its edges with soil to hold in place. Place a rock in the center of the plastic sheet.

(f) Lower the plastic sheet into the hole until it is about 18 inches below ground level. Make sure the cone's apex is directly over the container. Ensure the plastic does not touch the sides of the hole because the earth will absorb the moisture.

(g) Put more soil on the edges of the plastic to hold it securely and prevent the loss of moisture.

(h) Plug the tube when not in use so that moisture will not evaporate.

(i) Plants can be placed in the hole as a moisture source. If so, dig out additional soil from the sides.

(j) If polluted water is the only moisture source, dig a small trough outside the hole about 10 inches from the still's lip. Dig the trough about 10 inches deep and 3 inches wide. Pour the polluted water in the trough. Ensure you do not spill any polluted water around the rim of the hole where the plastic touches the soil. The trough holds the polluted water and the soil filters it as the still draws it. This process works well when the only water source is salt water.

(k) Three stills will be needed to meet the individual daily water intake needs.

UNITED STATES MARINE CORPS
Mountain Warfare Training Center
Bridgeport, California 93517-5001

WMC
04/07/02

STUDENT HANDOUT

FORAGING PLANTS AND INSECTS FOR SURVIVAL USES

PURPOSE The purpose of this period of instruction is to familiarize the Wilderness Medicine Student with the potential resources for survival sustenance.

OUTLINE

1. GENERAL CONSIDERATIONS. There are few places without some type of edible vegetation. Plants contain vitamins, minerals, protein, carbohydrates, and dietary fiber. Some plants also contain fats. The following are general considerations:

a. Do not assume that because birds or animals have eaten a plant, it is edible by humans.

b. Poor plant recognition skills will seriously limit your ability to survive.

c. Plant dormancy and snowfall make foraging plants difficult during the winter months.

d. Plants generally poison by:

(1) Ingestion: When a person eats a part of a poisonous plant.

(2) Contact: When a person makes contact with a poisonous plant that cause any type of skin irritation or dermatitis.

(3) Absorption: When a person absorbs the poison through the skin, which can interrupt a bodily function.

(4) Inhalation: Poisoning can occur through the inhalation of smoke that contains poisonous plant residue.

A7

e. Plant properties can change throughout the growing season. Plants can be edible during certain periods while poisonous in others.

2. PLANT RECOGNITION. Plant recognition can take years of study to learn. The following is used as a general guideline.

a. Types of plants that should be avoided are:

(1) All mushrooms.

(2) Plants with a milky sap.

(3) White and yellow berries should be avoided, as they are almost always poisonous.

(4) Plants with shiny leaves should be considered poisonous.

(5) Plants that are irritants to the skin should not be eaten.

(6) All beans and peas should be avoided.

b. Edible Plants: There are several edible plants found at the Marine Corps Mountain Warfare Training Center. All are easy to recognize and many of the same plants are located throughout the United States and abroad.

(1) Wild Onions - Complete plant.

(2) Dandelion - Roots and leaves.

(3) Mallow - Leaves, stem and flowers.

(4) Watercress -Complete plant.

(5) Cattail - Root, stalk and stem.

(6) Bull Thistle - The flower.

(7) Juniper Tree - Berries and cambium layer.

(8) All Conifers - Cambium layer, needles, and nuts within the cones.

(9) Currants - Berries.

c. Preparing an edible plant. All edible plants should be boiled twice with a change of water to remove any bitterness, acids, or bacteria that the plant may contain.

3. PLANT TESTING PROCEDURE. Always adopt the following procedure when trying unknown plants as food.

a. Wait 8 hours without eating.

b. Select a plant that grows in sufficient quantity in the local area. Separate the part of the plant you wish to test root, stem, leaf, or flower. Certain parts of plants are poisonous while the, other parts may be edible.

c. Rub a portion of the plant you have selected on your inner forearm. Wait 15 minutes and look for any swelling, rash, or irritation.

d. Boil the plant or plant part in changes of water. The toxic properties of many plants are water-soluble or can be destroyed by heat. Cooking and discarding two changes of water can lessen the amount of poisonous material or remove it completely. These boiling periods should last at least 5 minutes each.

e. Place 1 teaspoon of the prepared plant food in the mouth and chew for 5 minutes, but do not swallow. If unpleasant effects occur (burning, bitter, or nauseating taste), remove the plant from the mouth at once and discard it as a food source. If no unpleasant effects occur, swallow the plant material and wait 8 hours.

f. If after 8 hours no unpleasant effects have occurred (nausea, cramps, and diarrhea), eat two tablespoonfuls and wait 8 hours.

g. In no unpleasant effects have occurred at the end of this 8-hour period, the plant may be considered edible.

h. Completely document and sketch the plant in a logbook to refer to for future use. This will aid in future procurement of this plant. If plant properties have changed, you will have to repeat the plant testing procedure.

NOTE: Have charcoal ready to eat. If any side effects occur, attempt to induce vomiting and immediately eat charcoal. Charcoal will absorb toxins and reduce the chance of Death.

4. MEDICINAL PLANTS One of the many things that hinder a survivor's ability to survive is medical problems. Injuries incurred will reduce survival expectancy and the ability to evade.

a. Coyote Willow: Coyote Willow is a thick forming shrub with clustered stems and very narrow leaves. It habits wet soils, especially riverbanks, sandbars, and silt flats.

(1) Drinking tea made from the inner bark can reduce fever.

(2) The dried and powdered inner bark can be used to stop bleeding.

(3) Administering a half-cup of willow charcoal dissolved in water can treat diarrhea.

(4) The Coyote Willow roots can be mashed and applied to tooth aches.

b. Yarrow: Yarrow is a plant (tall, usually not branched, with many white, slat topped group flowers). Do not confuse Yarrow with Water Hemlock.

(1) Rubbing a handful of crushed Yarrow flowers and leaves on any exposed can make insect repellent skin.

(2) Placing a Yarrow leaf poultice on the wound can stop bleeding.

(3) Relief from many rashes, including poison oak and ivy, can be achieved by applying a Yarrow leaf compress to the effected area.

(4) The Yarrow root can be chewed to relieve the pain of a toothache or break a fever.

(5) A potent anesthetic can be made by scrubbing fresh Yarrow roots in water to clean them. Once the roots are clean, crush them into a spongy mass and apply gently to the wounded area.

c. Tea To make a tea, place the plants in a container and pour in two cups of boiling water. Let the mixture steep for 20 minutes and drink.

d. Poultice: Place plant (Coyote Willow, Yarrow, etc.) in gauze or other similar material and fold it so the gauze will hold the plant in place. Put gauze in a larger cloth, about 6"x8", and roll the sides inward. Fold the cloth over without losing the plant. While boiling water, dip the bottom portion of the cloth containing the plant into the hot water by holding the edges. Keep the plant submerged in the boiling water until it becomes saturated. Bring the cloth straight up and with a twisting motion, wring the excess water back into the pot. Apply the poultice to the affected area as soon as it has cooled enough to place on the wound. To be effective, the poultice should be as hot as you can tolerate it.

e. Compress: A compress is made just like a poultice; except it is cold when applied to the wound. NOTE: Poultice or Compress should be applied for 1 to 24 hours, as needed. When applying a poultice, you may experience a throbbing pain as it draws out the infection and neutralizes toxins. When the pain subsides, the poultice has accomplished its task and should be removed. Apply a fresh poultice as needed until the desired level of healing has been reached.

5. POISONOUS PLANTS. Rarely will a survivor have an ideal means of killing large game, however there are certain plants that can aid the survivor. The two plants that we will talk about are the Water Hemlock and Monkshood.

a. Preparing Poisonous Plants: Once a poisonous plant has been located, dig up the root of the plant. The roots of the Water Hemlock and Monkshood generally grow 8 to 10 inches deep. Extreme caution must be used when handling the root. Do not handle the root without a barrier between your hands and the root. A barrier can be gloves, socks, T-shirt, or even moss. Split the root lengthwise to expose the inside of the root where the toxin is located. With the root split, rub the tip of your spear/arrow inside of the root opening. In a slow controlled manner, work from the bottom of the tip to the top. Once the tip is thoroughly coated, allow the toxin to dry and apply another coat to the tip. Continue to apply coat after coat of toxin, until the root is completely drained of its' toxin. You are now finished and the tip is ready for use.

NOTE: Do not eat the meat in the vicinity of the poisoned wound.

CAUTION: Do not apply poison to the actual point and edges of tip. This avoids accidental puncture to you. Wash and boil all gloves and clothing used as a barrier before wearing

6. INSECTS. Insects are the most abundant life form on earth and are an excellent survival food. They are easy to catch and provide 65-80% protein; compared to 20% for beef. They aren't too appetizing, but personal bias has no place in a survival situation. The focus must remain on maintaining your health.

a. Insects to avoid.

(1) All adults that sting or bite.

(2) Hairy or brightly colored insects.

(3) All Caterpillars.

(4) Insects that have a pungent odor.

(5) All spiders.

(6) Disease carriers like ticks, flies, or mosquitoes.

b. Edible Insects.

(1) Insect Larvae.

(2) Grasshoppers.

(3) Beetles.

(4) Grubs.

(5) Ants.

(6) Termites.

(7) Worms.

c. Foraging for Insects. One must be careful not to expend more energy harvesting food than can be replaced. For example, catching insects such as grasshoppers can become exasperating and tiring.

(1) At night grasshoppers climb tall plants and clings to the stalks near the top. They can be picked from the plants in the early morning while they are chilled and dormant.

(2) Dig for worms in damp humus soil, under rocks/logs or look for them on the ground after it has rained.

(3) Carpenter ants are found in dead trees and stumps, which can be gathered by hand.

(4) Most other insects can be found in rotten logs, under rocks, and in open grassy areas.

d. Preparing Edible Insects.

(1) Insects with a hard outer shell have parasites. Remove the wings and barbed legs before cooking.

(2) Drop worms into potable water for at least a half-hour. They will naturally purge themselves. You can either cook or eat them raw.

(3) Most other insects can be eaten raw. Cooking insects will improve their taste. If the thought of eating insects is unbearable, grind them into a paste and mix with other foods.

UNITED STATES MARINE CORPS
Mountain Warfare Training Center
Bridgeport, California 93517-5001

WMC.A8.0
05/02/01

STUDENT HANDOUT

SURVIVAL TRAPS AND SNARES

PURPOSE: The purpose of this class is to familiarize the student with some of the techniques used to take game. Game can be useful for both food and clothing.

OUTLINE

1. GENERAL TIPS AND TECHNIQUES FOR TRAPPING

a. General Tips. Knowing a few general hints and tips will make the trapping of animals much easier and considerably more effective. The eight general tips for trapping are:

(1) Know your game: Knowing the habits of the animal you want to trap will help lure it into your trap. Such things as when and where they move, feed, and water will help you determine where the trap can be most effectively placed.

A8

(2) Keep things simple: You don't have time in a survival situation to construct elaborate traps and they do not necessarily do a better job.

(3) Set traps in the right place: Animals will travel and stop in certain locations. That is where to best employ traps.

(4) Cover up your scent: Animals will avoid an area, which smells strange to them. Smoke from your fire is the best cover to use.

(5) Use the right type of trap: Some traps work better than others do (i.e. a deadfall works better on squirrels than a spring pole).

(6) Use the right size trap: Adjust the size of your traps to the size of the animal.

(7) Check traps: Check your traps twice daily: morning and evening. Checking your traps less than twice a day can allow your game to escape, rot, or be taken by other predators.

(8) Bait your traps: Bait of any type will add to your chances of success.

b. General Techniques. A general technique is the method in which the trap is intended to kill or hold the animal.

(1) Strangulation: This method strangles the animal, such as a snare.

(2) Mangling: This method crushes the animal, such as a deadfall.

(3) Entanglement: This method entangles the animal, such as a net.

(4) Live: This method holds the animal, such as a box trap.

c. Types of Triggers. There are three basic triggers used for all traps and path guards. Depending on the situation, variations of these triggers can be used.

(1) Puite Figure 4.

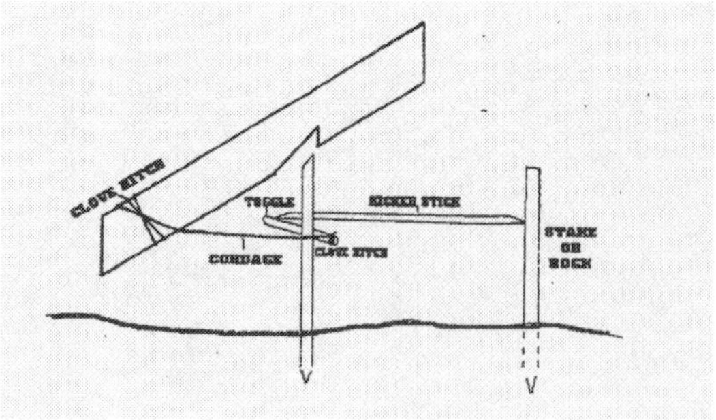

(2) Toggle.

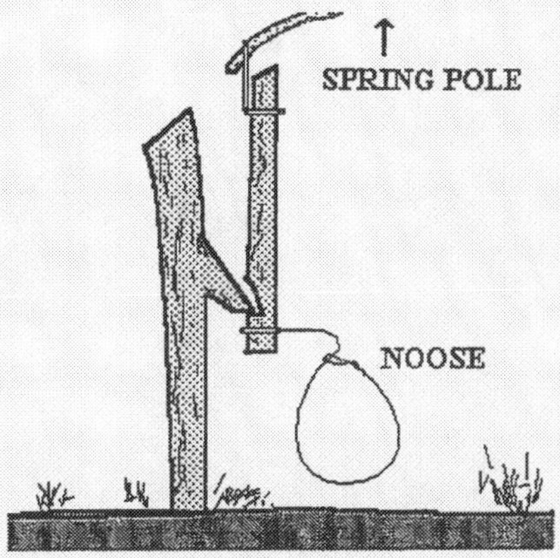

TOGGLE

(c) Universal.

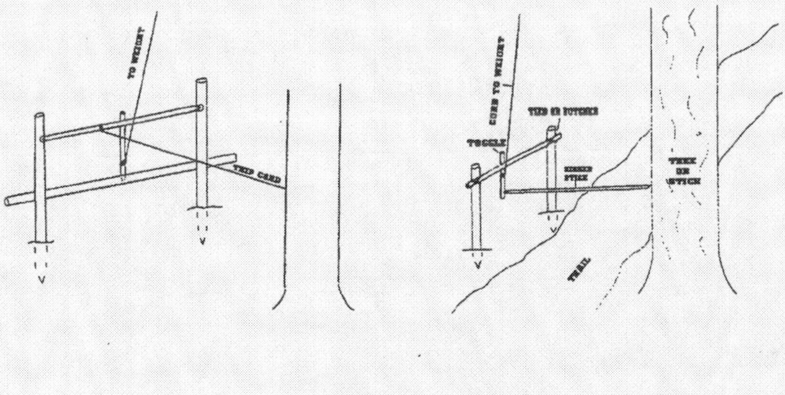

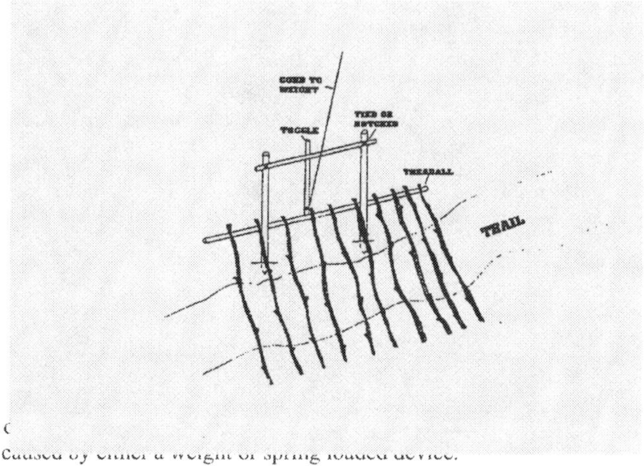

2. TRAPS. Traps are c
   action is generally caused by either a weight or spring loaded device.

a. Puite Deadfall: The Puite deadfall requires a knife and piece of cord to construct. It is designed to mangle small rodents. The trigger for the deadfall is the Puite figure 4.

b. Spring Pole: The spring pole requires a small sapling and cordage to construct. The trigger for the spring pole is the toggle. It is designed to lift the animal off the ground; not allowing predatory animals to take your game. Remember, the trigger can not be so tight that the intended game can not set it off.

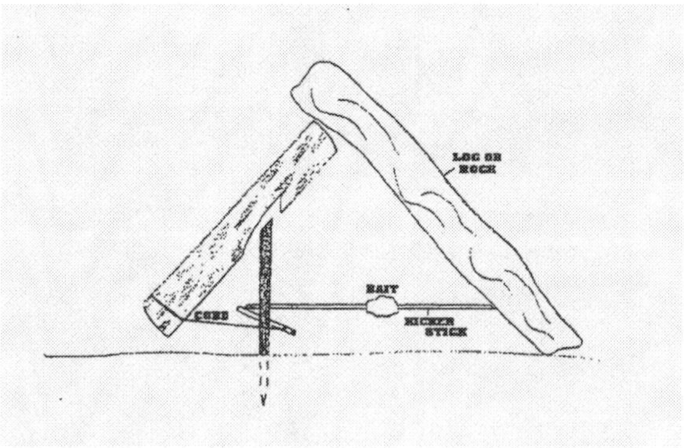

SPRING POLE

c. Box Trap: The box trap requires limited cordage to construct. It is designed to hold live small rodents and birds. The box trap trigger is the Puite figure 4.

3. PATH GUARDS Path guards is designed to protect and provide security for your shelter area against the enemy and predatory animals. They are classified into noise and casualty producing path guards.

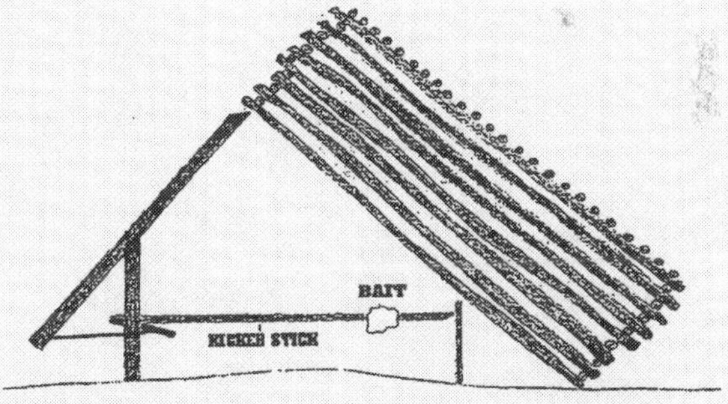

BOX TRAP

a. Noise producing path wards. Noise producing pathguards serve as an alarm for your shelter area. When triggered, it should produce some type of loud noise. Although construction can vary, depending on materials available, one example is as follows:

(1) Secure a young sapling to a universal trigger.

(2) At the end of the sapling, tie several pieces of metal to the sapling. Use what ever is available for metal.

(3) Camouflage the metal on the ground.

(4) When triggered, the sapling should swing back and forth, causing the metal rattle.

b. Casualty producing path wards. Casualty producing path guards, when triggered, should cause death or injury to the enemy or predatory animal. Tips should be poisoned as discussed in Survival Plant Uses class. Triggers for this type of path guard should be the universal triggers.

(1) Log Jerk

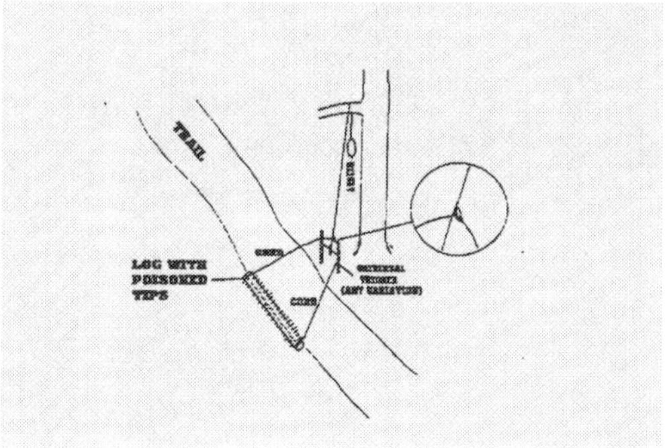

(2) Fish Hook Nightmare

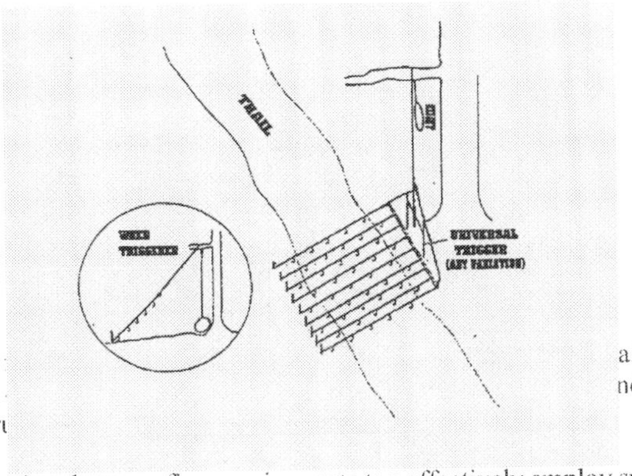

4. SNARING. A snare … a loop at one end, which tightens, … nd less time consuming to constr…

a. Requirements for snaring there are five requirements too effectively employ snares. They are as follows:

(1) Wire: Although snares can be used with rope or cord, they are less effective than wire. Wire should have memory and resist kinking. Aircraft cable type in size 1/32 to 3/16 inch should be used. This type of wire prevents animal chew out and resists breakage. Remember, you want the smallest diameter cable capable of holding the animal.

(2) Locking device: A locking device is imperative for a snare to work properly. Locking devices secure the snare around the animal's neck. There are several methods available for a locking device.

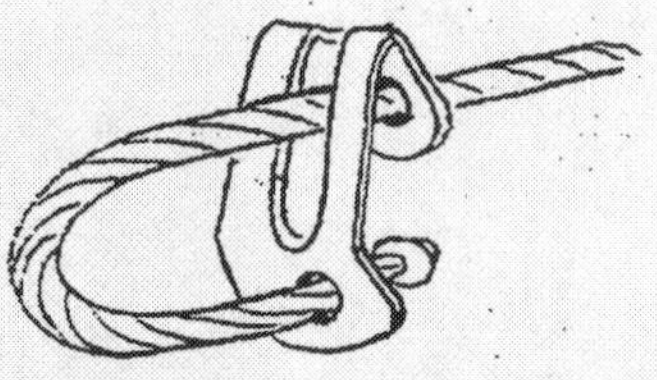

COMMERCIAL

BENT WASHER

(3) Channelization: The objective of channelization is to force the animal to move through the snare. Remember, animals will take the path of least resistance.

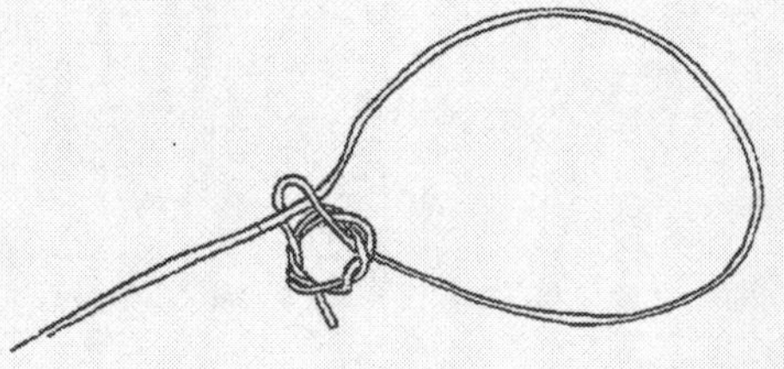

KNOT

(4) Loop size: Probably the single most important requirement in snaring. A correctly employed snare will have the snare holding the animal around the neck. Loop size is placed on the snare according to the intended animal. Too large will result in a body or leg catch, resulting in possible chew out or breakage. Too small will enable the animal to force the snare to the side, resulting in a miss. Additionally, the loop must be placed with specific ground clearance. The following chart will give a general guideline.

ANIMAL	NOOSE SIZE	GROUND CLEARANCE
SQUIRREL	2 1/2 TO 3 INCHES	1/2 TO 1 1/2 INCHES
RABBIT	4 TO 5 1/2 INCHES	1 1/2 TO 3 INCHES
RACCOON	6 INCHES	3 TO 4 INCHES
FOX	7 TO 10 INCHES	8 TO 10 INCHES
COYOTES	12 TO 14 INCHES	12 INCHES
BOBCAT	9 INCHES	8 INCHES

Note: (1) Noose size is the diameter of the snare loop.

(2) Ground clearance is measured from the bottom of the loop to the ground.

(5) Bait. In a survival situation, you will not be able to employ numerous snares. Baiting all snares increases your chances of success.

b. Snares although there are numerous ideas to employ snares, here are a few ideas.

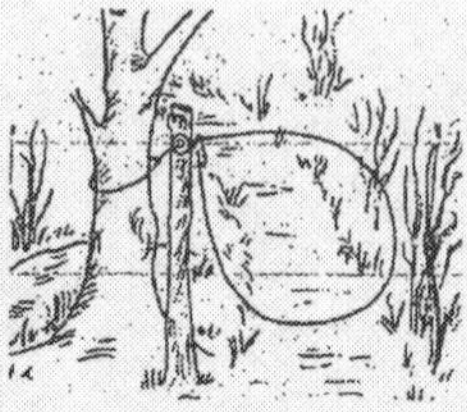

TRAIL SET

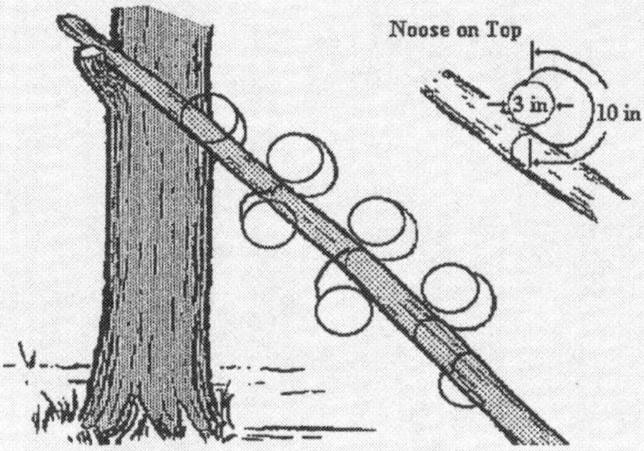

SQUIRREL POLE

DEN SET

CUBBIE SET

5. FISH. In a mountainous region, fish are normally an abundant resource. Not only are they a food source, the "left-overs" provide an excellent bait for traps and snares.

a. Fishing locations: Fishing in mountain streams is generally best done with a hand line. When fishing these streams, always look for these places to fish.

FISHING LOCATIONS

b. Expedient hooks: Although hooks should be carried in a survival kit, the survivor should be able to construct additional hooks if the situation arises. Expedient hooks are made to become lodged in the throat of the fish. Below are a few examples.

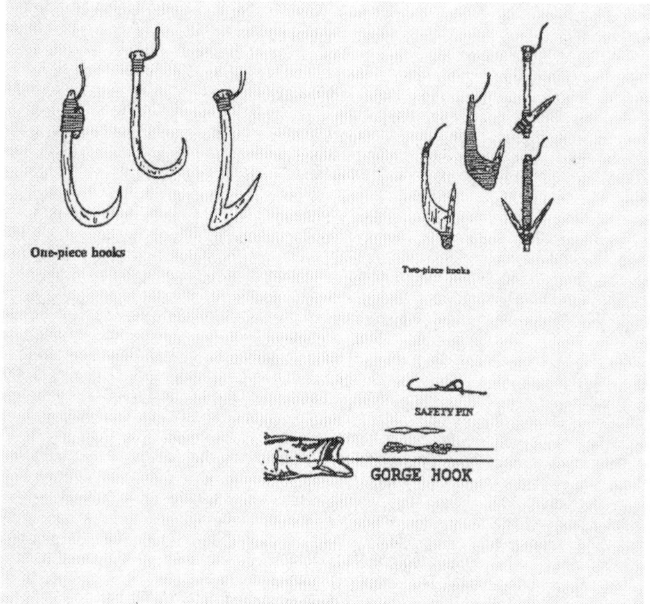

EXPEDIENT HOOKS

c. Fish traps: A fish trap can be effective if you have a shallow stream and time to construct it. A basic fish trap in nothing more than a barricade of rocks or sticks across a stream with another barricade using a funnel-type entrance which fish can be driven into but have a difficult time finding their way out. Once fish are trapped between these two barricades they may be speared, clubbed, or grabbed. This can be very effective when certain types of fish are moving in large groups to spawn. This type of trap is very effective in catching fish. A door can be constructed at the mouth of the trap so that excess fish can be kept live until needed.

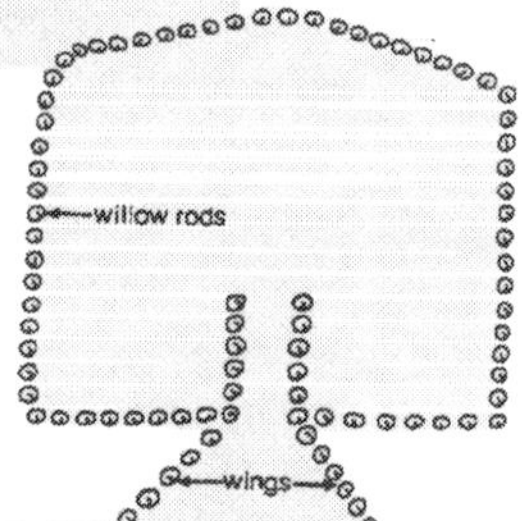

FISH TRAP

d. Set Lines: Set lines are an effective method of fishing which conserves energy. Put them out over night with several boated hooks attached. Place them with the hooks either on the bottom or suspended off the bottom, until you have determined where the fish are feeding.

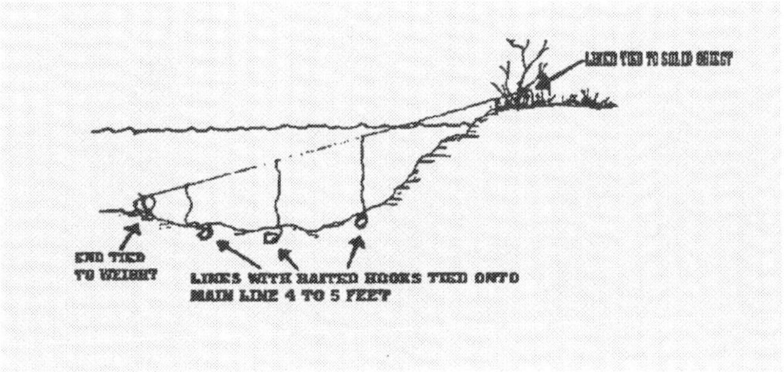

SET LINES

e. Netting: A gill net is most effective in still water, e.g., a lake (near the inlet and outlet are good locations) or back water in a large stream (for survival doesn't hesitate to block the stream). Nets can be constructed using the inner cords of parachute shroud lines. Place floats on top and weights on the bottom to keep the net vertical in the water. When the ice is on a lake, the fish are inclined to stay deeper. The smaller the mesh, the smaller the fish you can catch, but a small mest will still entangle a large fish. A mesh of 3 inches is a good standard. A gill net is very time consuming and requires allot of material to construct.

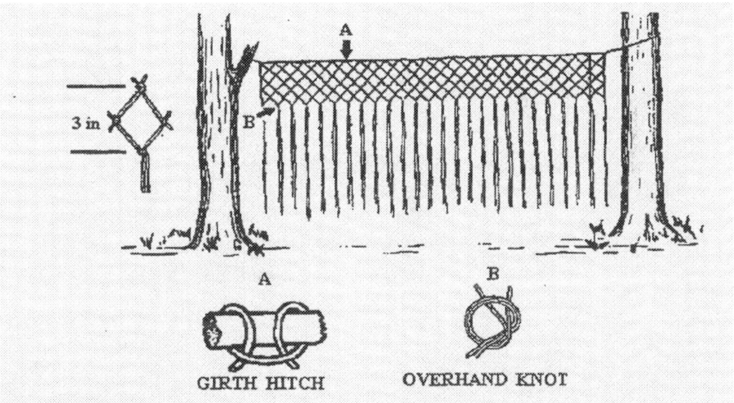

GILL NET

UNITED STATES MARINE CORPS
Mountain Warfare Training Center
Bridgeport, California 93517-5001

WMC
04/07/02

STUDENT HANDOUT

MOUNTAIN WEATHER

LESSON PURPOSE: The purpose of this period of instruction is to introduce you to weather patterns, the invisible aspects of weather, and some visible clues to help you forecast the weather. This lesson relates to all operations in the mountains at MWTC.

OUTLINE

1. GENERAL

a. The earth is surrounded by the atmosphere, which is divided, into several layers. The world's weather systems are in the troposphere, the lower of these layers. This layer reaches as high as 40,000 feet.

b. Dust and clouds in the atmosphere absorb or bounce back much of the energy that the sun beams down upon the earth. Less than one half of the sun's energy actually warms the earth's surface and lower atmosphere.

c. Warmed air, combined with the spinning (rotation) of the earth, produces winds that spread heat and moisture more evenly around the world. This is very important because the sun heats the Equator much more than the poles and without winds to help restore the balance, much of the earth would be impossible to live on. Where the air-cools, you can get clouds, rain, snow, hail, fog, frost, etc.

d. The weather that you find in any place depends on many things, i.e. how hot the air is, how moist the air is, how it is being moved by the wind, and especially, is it being lifted or not?

A10

2. PRESSURE

a. All of these factors are related to air pressure, which is the weight of the atmosphere at any given place. The lower the pressure, the more likely are rain and strong winds.

b. In order to understand this we can say that the air in our atmosphere acts very much like a liquid.

c. Areas with a high level of this liquid would exert more pressure on the Earth and be called a "high pressure area".

d. Areas with a lower level would be called a "low pressure area".

e. In order to equalize the areas of high pressure it would have to push out to the areas of low pressure.

f. The characteristics of these two pressure areas are as follows:

(1) High-pressure area. Flows out to equalize pressure.

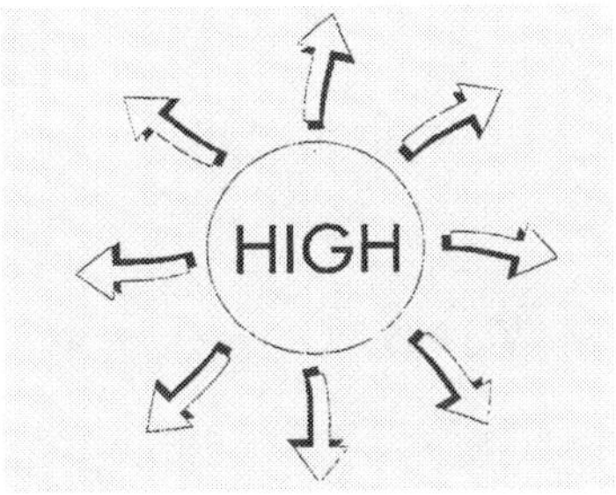

(2) Low pressure area. Flows in to equalize pressure.

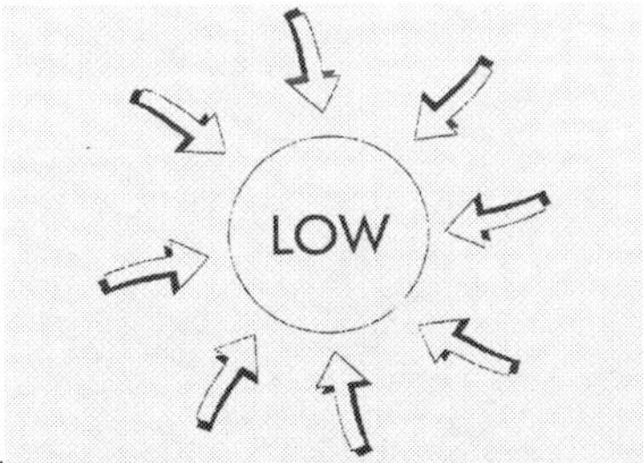

g. The air from the high-pressure area is basically just trying to gradually flow out to equalize its pressure with the surrounding air, while the low pressure is beginning to build vertically. Once the low has achieved equal pressure, it can't stop and continues to build vertically; causing turbulence, which results in bad weather.

NOTE: When looking on the weather map, you will notice that these differences in pressure resemble contour lines. These contour lines are called isobars and are translated to mean equal pressure area.

h. Isobars: Pressure is measured in millibars or another more common measurement is inch mercury.

i. Fitting enough, areas of high pressure are called ridges and areas of low pressure are called troughs.

NOTE: The average air pressure at sea level is:

29.92 inches mercury.
1,013 millibars.

j. As we go up in altitude, the pressure (or weight) of the atmosphere decreases.

EXAMPLE: At 18,000 feet in elevation it would be 500 millibars vice 1,013 millibars at sea level.

3. HUMIDITY Humidity is the amount of moisture in the air. All air holds water vapor, although it is quite invisible.

a. Air can hold only so much water vapors, but the warmer the air, the more moisture it can hold. When the air has all the water vapor that it can hold, the air is said to be saturated (100% relative humidity).

b. If the air is then cooled, any excess water vapor condenses; that is, it's molecules join to build the water droplets we can see.

c. The temperature at which this happens is called the "condensation point". The condensation point varies depending on the amount of water vapor and the temperature of the air.

d. If the air contains a great deal of water vapor, condensation will form at a temperature of 20°C (68°F). But if the air is rather dry and does not hold much moisture, condensation may not form until the temperature drops to 0°C (32°F) or even below freezing.

e. Adiabatic Lapse Rate: The adiabatic lapse rate is the rate that air will cool on ascent and warm on descent. The rate also varies depending on the moisture content of the air.

(1) Saturated Air = 2.2°F per 1,000 feet.

(2) Dry Air = 5.5°F per 1,000 feet.

4. WINDS. As we stated earlier, the uneven heating of the air by the sun and rotation of the earth causes winds. Much of the world's weather depends on a system of winds that blow in a set direction. This pattern depends on the different amounts of sun (heat) that the different regions get and also on the rotation of the earth.

a. Above hot surfaces, air expends (air molecules spread out), and move to colder areas where it cools and becomes denser, and sinks to the earth's surface. This forms a circulation of air from the poles along the surface or the earth to the Equator, where it rises and moves towards the poles again.

b. Once the rotation of the earth is added to this, the pattern of the circulation becomes confusing.

c. Because of the heating and cooling, along with the rotation of the earth, we have these surface winds. All winds are named from the direction they originated from:

(1) Polar Easterlies: These are winds from the polar region moving from the east. This is air that has cooled and settled at the poles.

(2) Prevailing Westerlies: These winds originate from approximately 30 degrees North Latitude from the west. This is an area where prematurely cooled air, due to the earth's rotation, has settled back to the surface.

(3) Northeast Tradewinds: These are winds that originate from approximately 30 degrees North from the Northeast. Also prematurely cooled air.

d. Jet Stream: A jet stream can be defined as a long, meandering current of high speed winds near the tropopause (transition zone between the troposphere and the stratosphere) blowing from generally a westerly direction and often exceeding 250 miles per hour. The jet stream results from:

(1) Circulation of air around the poles and Equator.

(2) The direction of airflow above the mid latitudes.

(3) The actual path of the jet stream comes from the west, dipping down and picking up air masses from the tropical regions and going north and bringing down air masses from the Polar Regions.

NOTE: The average number of long waves in the jet stream is between three and five depending on the season. Temperature differences between polar and tropical regions influence this. The long waves influence day to week changes in the weather; there are also short waves that influence hourly changes in the weather.

5. AIR MASSES. As we know, all -of these patterns move air. This air comes in parcels known as "air masses". These air masses can vary in size from as small as a town too as large as a country. These air masses are named for where they originate:

a. Maritime: Over water.

b. Continental: Over land.

c. Polar: Above 60 degrees North.

d. Tropical: Below 60 degrees North.

e. Combining these give us the names and description of the four types of air masses:

(1) Continental Polar: Cold, dry air mass.

(2) Maritime Polar: Cold wet air mass.

(3) Continental Tropical: Dry, warm air mass.

(4) Maritime Tropical: Wet, warm air mass.

ABBREVIATED MEANINGS
CP = Continental Polar.
MP = Maritime Polar.
CT = Continental
Tropical. MT =
Maritime Tropical.

f. The thing to understand about air masses; they will not mix with another air mass of a different temperature and moisture content. When two different air masses collide, we have a front, which will be covered in more detail later in this period of instruction.

g. Here are some other types of winds that are peculiar to mountain environments but don't necessarily affect the weather:

(1) Anabolic wind: These are winds that blow up mountain valleys to replace warm rising air and are usually light winds.

(2) Catabolic wind: These are winds that blow down mountain valley slopes caused by the cooling of air and are occasionally strong winds.

6. LIFTING/COOLING. As we know, air can only hold so much moisture depending on its temperature. If we cool this air beyond its saturation point, it must release this moisture in one form or another, i.e. rain, snow, fog, dews, etc. There are three ways that air can be lifted and cooled beyond its saturation point.

a. Orographic uplift: This happens when an air mass is pushed up and over a mass of higher ground such as a mountain. Due to the adiabatic lapse rate, the air is cooled with altitude and if it reaches its saturation point we will receive precipitation.

OROGRAPHIC UPLIFT

b. Convection effects: This is normally a summer effect due to the sun's heat radiating off of the surface and causing the air currents to push straight up and lift air to a point of saturation.

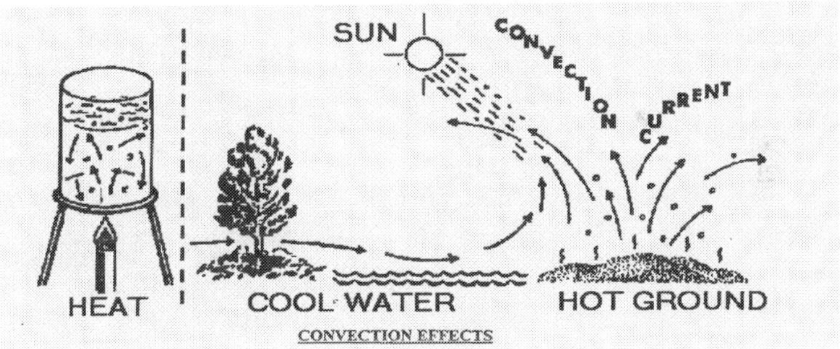

CONVECTION EFFECTS

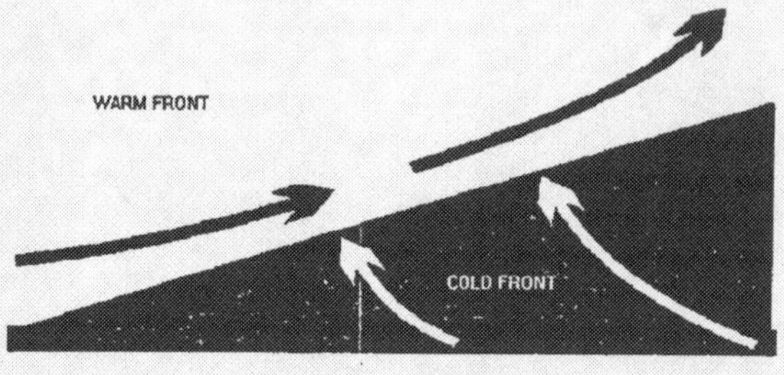

FRONTAL LIFTING

c. Frontal ... rature content ... is forced aloft; fi ... where we receive ... ifting is not uncom...

7. CLOUDS. Anytime air is lifted or cooled beyond its saturation point (100% relative humidity), clouds are formed. Clouds are one of our signposts to what is happening. Clouds can be described in many different ways; they can also be classified by height or appearance, or even by the amount of area covered, vertically or horizontally.

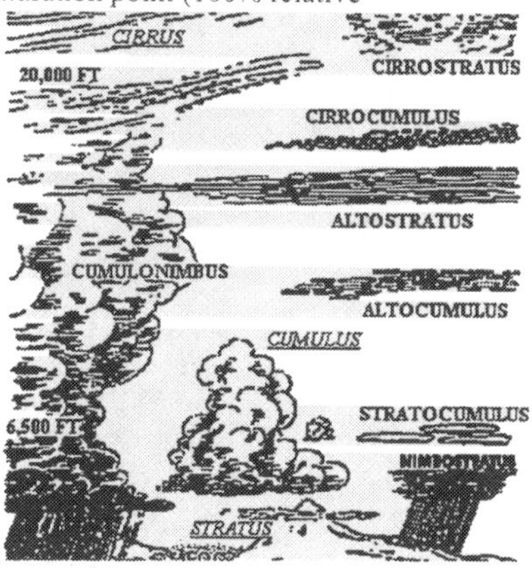

a. Cirrus: These clouds are formed of ice crystals at very high altitudes (usually 20,000 to 35,000 feet) in the mid-latitudes and are thin, feathery type clouds. These clouds can give you up to 24 hours warning of approaching bad weather, hundreds of miles in advance of a 'warm front. Frail, scattered types, such as "mare-tails" or dense cirrus layers, tufts are a sign of fair weather but predictive may be a prelude to approaching lower clouds, the arrival of precipitation and the front.

b. Cumulus: These clouds are formed due to rising air currents and are prevalent in unstable air that favors vertical development. These currents of air create cumiliform clouds that give them a piled or bunched up appearance, looking similar to cotton balls. Within the cumulus family there are three different types to help us to forecast the weather:

(1) Cotton puffs of cumulus are Fair Weather Clouds but should be observed for possible growth into towering cumulus and cumulonimbus.

(2) Towering cumulus are characterized by vertical development. Their vertical lifting is caused by some type of lifting action, such as connective currents found on hot summer afternoons or when wind is forced to rise up the slope of a mountain or possibly the lifting action that

may be present in a frontal system. The towering cumulus has a puffy and "cauliflower-shaped" appearance.

(3) Cumulonimbus clouds are characterized in the same manner as the towering cumulus, form the familiar "thunderhead" and produce thunderstorm activity. These clouds are characterized by violent updrafts, which carry the tops of the clouds to extreme elevations. Tornadoes, hail and severe rainstorms are all products of this type of cloud. At the top of the cloud, a flat anvil shaped form appears as the thunderstorm begins to dissipate.

c. Stratus: Stratus clouds are formed when a layer of moist air is cooled below its saturation point. Stratiform, clouds lie mostly in horizontal layers or sheets, resisting vertical development. The word "stratus" is derived from the Latin word "layer". The stratus cloud is quite uniform and resembles fog. It has a fairly uniform base and a dull, gray appearance. Stratus clouds make the sky appear heavy and will occasionally produce fine drizzle or Very light snow with fog. However, because there is little or no vertical movement in the stratus clouds, they usually do not produce precipitation in the form of heavy rain or snow.

d. As previously stated, clouds are formed when air is lifted to a point where it cools to its saturation point. We also know that frontal lifting affects our fronts, which produce the largest portion of our precipitation.

8. FRONTS. As we know, fronts often happen when two air masses of different moisture and temperature content interact. One of the ways we can identify that this is happening is by the progression of the clouds.

a. Warm Front: A warm front occurs when warm air moves into and over a slower (or stationary) cold air mass. Since warm air is less dense, it will rise naturally so that it will push the cooler air down and rise above it. The cloud you will see at this stage is cirrus. From the point where it actually starts rising, you will see stratus. As it continues to rise, this warm air-cools by the cold air and, this, receiving moisture at the same time. As it builds in moisture, it darkens becoming "nimbus-stratus", which means rain of thunderclouds. At that point some type of moisture will generally fall.

b. Cold Front. A cold front occurs when a cold air mass (colder than the ground that it is traveling over) overtakes a warm air mass that is stationary or moving slowly. This cold air, being denser, will go underneath the warm air, pushing it higher. Of course, no one can see this "air", but they can see clouds and the clouds themselves can tell us what is happening. The cloud progression to look for is cirrus to cirrocumulus to cumulus and, finally, to cumulonimbus.

c. Occluded Front. Cold fronts move faster than warm ones so that eventually a cold front overtakes a warm one and the warm air becomes progressively lifted from the surface. The zone of division between cold air ahead and cold air behind is called a "cold occlusion". If the air behind the front is warmer than ahead, it is a warm occlusion. Most land areas experience more occlusions than other types of fronts. In the progression of clouds leading to fronts, orographic uplift can play part in deceiving you of the actual type of front, i.e. progression of clouds leading to a warm front with orographic cumulus clouds added to these. The progression of clouds in an occlusion is a combination of both progressions from a warm and cold front.

9. USING PRESSURE AS AN INDICATOR. A very important factor of telling us what might happen is the pressure. As we know, low pressure or dropping pressure normally indicates deteriorating weather whereas high pressure usually gives us more good weather or clearing of bad weather. There are a couple of ways to monitor our pressure and are as follows:

a. Barometer: A barometer could be described as a pan of mercury with a tube leading out of the pan. Pressure from the atmosphere causes the mercury to rise in the tube.

(1) The tube is marked in millibars and the station that's reading these millibars will know how much it should rise for that location. Once again, if it rises more than normal, it would be considered a high-pressure reading.

b. Altimeter: Another means that is used to measure pressure is an altimeter, which is commonly used by mountaineers. It works like this:

(1) As you rise in elevation the pressure becomes less, thus allowing the needle in the altimeter to rise. If the needle rises without you rising with it, there is less pressure in the atmosphere than before and thus, a low-pressure area.

c. Contrail Lines: A basic way of identifying a low-pressure area is to note the contrail lines from jet aircraft. If they don't dissipate within two hours, that indicates a low-pressure area in your area. This usually occurs about 24 hours prior to an oncoming front.

d. Lenticulars: These are optical, lens-shaped cumulus clouds that have been sculpted by the winds. This indicates moisture in the air and high winds aloft. When preceding a cold front, winds and clouds will begin to lower.

10. USING SIGNS FROM NATURE. These signs will give you a general prediction of the incoming weather conditions. Try to utilize as many signs together as possible, which will improve your prediction. All of these signs have been tested with relative accuracy, but shouldn't be depended on 100%. But in any case you will be right more times than wrong in predicting the weather. From this we can gather as much information as needed and compile it along with our own experience of the area we are working in to help us form a prediction of incoming weather. The signs are as follows:

a. A spider's habits are very good indicators of what weather conditions will be within the next few hours. When the day is to be fair and relatively windless, they will spin long filaments over which they scout persistently. When precipitation is imminent, they shorten and tighten their snares and drowse dully in their centers.

b. Insects are especially annoying two to four hours before a storm.

c. If bees are swarming, fair weather will continue for at least the next half day.

d. Large game such as deer, elk, etc., will be feeding unusually heavy four to six hours before a storm.

e. When the smoke from a campfire, after lifting a short distance with the heated air, beats downward, a storm is approaching. Steadily rising smoke indicates fair weather.

f. A red sun or sky in the morning indicates nearing rain due to the excessive amount of moisture in the air.

g. A red evening sky indicates that the air contains so little moisture that rain within the next 24 hours is highly improbable.

h. A gray, overcast evening sky indicates that moisture carrying dust particles in the atmosphere have become overloaded with water; this condition favors rain.

i. A gray morning sky indicates dry air above the haze caused by the collecting of moisture on the dust in the lower atmosphere; you can reasonably a fair day.

j. When the setting sun shows a green tint at the top as it sinks behind clear horizon, fair weather is probable for most of the next 24 hours.

k. A rainbow in the late afternoon indicates fair weather ahead. However, a rainbow in the morning is a sign of prolonged bad weather.

l. A corona is the circle that appears around the sun or the moon. When this circle grows larger and larger, it indicates that the drops of water in the atmosphere are evaporating and that the weather will probably be clear. When this circle shrinks by the hour, it indicates that the water drops in the atmosphere are becoming larger, forming into clouds, rain is almost sure to fall.

m. In fair weather, air currents flow down streams and hillsides in the early morning and start drifting back up towards sunset. Any reversal of these directions warns of a nearing storm.

n. When the breeze is such that the leaves show their undersides, a storm is likely on the way.

o. It is so quiet before a storm, that distant noises can be heard more clearly. This is due to the inactivity of wildlife a couple of hours before a storm.

p. When in the mountains, the sight of morning mist rising from ravines is a good sign of clear weather the rest of the day.

q. A heavy dew or frost in the morning is a sign of fair weather for the rest of the day. This is due to the moisture in the atmosphere settling on the ground vice in the form of precipitation such as rain, snow, etc.

UNITED STATES MARINE CORPS
Mountain Warfare Training Center
Bridgeport, California 93517-5001

FMST.07.42
03/14/02

STUDENT HANDOUT

RESCUE PHILOSOPHY FOR SWIFTWATER RESCUE TEAMS

TERMIANL LEARNING OBJECTIVE In a swift water rescue scenario, organize a swift water rescue, in accordance with the reference. (FMST.07.42)

ENABLING LEARNING OBJECTIVES.

(1) With out the aid of reference, from a given list select the correct components of rescue preparation, in accordance with the reference. (FMST.07.42a)

(2) With out the aid of reference, from a given list select the correct role description of the rescue team members, in accordance with the reference. (FMST.07.42b)

(3) With out the aid of reference, from a given list select the correct methods of rescue, in accordance with the reference. (FMST.07.42c)

OUTLINE.

1. **COMPONENTS OF RESCUE PREPARATION**. (FMST.07.42a) These components comprise the backbone of both technical skills and the proper mental attitude required of a rescue professional. Remember that this course only provides you with the first component.

a. Training. Provides the students with the basic technical skills required to perform a safe rescue, these can be gained by attending an approved SWIFTWATER RESCUE COURSE.

b. Practice. Use of the skills learned in training.

c. Experience. Gained through advanced training, practical exercises and real-world situations.

d. Judgment. The combination and application of skills for a given situation. Remember that every rescue situation is different. Don't get into a SOP mindset.

2. **PRIORITIES OF SAFETY**. Most of you have heard the saying that "It is better to be judged by twelve then carried by six". With that in mind you must realize that, at times, there is no way to affect a safe rescue. If you place yourself or your team in a hostile environment, from which you cannot escape, you are adding to the problem, not assisting in the solution. For this reason we set priorities of safety.

a. Self-aid. Be able to rescue yourself.

b. Buddy-aid. Ensure all safety measures are in force and that team members can rescue each other in an emergency.

c. Victim-aid. Only after self and buddy-aid are ensured should you attempt to rescue the victim. This will enable you to concentrate and focus your entire effort on the rescue.

3. **ROLES OF THE TEAM MEMBERS**. (FMST.07.24b) Often when dealing with rescues, you will find yourself working with outside agencies and organizations. In order to successfully coordinate the entire effort the Incident Command System (ICS) was developed. In swiftwater rescue situations there are four specific roles that must be filled.

a. Incident Commander. Is in overall command of the rescue efforts. Does not necessarily have the technical knowledge to conduct the rescue, but must be able to coordinate the rescue effort.

b. Rigger. Personnel responsible for equipment and rope systems, i.e. establishing the tension diagonal or the tyrolean high line.

c. Gofer. This group of individuals fills all other supporting roles, i.e. Communications, logistics, medevac, accounting, equipment accountability etc.

d. Rescuer. The individuals who are actually conducting the rescue.

4. **COMPONENTS OF SEARCH AND RESCUE**. In order to successfully complete a rescue there is four objectives that must be met. In order to remember them we use the acronym LAST.

a. Locate. This is the search in the search and rescue business.

b. Access. The determination of the assets, equipment, technical skills, and personnel required safely reaching and extracting the victim.

c. Stabilization. Providing the on scene medical attention required in order to extract the victim without causing further injury or aggravating an existing condition.

d. Transport. Removing the victim from the hostile environment to a safe one where further treatment can be provided.

5. **METHODS OF RESCUE**. (FMST.07.42c) In any rescue, we must access the situation to determine the best techniques to apply. Bearing in mind the self-aid, buddy-aid, victim-aid concept as well as time, weather, victim's condition, etc. In all cases we will attempt the lowest risk methods first, while preparing for a higher risk option. In all cases remember the KISS principle.

a. Reach. Use a ladder, fire hose, stick, paddle, rifle or any other sturdy object to extend your reach in order to reach the victim.

b. Throw. Get some type of floatation device to the victim. Or throw a line to the victim in order to pendulum or pull him/her to the shore.

c. Row. Putting a boat in the water to reach the victim.

d. Go and Tow. These are the contact rescues that, while look good on film, put you at the extreme risk of becoming a second victim.

e. Helo. Most civilian helicopter crews are not trained in SAR evolutions. Military aircraft and crews are a better bet. Several uses for helos are in the support role. Locating the victim, inserting and extracting teams and moving equipment are some of these roles. If using the helo in a direct role it is better to use a qualified SAR swimmer/diver to hook-up or pick-up victims.

UNITED STATES MARINE CORPS
Mountain Warfare Training Center
Bridgeport, California 93517-5001

FMST.07.25
03/14/02

STUDENT HANDOUT

PERSONAL AND TEAM EQUIPMENT

TERMINAL LEARNING OBJECTIVE As a member of a swiftwater rescue team, maintain swiftwater rescue equipment, in accordance with the reference. (FMST.07.25)

ENABLING LEARNING OBJECTIVES.

(1) Without the aid of references, from a given identify the correct items of personal equipment, in accordance with the reference. (FMST.07.25a)

(2) Without the aid of references, from a given identify the correct items of team equipment, in accordance with the reference. (FMST.07.25b)

OUTLINE.

1. **PERSONAL EQUIPMENT**. (FMST.07.25a) The following items of personal equipment should be in each man's possession for a swiftwater rescue.

   a. Dry Suit. Warmest, most versatile, full body protection available. Extra thermal layers can be added for varying water temperatures. Provides wind protection out of the water and provides protection against contaminated water.

   b. Wet Suit. Excellent protection from flotsam. Variety of thick nesses for thermal protection. Insulating quality is degraded in moving water due to the flushing effect.

   c. Footwear. Primary is full sole scuba booties. You can use tennis shoes or jungle boots with heavy wool socks.

   d. Gloves. Neoprene is recommended.

   e. Hoods. Wetsuit hood or pile hood for insulation. Remember 30-50% of your body heat losses can occur through the head and neck region of your body if left unprotected.

B2

f. Helmet. Protects you from flotsam and impact injuries. Should be padded and ventilated to allow water to flow through.

g. Knife. Fixed or locking blade attached to the vest, handle down.

h. Whistle. Required for communication. Best without "pea".

i. Eye Protection. Protects subject from debris.

j. Personal Floatation Device. This is the single most important piece of equipment. Jackets should provide at least 15 lbs. of floatation. Proper fit is a priority. Rescue vests with a "Blow-Out" belt should only be used by trained professionals.

2. **TEAM EQUIPMENT**. (FMST.07.25b) The following is a list of some of the team equipment for a typical four-man team. Of course the situation will dictate which equipment will be required for the mission on hand, but all hands should be familiar with this equipment. This is only a list. The uses and application of this gear, along with practical application exercises later in this course will provide you with the hands-on training and working knowledge to properly and safely use this equipment.

QTY	ITEM
1	300 ft, l Static Rope
4	150 ft, 11 mm Static Rope
25	85 Locking Stubai Carabiners
4	2" Pulleys
2	Rescue Figure 8
6	3 ft, 7mm Prussic
20	15 ft, 1" Tubular Nylon Runners
2	Ascenders
3	Throw Bags
2 pr	Fins
1	Rescue PFD
1	Self Bailing Raft (13 or 14 foot)
1	Oar Frame w/ Oars, or Seven Paddles
3	Radios (waterproofed)
1	Medical Kit
1	Litter

NOTE: This is not a complete list of rescue equipment. However, it is a good basic load.

UNITED STATES MARINE CORPS
Mountain Warfare Training Center
Bridgeport, California 93517-5001

FMST.07.41
03/14/02

STUDENT HANDOUT

SWIFTWATER TERMINOLOGY AND DYNAMICS

TERMINAL LEARNING OBJECTIVES In a swift water rescue scenario, evaluate swift water river conditions in accordance with the reference. (FMST.07.41)

ENABLING LEARNING OBJECTIVES.

1) Without the aid of reference, from a given list correctly identify the swiftwater terminology, in accordance with the reference. (FMST.07.41a)

2) Without the aid of reference, from a given list correctly identify the classifications of rivers, in accordance with the reference. (FMST.07.41b)

OUTLINE.

1. **SWIFTWATER TERMINOLOGY**. (FMST.07.41a)

a. Hydraulic or Hydraulic effect: Movement of water caused by pressure.

b. Eddy: Horizontal reversal. Current is blocked by an obstacle creating a vacuum which, when filled, causes water to flow upstream.

c. Eddy Fence: Reversal line between downstream current and the eddy. This fence can be very strong and the water flow may vary in height by several inches.

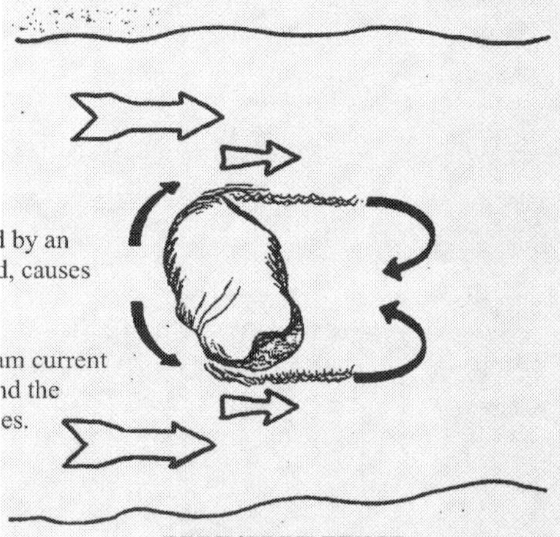

EDDY/EDDY FENCE

B3

d. Hole, Stopper, Keeper: Vertical reversal as water flows over an obstacle the main current follows the base of the river causing a surface vacuum. Downstream water flows back upstream to fill this void. This sets up a recirculating effect.

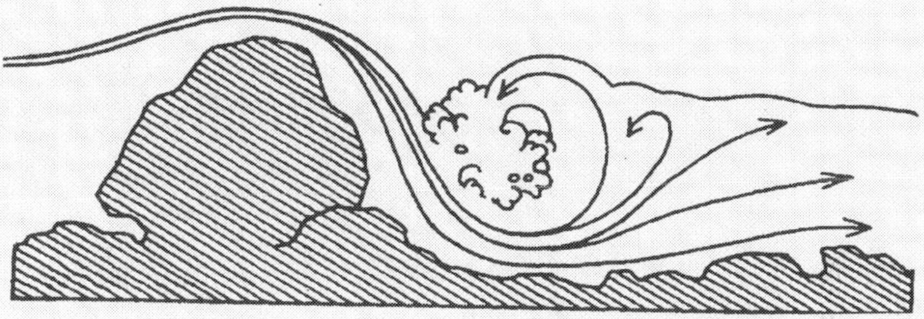

HOLE/STOPPER/KEEPER

e. Standing Wave. Haystack: Series of waves in the main channel caused by obstacles, ledges, or increases in the flow rate, gradient, or volume.

STANDING WAVES/HAYSTACKS

f. Downstream V. Tonal: Represent largest volume of water or main channel.

g. Upstream V: Caused by obstacles submerged just below the surface.

h. Strainer: Debris, logs, small trees on flooded banks, which allow water to flow through but will not allow a body to pass.

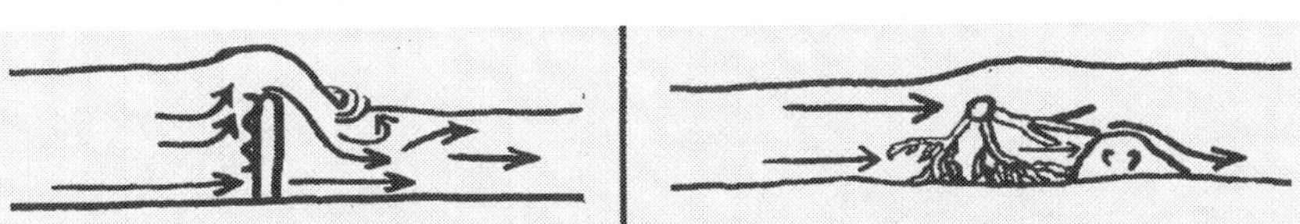

STRAINER / GUARD RAIL

STRAINER / TREE

i. Cushion, Pillow: Build-up of water on the upstream side of obstacles or the outside of curves.

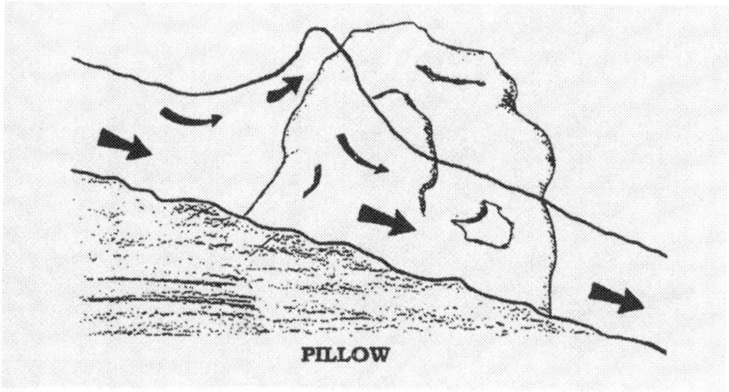

j. Confluence: Merging of streams.

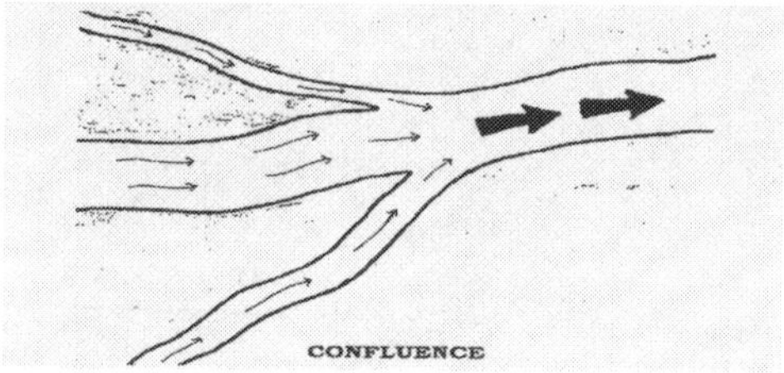

k. River Right. On the right side of the river looking downstream.

l. River Left. On the left side of the river looking downstream.

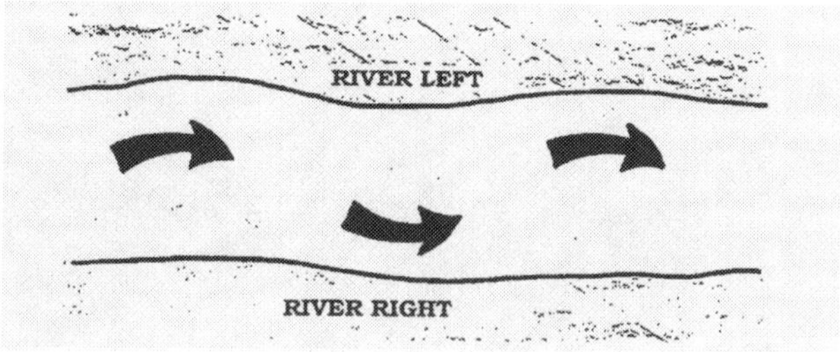

2. **CLASSIFICATIONS OF RIVERS**: (FMST.07.41b)

a. Class 1. Moving water, few obstacles, small ripples or waves.

b. Class 2. Moving water, clear channels, wide waves to 3', no scouting required.

c. Class 3. Rapids, narrow channels that require complex maneuvering, waves capable of flipping rafts or swamping canoes, scouting required.

d. Class 4. Long hazardous rapids, precise boat handling in rough water, scouting required, rescues hazardous.

e. Class 5. Severe, extended, extremely violent rapids, professional boat handlers and scouting required. Rescues extremely hazardous, situations are life threatening.

f. Class 6. Only Expert/Professional River Runners. Rescue may be impossible.

g. CLASSES U DON'T EVEN THINK ABOUT IT! Waters are not navigable.

3. **RIVER STRENGTH**.

a. Volume of Water. The volume of water in a given stretch of the river between two points is based on a simple formula and is measured in Cubic Feet Per Second (CFS). Simply take the river depth x river width x river velocity = volume of water in CFS.

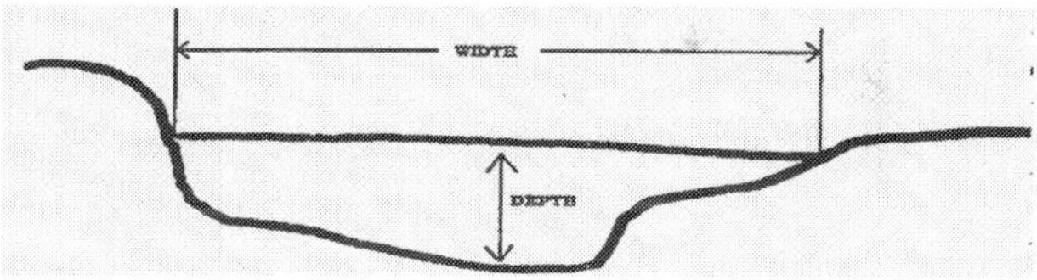

For example: 6 feet deep x 100 feet wide x 2 feet per second = 1,200 CFS

b. River Velocity. In order to determine the velocity or current speed measure a 100-foot section of the river and drop a float at the upstream end. Then, time its movement to the lower end and use the chart below.

TIME IN SECONDS	VELOCITY FT PER SEC	VELOCITY MPH
5	20	13
10	10	7
15	6.7	5
17	5.9	4
20	5	3.5
25	4	3
29	3.5	2.5
37	2.7	2
50	2	1.5
80	1.3	1
110	0.7	0.5

NOTE: Should the shape of the river change by becoming wider or narrower, once you have the flow in CFS, you can determine the new velocity by using the following formula:

$$\text{Velocity} = \frac{\text{Volume in CFS}}{\text{Width x Depth}}$$

c. Force Exerted. The velocity of the water will determine the force exerted X's the amount of exposed area under pressure. For simplicity we will use the following chart:

CURRENT VELOCITY	FORCE EXERTED ON OBJECT		
MPH	LEGS RAFT	BODY	WRAPPED
3	17	34	168
6	67	134	672
9	151	302	1,512
12	269	538	2,688

4. **LOW HEAD DAMS**. Low head dams are used throughout the world for irrigation, flood control, at hydroelectric plants and for many other uses. Basically they are pouring over obstacles that create hydraulics, keeper holes, and recirculating currents at their bases. Unlike natural obstacles, that rarely span the entire river and that allow you to be "flushed" out of the bottom or sides, low head dams create a hydraulic that extends from shore to shore.

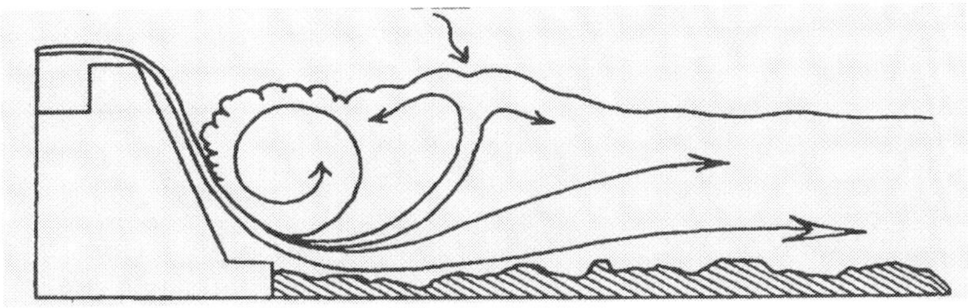

a. The size of the hydraulic created is based on several denominators:

(1) Height of drop.

(2) Angle of drop.

(3) Volume of water CFS.

(4) Velocity of water MPH.

(5) Bottom composition of runout.

b. The degree of difficulty for the rescuer is also compounded by:

(1) Water depth below dam.

(2) Debris build-up.

(3) Dam construction/conformation.

(4) Lack of access.

(5) Lack of natural anchors.

(6) Distance between riverbanks.

c. The basic hydrology is the same as any other pour over obstacle. As the river flows over the dam, the water follows the face of the dam to the river bottom. When sufficient speed diminishes the water rebounds to the surface and continues downstream. This flow has caused a vacuum or pocket directly below the dam that must be filled. Therefore, when the water rebounds to the surface, some of it flows back upstream to form a vertical reversal or pour-over eddy. The line of demarcation for the upstream/downstream flow is called the boil line. The shape and angle of the dam face and the volume and velocity of the water flowing over the dam create the distance and width of the boil line.

5. **REVIEW**.

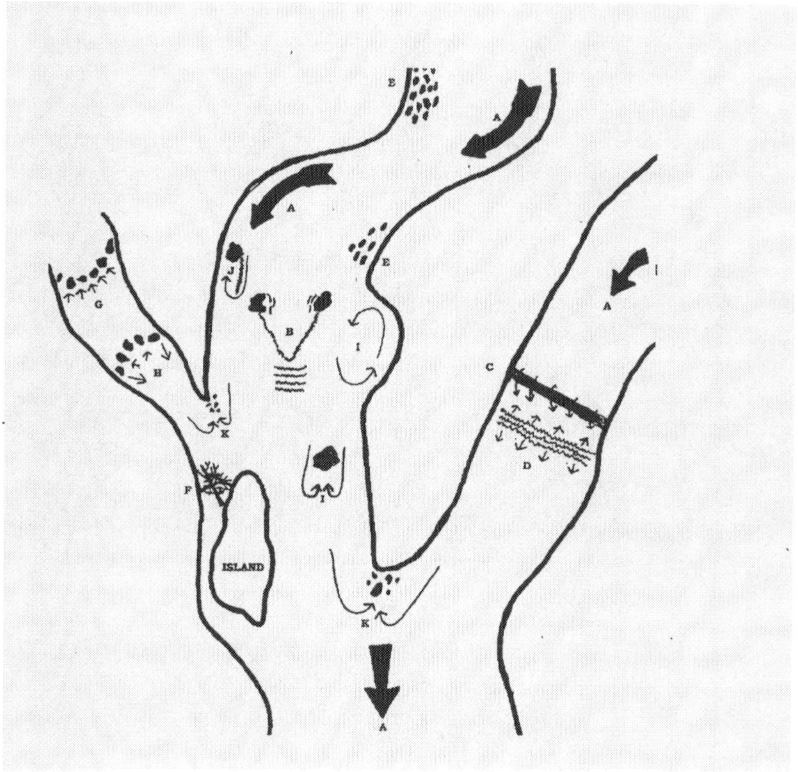

A. Main Current
B. Downstream V
C. Low head Dam
D. Boil Line
E. Rock Garden
F. Debris Pile/Strainer
G. Frowning Hole
H. Smiling Hole
I. Eddy
J. Eddy Fence
K. Confluence

UNITED STATES MARINE CORPS
Mountain Warfare Training Center
Bridgeport, California 93517-5001

FMST.07.24
04/14/02

STUDENT HANDOUT

COMMUNICATIONS

TERMINAL LEARNING OBJECTIVES Given a casualty in a swift water environment, assignment as a member of a rescue team, and a whistle, employ swift water rescue communication signals to support mission requirements per the reference. (FMST.07.24)

ENABLING LEARNING OBJECTIVES.

(1) In a summer mountainous environment, prior to swift water rescue operations, when the instructor verbally states a hand and arm signal, the student will demonstrate that signal, in accordance with the references. (FMST.07.24a)

(2) In a summer mountainous environment, standing at the edge of a swift water stream, when the instructor blasts a whistle a varied amount of times, the student will focus their attention in the proper direction, in accordance with the references. (FMST.07.24b)

OUTLINE.

1. **RELIABLE COMMUNICATION**. In many cases time is critical in a SWR scenario. Therefore immediate, understandable communication is a must. Radios, bull horns, and voice work well in some situations, but wet equipment, dead batteries, and ambient noise levels can render these methods ineffective. Therefore, rescue professionals have developed a series of simple hand and arm, and whistle signals. Any rescue team can develop their own signals. The only requirement is that all hands understand the signals and that they are briefed to any other agency you are working with.

B4

2. **HAND AND ARM SIGNALS**. (FMST.07.24a)

ACTION	MEANING
1 hand extended over head	Require Assistance
Both hands extended forming an "O" with one hand touching The head	Okay
Extend both hands above head, then point left/right	Move in direction indicated
Extend both hands above head, wave, then point left/right	Eddy out in direction indicated
Cross both arms over chest	Require medical assistance or Corpsman

3. **WHISTLE BLASTS**. (FMST.07.24b)

ACTION	MEANING
One Blast	Stop/Attention
Two Blasts	Upstream
Three Blasts	Downstream
Three Blasts Repeated	Emergency

UNITED STATES MARINE CORPS
Mountain Warfare Training Center
Bridgeport, California 93517-5001

FMST.0
7.27
04/14/02

STUDENT HANDOUT

THROW BAGS AND FOOT ENTANGLEMENTS

TERMINAL LEARNING OBJECTIVES In a swiftwater rescue scenario employ a throw bag, in accordance with the reference. (FMST.07.27)

ENABLING LEARNING OBJECTIVES.

1) Without the aid of reference, provided with a throw bag and swimming victim, rescue the victim, in accordance with the reference. (FMST.07.27a)

OUTLINE.

1. **THE THROW BAG**: The throw bag is very simply a cloth bag with a small float insert, packed with 50 to 70 feet of 1/4" to 3/8" floating polypropylene or spectra/poly line. The concept is to hold the exposed end of the line and throw the packed bag to a swimmer. As the bag flies, the rope deploys enabling you to reach a victim 50 to 70 feet from shore. It is used either from a standing position, or a moving position on open shores.

a. Standing Procedures. (FMST.07.27a)

1. Assume a stable belay position and ensure your footing is solid. Avoid loose, sloped or wet rock.

2. Attract victim by blowing your whistle (1 blast). Show him/her the throw bag and yell, "grab the rope". NOTE: If the victim grabs the bag rather than the rope, the rope will continue to deploy to its full length.

3. Grasping the exposed end of the rope, throw the bag underhand in a smooth motion directly at the victim. You may need to lead them in order to score a direct hit.

4. Tell the victim to place the rope over his/her upstream shoulder. This will establish a good ferry angle.

B5

5. Do not wrap the rope around yourself or tie it off at anytime. You may have to release the victim to avoid an obstacle or strainer. Use a dynamic belay and pull the victim to shore or pendulum the victim to other rescuers downstream.
6. If available have other rescuers downstream to recover the victim or repeat the procedure if you miss!

b. Moving Procedures. The moving procedures are the same as the standing procedures except you are on a smooth, open shoreline, so you move slightly downstream of the victim as you pull them gently ashore.

2. **FOOT ENTRAPMENTS AND VERTICAL PINNINGS**. These scenarios are caused by a variety of actions. Swimmers put their feet down or fishermen, in waders, get their feet stuck between rocks and are pushed over by the current. Swimmers are stuck in a strainer. Rafters get wrapped on a rock with victims pinned between the raft and rock or kayakers drop down a vertical shoot and get lodged in the rocks at the bottom. Whatever way it occurs you must respond quickly in order to save the victim. Initially take the following precautions:

   1. Position upstream spotters, to provide warning of boats or debris.

   2. Plan alternate methods if initial try doesn't work.

   3. Have back-up rescuers downstream in order to retrieve the victim.

   a. Once these actions are complete conduct the rescue using one of the five following methods.

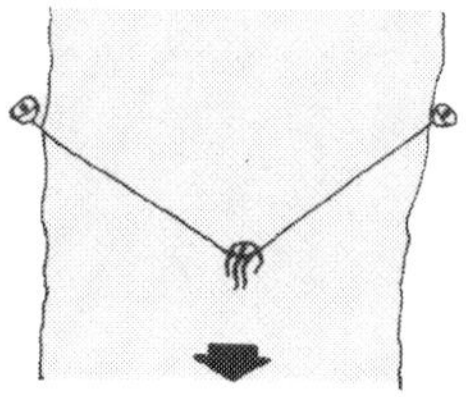

   1. Bechel-Ray One Line Method. The simplest and safest method. Get a rescuer on both sides of the river. Establish a rope between the two rescuers and downstream of the victim. Walk upstream until the rope hooks the victim. If you can position the line near the victims ankles you may be able to pop him/her loose and the downstream safety can pick him/her up. If not you can haul the victim upstream until he/she is clear of the obstacle, then by one rescuer playing out line and the other retrieving it you can bring the victim ashore. Remember to bring the victim to the shore with the medevac capability.

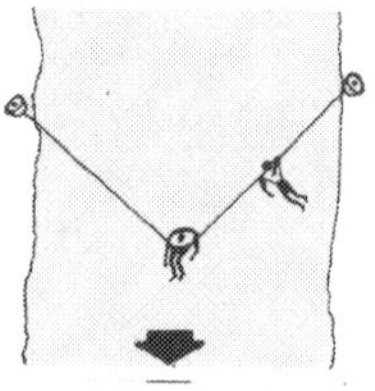

   2. The Strong Swimmer-variation. If the previous method doesn't work, leave the line in place and anchor it to stabilize the victim above the surface. Then have a "strong swimmer" go down one side of the rope, as would a tension diagonal. This rescuer is not tied in. Upon arrival the rescuer can readjust the line on the victim, attempt to pull them free, give them a PFD and helmet, or if a debris pile or rock is near, get on it and assist in further efforts.

3. Two Line Loop. There are two variations on this method both of which are applied using the initial rope used in the Bechel-Ray attempt.

(1) Deploy a second rope across the river and hook it onto the first line. Allow it to go one-half way down the rope to the victim. Pull at a 90° angle to rope #1. This will establish a mechanical advantage and pull the victim loose.

(2) Once the victim is free walk the line downstream trapping the victim in a loop. Then by releasing slack on line one, you can bring him/her ashore.

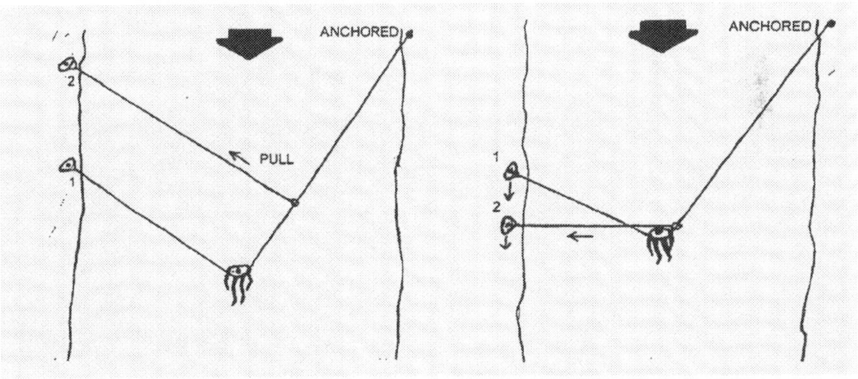

4. European Tethered Rescuer Lower. Using a rescue vest with a "blow-out" belt establish an upstream belay and lower the tethered rescuer to the victim. The rescuer's body will block the water forming an eddy around the victim enabling him to be pulled free. Ensure that the rescuer has a working rescue vest, as this will enable him to cut away if the situation dictates.

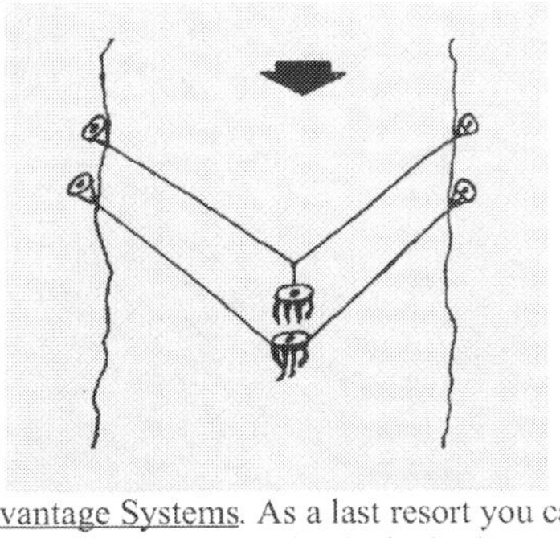

5. Mechanical Advantage Systems. As a last resort you can use a mechanical advantage system. This method is usually used only for body recoveries, as the amount of force exerted on the victim is tremendous and can cause severe injuries.

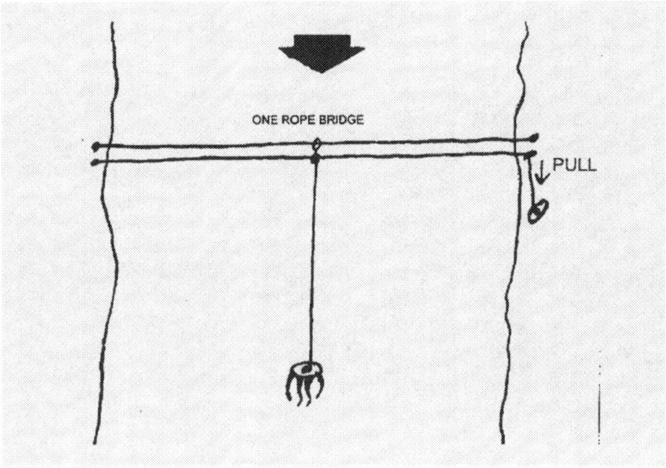
ONE ROPE BRIDGE
PULL

UNITED STATES MARINE CORPS
Mountain Warfare Training Center
Bridgeport, California 93517-5001

FMST.07.28
04/14/02

STUDENT OUTLINE

DEFENSIVE SWIMMING AND FERRY ANGLE

TERMINAL LERNING OBJECTIVES Given a swift water river and throw bags, perform aggressive swift water swimming to meet mission requirements per the reference. (FMST.07.28)

ENABLING LEARNING OBJECTIVES.

(1) Without the aid of reference, demonstrate the defensive swimming position, in accordance with the reference. (FMST.07.28a)

(2) Without the aid of reference, demonstrate use of the ferry angle to cross the river, in accordance with the reference. (FMST.07.28b)

OUTLINE

1. **THE DEFENSIVE SWIMMING POSITION (DSP)**. (FMST.07.28a) The DSP is assumed by lying flat on your back, chin on chest (to read the river & select a route), with your feet up, heels just below the level of your buttocks. Holding your feet in this position will enable you to fend off of shallow rocks and will reduce the likelihood of a foot entrapment or pinning. Your arms and hands should be kept close to the body but may be used like oars on a boat to maneuver. Your head will remain upstream except to cross a strainer or to over take a victim in a swimming rescue.

2. **FERRY ANGLE**. (FMST.07.28b) A ferry angle is used to maneuver diagonally across a river. By placing your body at a 45° angle to the current, the river will push you across the river. This is because the force of the current will travel down the exposed, upstream, side of the body moving it laterally more rapidly then the current pushing straight. This is like traversing a slope with edged skis vice sideslipping down on flat skis. In order to cross faster or to reach an eddy you may roll to your belly and execute the rescue crawl stroke or breaststroke. Remember to maintain your ferry angle with your head upstream this will enable you to travel diagonally across the river.

3. **STRAINER CROSSING**. The first rule is to avoid the strainer or shallow submerged obstacle. Read the river and maneuver to the main channel by using downstream "V's" or tongues. If the obstacle cannot be avoided, roll on your belly, head downstream and use the rescue crawl stroke. You must get your body moving faster than the current. Upon arrival at the obstacle use both hands to "vault" yourself over. A well-timed dolphin kick will assist in keeping your legs clear of entanglements. Once clear of the obstacle, resume the defensive swimming position. Bear in mind that it is usually safer to remain in the main current or to eddy hop downstream than it is to attempt to get out on a flooded riverbank, the outside of a bend or on slippery rocks. Also, use the dolphin kick rather than placing your feet on the bottom and kicking off. This will reduce the likelihood of foot entrapments.

UNITED STATES MARINE CORPS
Mountain Warfare Training Center
Bridgeport, California 93517-5001

FMST.07.25
03/14/02

STUDENT HANDOUT

FEET-WET RESCUES AND C-SPINE ROLLS

TERMINAL LEARNING OBJECTIVES In a swift water rescue scenario, conduct a swimming rescue, in accordance with the reference. (FMST.07.25)

ENABLING LEARNING OBJECTIVES.

(1) Without the aid of reference, while in a river and provided with a victim, perform a contact rescue, in accordance with the reference. (FMST.07.25a)

(2) Without the aid of reference, while in a river and provided with a simulated unconscious victim, perform a C-spine roll, in accordance with the references. (FMST.07.25b)

OUTLINE.

1. **FEET WET RESCUE GENERAL CONSIDERATIONS**. (FMST.07.25a)

   a. Always wear the proper personal equipment, i.e. PFD, helmet, drysuit, etc.

   b. Don't make physical contact if it can be avoided. Make him/her "chase" you ashore. Use a life ring, can, or boogie board, and tow it along with the victim. Toss him a throw bag and swim/tow him/her until you reach the shore, then pendulum or haul the victim in. Remember, to a panicked victim YOU are a floatation device.

2. **CONTACT RESCUE**.

   a. Approach the victim from upstream. Use the rescue crawl stroke to catch-up to him/her. Talk to the victim and attempt to get him/her to swim to shore or "chase" you to the shore.

B7

b. Explain to the victim the defensive swimming position and attempt to get them in this position. Assume this position yourself, in contact rescues this is called the "Reverse and Ready" position. Tell the victim what you are going to do prior to grabbing them.

c. Gain contact by grabbing the shoulders of the life jacket or utilize a cross-chest carry. If the victim is unconscious and face down use a C-Spine roll down and maintain stability as much as possible.

d. Assume a good ferry angle with the victim downstream of you and use his/her body to protect you from obstacles.

e. If the victim panics and attempts to climb on top of you, break away or lock the victim up and go for an alligator roll. Reestablish control and calm the victim as soon as possible.

f. Once contact is made remember that you may be held accountable under the "abandonment" laws if you quit.

3. **TOWED VICTIM SWIM**. When towing a victim it is essential that you establish a ferry angle. If the victim is on a boogie board or float, explain to him/her what you are doing and have them help by kicking their feet. If you are towing by the PFD grasp the shoulder towards the far bank and remain in the DSP (defensive swimming position) in case your victim panics.

4. **LIVE BAIT RESCUES**. The final go and tow method is the "live bait" rescue. Using a rescue vest with a "blow out" belt and a mule team on the shore to belay you, swim to the victim as explained in contact rescues. Once you establish contact you are, in effect, a throw bag with hands. The belayer will then either pendulum you ashore or hauls you in. In an emergency you can cut away and conduct a normal contact rescue.

5. **C-SPINE ROLLS AND STABILITY**.

a. Basic Rules: (FMST.07.25b)

(1) Suspect C-Spine involvement in any head injury.

(2) Stabilize and immobilize, use C-Collar, Sam Splints, or a field expedient method.

(3) Keep the victim in the water until a spine board or equivalent is available, or there are enough rescuers to move the victim safely.

(4) Observe for other signs, i.e. cranial bleeding, bleeding from ears, eyes, nose (Do not attempt to stop this bleeding) or unequal/unresponsive pupil response.

b. Methods:

(1) First, approach the victim from behind. Reach under the arms and grasp the victim's head with both hands, then roll the victim. This is the least desirable method as you cannot swim properly or evaluate the ABC's of the victim.

(2) Second, approach the victim from the side, extend his/her arms above and alongside his/her head, reach across and roll the victim onto your arm maintaining the victims arms in place to ensure stabilization. Evaluate the ABC's, provide rescue breathing and use your free arm to swim.

(3) Finally, approach a victim wearing a PFD from the head. If face down, cross your arms thumbs inboard and use your forearms to immobilize the neck. Then roll the victim over. If face up, simply approach as if gaining contact on a contact rescue and maintain control of the C-Spine.

UNITED STATES MARINE CORPS
Mountain Warfare Training Center
Bridgeport, California 93517-5001

WMC

04/14/02

STUDENT HANDOUT

BOATS AND BOAT HANDLING

LESSON PURPOSE. The purpose of this period of instruction is to introduce the student to the boats available at MWTC for river rescue evolutions.

OUTLINE

1. **GENERAL**. There are a wide variety of boats and hybrid river craft available to rescue crews. Whitewater rafts, rigid raiders, zodiacs, airboats, and regular pleasure boats can all be used in the proper situations. All have been used both safely and have caused deaths and injuries when used improperly. Wherever you are, learn the equipment available, both its assets and its limitations.

2. **WHITEWATER RAFTS**. The whitewater raft of today is an inflatable, self-bailing, multi-chambered vessel with several tie rings and foot cups. Additionally, the materials used in construction resist rips, tears, and punctures much better than their predecessors. The size most often found on local rivers is 12 to 14 feet long due to the size of the rivers. In the MWTC SAR locker, we have a 13-foot self-bailer. This will be our primary rescue boat.

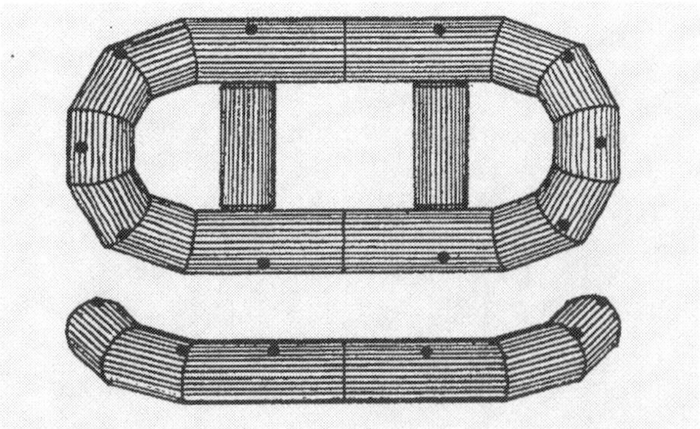

B8

3. **RAFTING COMMANDS**. In order to control the direction of a peddler powered boat the coxswain or guide must use a series of simple, concise, and understandable commands.

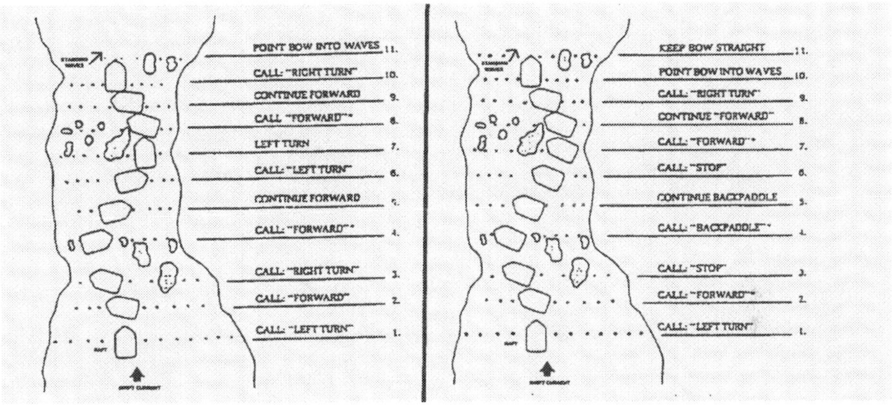

NORMAL TACTICS APPROACH | CONSERVATIVE/FAST CURRENT

a. Forward - All hands paddle forward together.

b. Back Paddle - All hands back paddle together.

c. Stop - Stop paddling - Remove blades from water.

d. Left Turn - Left side back paddles, right side forward.

e. Right Turn - Right side back paddles, left side forward.

f. High Side - Move quickly to the down streamside of the boat.

4. **PADDLE STROKES**. In order to control the raft from the coxswain position, either with a crew or alone on a tether system, you will need to apply the following strokes:

a. Sweep - Stroke in a wide arc around the fantail of the boat.

b. Draw - Reach out 90° away from the boat and "draw" it straight back to you.

c. Pry - Put paddle blade beside raft and "pry" the water away from the boat.

d. "J" - Used by lone guide to keep boat straight. Regular straight stroke finished with an outboard stroke or tail.

UNITED STATES MARINE CORPS
Mountain Warfare Training Center
Bridgeport, California 93517-5001

WMC
04/14/02

STUDENT HANDOUT

ROPE SYSTEMS AND BOAT ANCHORS

LESSON PURPOSE: The purpose of this period of instruction is to introduce the student to the rope systems and boat anchors used in swiftwater rescue situations.

OUTLINE

1. **TWO AND FOUR POINT TETHERING SYSTEMS**.

a. With all rope systems the first order of business is to get a rope across the river. Best case is to throw a rope bag across to someone already on the other side. Next best is to use another boat or raft to ferry the rope across. Last we use a swimmer to drag the rope across. Remember to use a smaller line on the swimmer and not to tie it to him/her. When swimming, start high and pendulum across using a good ferry angle.

b. Clip in on the side "D" rings forward for a two point or on the two forward and two aft "D" rings for a four-point tether system. For a two point you now need to simply load in a rescuer (if used) and move the boat into the river by taking slack on the far side and giving slack from the near side. For a four point man the two nearside lines and load a far side aft lineman and a rescuer (if used) into the boat and ferry the boat across. The far side aft belayer will get out and the boat is on a four point system ready for action.

c. This system is good for Class II Rivers with clear banks, few mid stream obstacles, and little floating debris. Use more belayers as required by river speed and boatload.

d. In all rope systems the whistle and hand and arm signals used to communicate are those taught in the communications class earlier in this course.

e. To break down the system simply reverse the procedures. The only difference is that the last man rides in the boat instead of swimming.

B9

2. **THREE POINT AND BOATMEN'S SELF-EQUALIZING ANCHORS.**

a. These anchors have in common the fact that they are self equalizing. This means that, regardless of the direction of pull, all anchor points ("D" rings, thwarts, etc) will remain under equal tension.

b. Using ropes the knots required are the double figure 8 on a bight, a directional figure 8, and a figure 8 on a bight.

c. Using webbing (one inch tubular nylon) a water tape knot is required.

d. If you have an abundance of carabineers, simply clip into the anchor points and back into the small loop on the double figure 8 on a bight, or into the directional figure 8. If not run the rope through the small loop. Remember that this is not preferred as it causes rope on rope stress and rubbing that can melt the rope.

e. When using webbing the same concept applies. Ensure you put a twist in the webbing at the contact point to ensure that a single blow-out will not collapse the entire system.

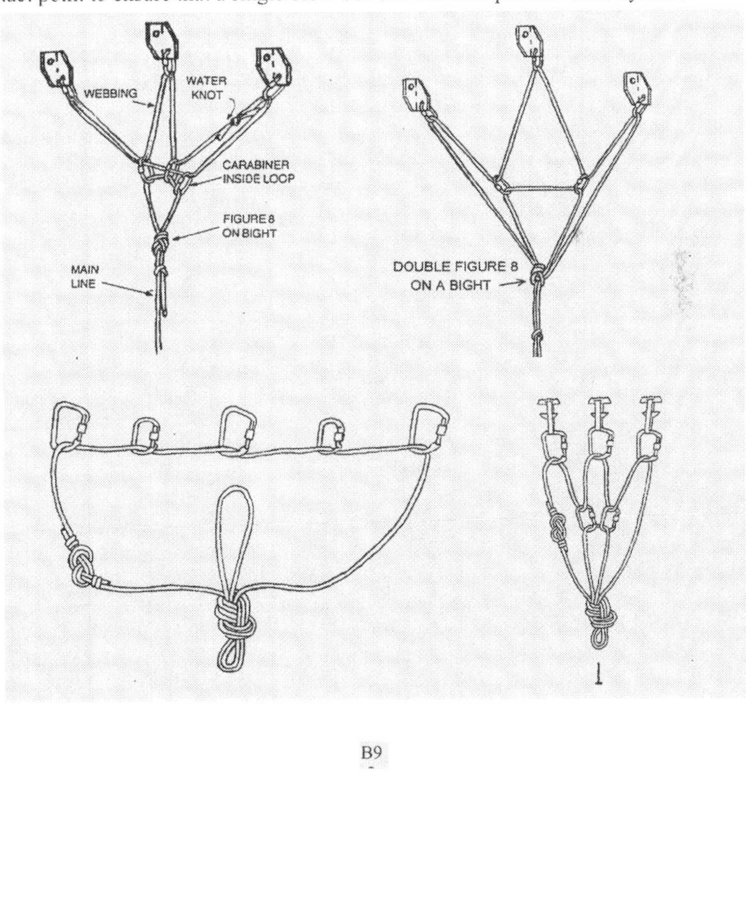

3. **PIG-RIG FOR MECHANICAL ADVANTAGE.** At times you will find yourself with ropes that are long enough to span the river, but not long enough to establish a mechanical advantage system. In these situations, you can use a piggy-back line or "pig-rig" by following the procedures below:

a. Tie off the mainline on either ends, or tie off the far side and set up a belay device on the near side using a munter hitch or rescue 8.

b. Establish an anchor for the pig-rig below the near side anchor.

c. Tie a prussic on the mainline and clip in a carabiner. Tie a figure 8 on a bight into the end of the pig-rig line and clip it into the prussic's carabineer.

d. Run the pig-rig line back to the anchor carabineer and then back through the prussic carabiner.

e. Pull on the running end towards the anchor. This will give you a 4:1 mechanical advantage on the main line.

f. If more tension is required you can either retie the nearside anchor (if tied), or use your belaying device to tension the mainline before moving the prussic.

g. If no prussic cord is available, you can accomplish this with a directional figure 8 on the mainline. However, this will only allow you to tighten the system once. A better method is to use the pig-rig line to tie a prussic directly on to the main line and run your system off of a directional figure 8 tied below the prussic on the pig-rig line. Remember that if the lines are of the same diameter this system will not work as the prussic will slip, therefore, you should use a kragur knot with ropes of the same diameter.

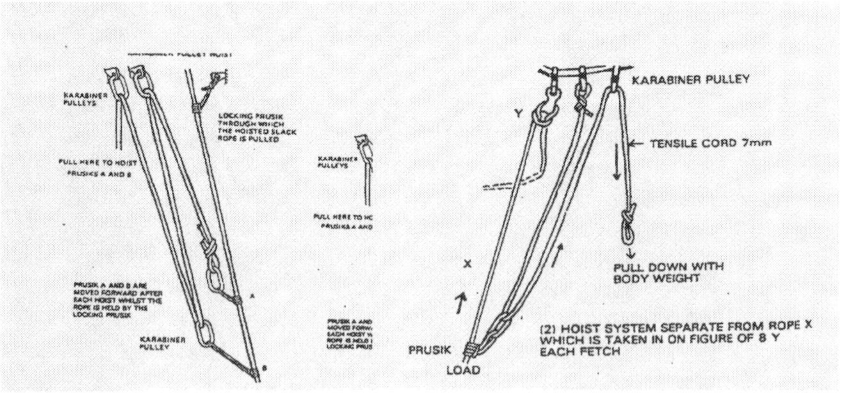

4. **TYROLEAN HIGHLINE AND TELFER LOWERING SYSTEMS** Both of these systems are used to maintain a boat as a stable rescue platform. The lateral movement (side to side/shore to shore) in both systems is controlled by mule teams on the shore. The main difference is that in the tyrolean highline the movement up and down-stream is also controlled from the shore, while in the telfer lowering system, the rescuer in the boat controls that aspect of movement. Both systems have good and bad points and both should be mastered. The situation will dictate which system you will use. Follow the steps below to establish these systems:

a. First you must establish TWO lines across the river. You can use any method taught to accomplish this.

b. Establish one line as the anchor line and tighten it as taught in the "one rope bridge" class or use a pig-rig, if required. If you have uneven mule teams set it at a slight down-stream angle towards the weak team's shore (no more then 10 to 15 degrees) to assist the weak team.

c. Rig the boat with a self-equalizing anchor system.

d. Rig the lowering systems as shown in the diagrams. Either tyrolean or telfer.

e. Load a rescuer or rescue team into the boat and execute. Remember the rescuer can act as coxswain and assist in maneuvering the boat by establishing a proper ferry angle.

f. Communications will be as taught earlier. Ensure you have a shore commander who maintains eye contact with the rescuers and mule teams on both the near and far shores.

g. When using the telfer system, a counter hitch, stitch plate or rescue 8 can be used as a belay device. This enables you to tie-off when you reach the victim so that you can affect the rescue without worrying about controlling the boat.

h. When using either of these systems it is a good idea to set-up a mechanical advantage system between the control (central) pulleys aid the raft/rescue platform.

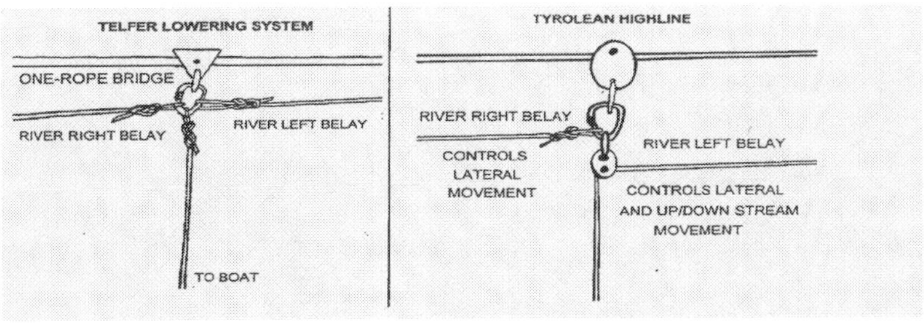

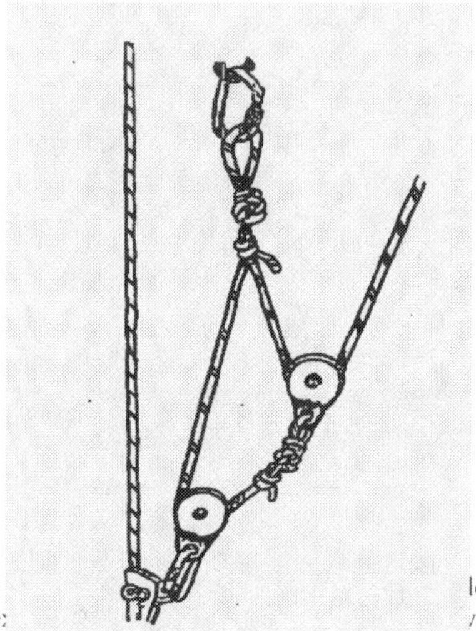

5. **BOAT WRAPS**. As [illegible] logy and Dynamics" class, the river can put an e[illegible] vrapped around an obstacle in the river. You first job is to ensure that NO ONE is trapped between the boat and the obstacle. If you do not have full accountability of personnel, don't hesitate! CUT THE BOAT! The easiest method to do this is to cut the floor out of the older "bucket" boats, or to cut the laces holding in the floor of a "self bailing" raft. Depending on how tightly wrapped the boat is you have several options available:

   a. First ensure the crew is accounted for and treated for any injuries.

   b. Establish a tag-line and attempt to bounce and pull the raft free.

   c. Set a Z-drag inside the boat to dump water.

   d. Use an oar or tree to lever the boat off the obstacle.

e. Set up a mechanical advantage system and attempt to "peal" the boat free.

f. Release air from the tubes. Be aware that this can cause more harm than good by reducing tube strength and exposing more surface area to the water pressure.

g. If all else fails, consider cutting out the floor as discussed earlier.

UNITED STATES MARINE CORPS
Mountain Warfare Training Center
Bridgeport, California 93517-5001

WMC
04/14/02

STUDENT HANDOUT

VEHICLE RESCUES

LESSON PURPOSE. The purpose of this period of instruction is to train the students in the procedures used to rescue personnel from vehicles trapped in swift water.

1. **VEHICLE BEHAVIOR**. There are three types of hazards in a river. Floating objects such as trees and branches, suspended objects or tumbling objects such as sawyers, rocks, and other debris, and stationary objects such as live trees on flooded banks and boulders. A vehicle will become all of these "loads". For the first minute, or so, they will float, as they become partially filled with water, they will be suspended. This is normally when they tumble or roll, and finally they will become stationary as they lodge against another stationary object or fill completely. Once they become stationary, depending on the water depth, they may form an eddy, a pour through or pour over, creating a hydraulic or a standing wave if fully covered. Also, the straighter a vehicle is (head-on) to the water flow the more stable it is. Finally, bear in mind that changes in water flow or displacement (weight) of the vehicle (taking a victim off it) may cause the vehicle to once again become a suspended load and tumble.

2. **RESCUE PROCEDURES**

a. The initial action is to secure the vehicle. Ensure that you have adequate anchors. Trees, guardrails, and bridge abutments are good. Avoid using another vehicle if possible.

b. Approach the vehicle on the downstream side by using the eddy created by the vehicle. Use a boat, either tethered or under power, a strong swimmer or boogie board. Another method is to use the line securing the vehicle as a tension diagonal to reach the vehicle. Then establish a second tension diagonal or endless rope to move the victims ashore. Ensure your upstream spotters are in place to warn of other suspended debris coming your way.

B10

c. If the vehicle is sealed tight, crack one window slightly to allow water to get in the vehicle slowly and equalize the pressure. Warning: If you pop or shatter a window you may cause a sympathetic detonation of the remaining windows.

d. Equip the victims with PFD's and remove to the boat or onto the rope system. Consider the following points:

(1) The victims weight might be the only thing stopping it from rolling. You might need to transfer a rescuer in a one for one exchange.

(2) On hard pack or asphalt the vehicle will shift more easily.

(3) If the vehicle is sideways keep the victims on the upstream side. The high side rule is reversed.

(4) Vehicles roll slowly.

(5) Helo's are a last resort.

UNITED STATES MARINE CORPS
Mountain Warfare Training Center
Bridgeport, California 93517-5001

FMST.07.3
0
04/14/02

STUDENT HANDOUT

STREAM CROSSING

TERMINAL LEARNING OBJECTIVE: Given a mountain swift water environment, and a staff, conduct a mountain stream crossing to meet mission requirements per the reference. (FMST.07.30)

ENABLING LEARNING OBJECTIVES:

(1) Without the aid of references, choose from a given list the considerations in site selection for a stream crossing, in accordance with the references. (FMST.07.30a)

(2) Without the aid of the reference, choose from a given list the safety precautions taken while conducting a stream crossing, in accordance with the reference. (FMST.07.30b)

(3) In a summer mountainous environment, execute the individual preparations taken prior to crossing a stream, in accordance with the reference. (FMST.07.30c)

(4) In a summer mountainous environment and given a staff, cross a mountain stream using the staff method, in accordance with the reference. (FMST.07.30d)

(5) In a summer mountainous environment, cross a mountain stream using team methods, in accordance with the reference. (FMST.07.30e)

1. **SITE SELECTION** (FMST.07.30a)

   a. Mountain streams and rivers are military obstacles and therefore are danger areas for units crossing them. In order to reduce the time in the vicinity of the danger area, a recon team should precede the main body and select the best crossing site. The site selection for a stream crossing should include these eight considerations:

B11

(1) Look for logjams, rocks or fallen trees that will provide a dry crossing if possible.

(2) If a dry crossing is not possible, select a crossing point at a wide and shallow point where the current is slower.

(3) Avoid sharp bends. They can be deep with a strong current on the outside of the bend.

(4) Look for a firm, smooth bottom. This is because large rocks and boulders provide poor footing and cause a great deal of turbulence in the water.

(5) It may be easier to cross several small channels of water rather than one large one.

(6) Do not cross just above rapids, falls, or logjams, taking a fall or slipping could have serious consequences.

(7) Cross in the early morning. The water level will be lower since there has been less daylight for the snow to melt. Also, on sunny days, you will have more time to dry clothing and equipment.

(8) There should be a suitable spot downstream for safety swimmers.

2. **SAFETY PRECAUTIONS** (FMST.07.30b). The following two safety precautions must be taken while conducting stream crossings.

   a. There must be a safety line at a 45-degree angle downstream across the stream. This is for anyone who slips and is swept downstream to grab so that he is capable to stopping himself.

   b. There must be safety swimmers downstream. These are strong swimmers who are positioned downstream to help anyone who is swept downstream. The safety swimmers will use throw bags.

NOTE: The safety line, as well as any other lines that must be taken across the stream, will be taken across by using the lead swimmer method taught in ONE-ROPE BRIDGE.

3. **INDIVIDUAL CROSSING PREPARATIONS.** (FMST.07.30c) Prior to beginning to cross a stream, there are certain preparations that each individual should take. The five preparations are as follows:

   a. Wear pack with shoulder straps fastened snugly. Waterproof the pack for buoyancy if possible.

   b. Weapons will be slung diagonally over the shoulder with the weapon itself being between the pack and the individual's back.

c. Button all pockets and remove blousing garters. This prevents the water from creating added drag against the individual, which it would do if it could flow into open pockets.

d. Wear boots to protect the feet, but remove socks and insoles to keep them dry:

e. Wear the minimum amount of clothing. This reduces the amount of clothing that must be dried after the crossing.

f. Wear helmet in slow moving, shallow water.

NOTE: If you are in a tactical situation, the actual situation will dictate which of the above precautions will be taken.

4. **INDIVIDUAL CROSSING METHODS**. There are three individual methods that may be used.

a. Staff Method. (FMST.07.30d) A strong staff or pole about 6 feet long is used as a crossing aid. It should be strong enough to support the Marine's weight and trimmed clean of any branches. Placing both hands on the pole, the Marine should place the staff just upstream of his intended path. He should use the staff as the third leg of a tripod and should move only one leg or the staff at a time. He should face upstream using the staff to retain his balance. The staff is also used as a probe to discover bottom irregularities that could trip the Marine. The Marine should drag his feet instead of picking them up.

STAFF CROSSING METHOD

b. Swimming. This is an obvious method, if your Marines are good swimmers. This is not always the case, so usually this method is not a preferred one.

(1) In fast, shallow water, the Marine should angle across on his back with his feet downstream and his head up. He should use his hands to tread water and his feet to fend off obstructions.

(2) In fast, deep water, the Marine should angle across the stream on his stomach with his head upstream, to establish a proper ferry angle.

c. Belayed Method. In chest deep or water with a strong current, a rope can assist greatly. A rope will be secured from bank to bank, with the far anchor slightly downstream from the near anchor. The rope will be anchored off so that it lays at a minimum of 45 degrees. The Marine attaches himself to the rope by using a sling rope as safety line, tying a bowline around his waist and a figure-of-eight loop with a steel locking carabiner inserted. He then attaches the carabiner to the crossing line and crosses using the current to assist him.

5. **TEAM CROSSING METHODS** (FMST.07.30e)

a. Line Abreast. Small units (squad to platoon) can cross in moderate currents up to chest deep, by lining arms in a line abreast or chain method. The largest man of the chain is placed on the upstream side of the group. The group will enter the stream parallel to the flow of the stream. The middleman of the chain will control the group's movement and give the command when to step.

LINE ABREAST

b. Line Astern Method. Three or more men can line up facing the current. The upstream man, who should be the largest man in the group, breaks the current while the downstream men hold him steady. The upstream man may use a staff, similar to the individual staff method, to steady himself. All men side step at the same time with one man calling the cadence..

LINE ASTERN

c. Huddle Method. Between three and eight men can face inboard as in a football huddle. They will wrap their arms around each other's shoulders and cross the stream in this formation. The upstream man should change position as they cross because the entire formation will rotate. This prevents one man from becoming exhausted in the upstream position.

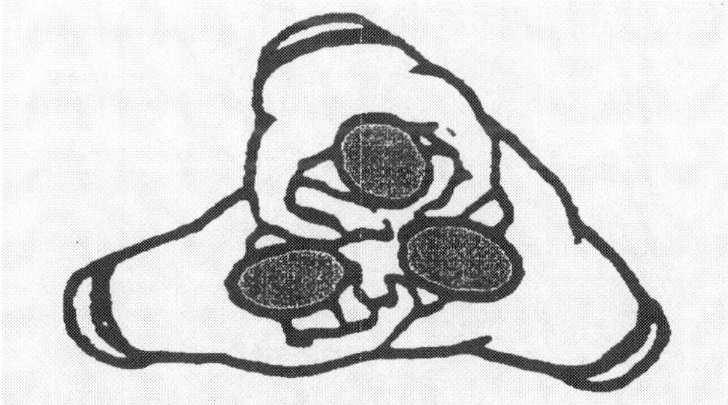

HUDDLE METHOD

UNITED STATES MARINE CORPS
Mountain Warfare Training Center
Bridgeport, California 93517-5001

WMC
04/09/02

STUDENT OUTLINE

ROPE MANAGEMENT

LESSON PURPOSE: To familiarize the student with some of the terms and procedures for working with a rope.

BODY

1. **TERMS USED IN ROPE WORK**

a. Bight. A simple bend in the rope in which the rope does not cross itself.

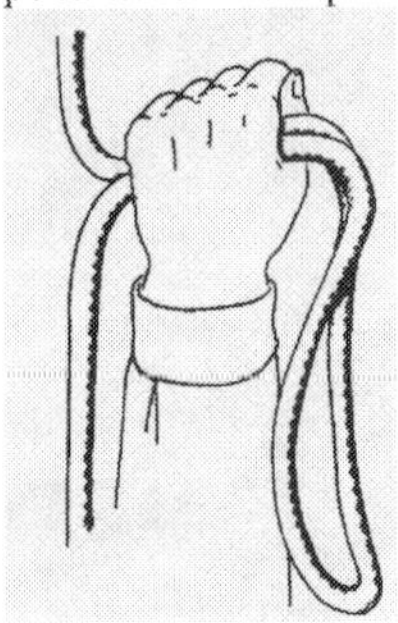

BIGHT OF ROPE

b. Loop. A simple bend in the rope in which the rope does cross itself.

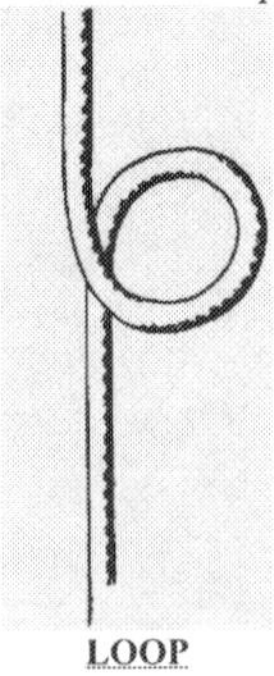

LOOP

c. Half Hitch. A loop which runs around an object in such a manner as to bind on itself.

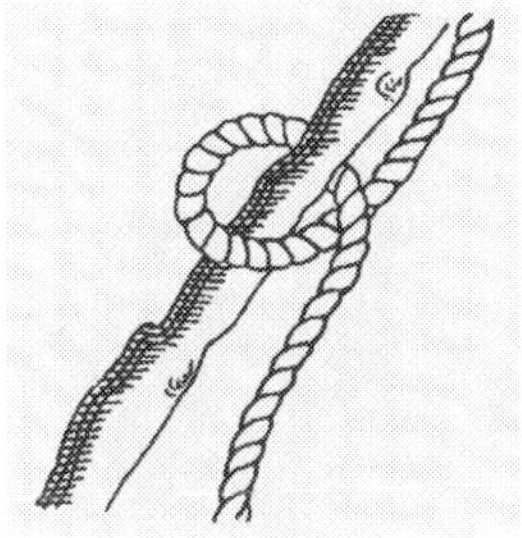

HALF HITCH

d. Standing End. The part of the rope which is anchored and cannot be used, also called the static end.

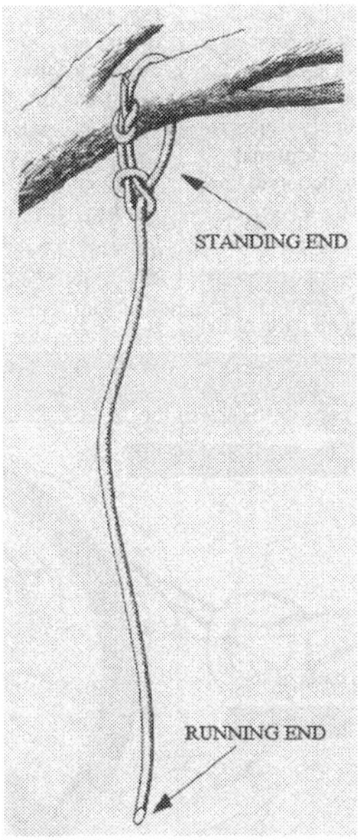

e. Running End. The free end of the rope which can be used.

f. Lay. The same as the twist of the rope. (Applies only to hawser laid ropes, such as manila.)

g. Pigtail. The short length left at the end of a rope after tying a knot or coiling a rope. It may or my not be tied off with a secondary knot, depending on the circumstance.

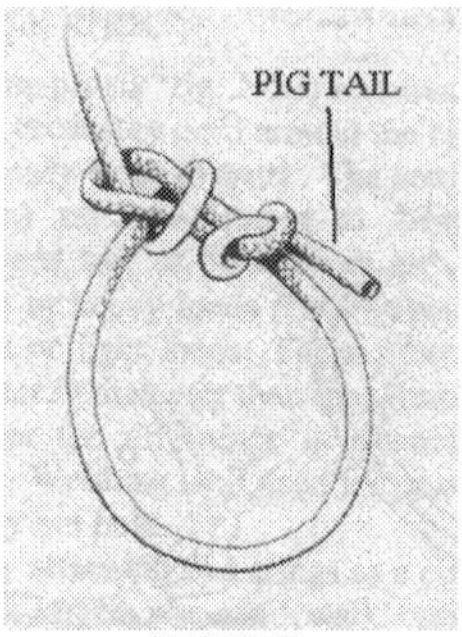

PIG TAIL

h. Stacking (or Flaking). Taking off one wrap at a time from a coil, and letting it fall naturally to the ground.

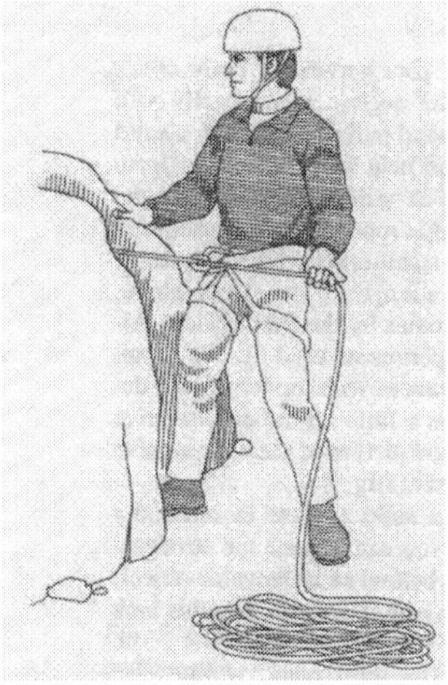

FLAKED OUT ROPEON THE GROUND

i. Dressing the knot. This involves the orientation of all of the knot parts so that they are properly aligned, straightened, or bundled, and so the parts of the knot look like the accompanying pictures. Neglecting this can result in an additional 50% reduction in knot strength.

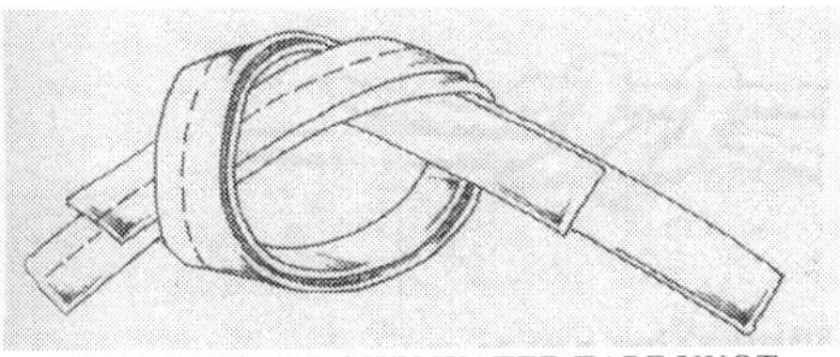

NON-DRESSED DOWN WATER TAPE KNOT

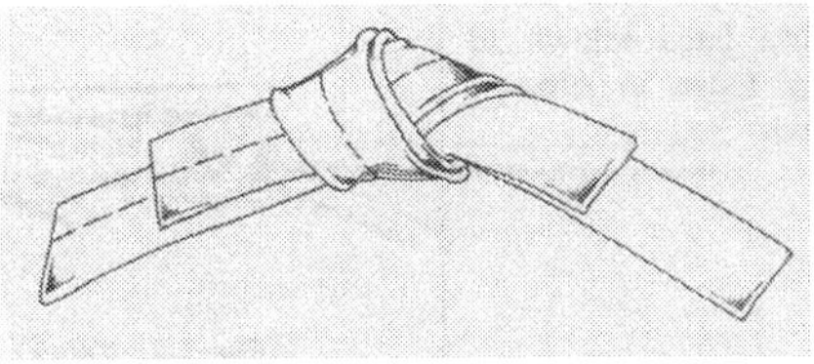

DRESSED DOWN WATER TAPE KNOT

j. Setting the knot. This involves tightening all parts of the knot so that all of the rope parts bind upon other parts of the knot so as to render it operational. A loosely tied knot can easily deform under strain and change character.

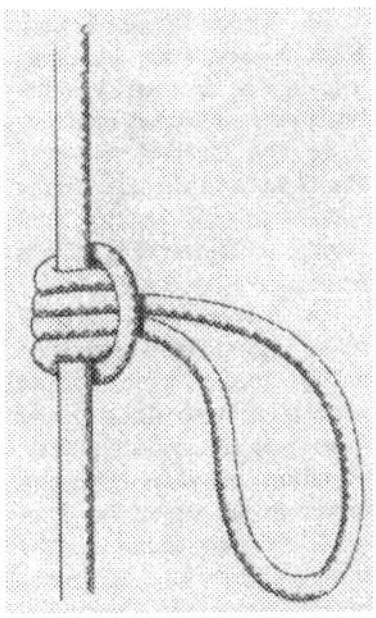

A FULLY SET PRUSIK KNOT

2. CONSIDERATIONS FOR THE CARE OF ROPE

a. The rope should not be stepped on or dragged on the ground unnecessarily. Small particles of dirt will get into and through the sheath causing unnecessary wear to the rope within.

b. The rope should never come in contact with sharp edges of any type. Nylon rope is easily cut, particularly when under tension. If a rope must be used around an edge which could cut it, then that edge must be padded or buffed using fire hose if available, or several small sticks.

c. Keep the rope as dry as possible. If it should become wet, hang it in large loops, above the ground, and allow it to dry. A rope should never be dried out by an open flame, or be hung to dry on metal pegs, as this will cause rust to get in the rope thus rendering it unserviceable.

d. Never leave a rope knotted or tightly stretched longer than necessary.

e. When using rope installations, never allow one rope to rub continually against another.

NOTE: With manila ropes this will cause the rope to fray, whereas nylon ropes can melt under the friction that this causes.

f. The rope should be inspected prior to each use for frayed or cut spots, mildew, rot or defects in construction.

g. Mark all climbing ropes at their midpoints to facilitate establishing the midpoint for a procedure requiring you to use the middle of the rope. The rope should be marked with a bright colored adhesive tape.

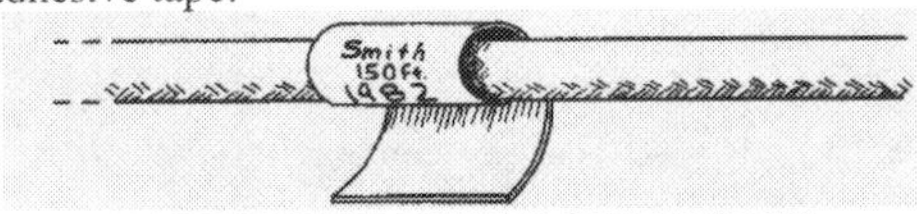

MARKING THE MIDDLE OF THE ROPE

h. The rope should not be marked with paints or allowed to come in contact with oils or petroleum products for these products will weaken it.

i. A climbing rope should NEVER be used for any other purpose except for mountaineering, i.e., towing vehicles.

j. The ends of a new rope or ends caused by a cut should be cut with the rope cutter contained in the MACK and marked with a serial number.

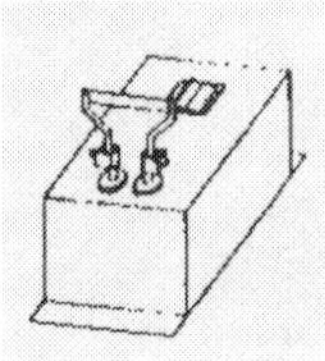

ROPE CUTTER

k. The rope should never be subjected to high heat or flame as this can significantly weaken it.

l. To clean rope use mild soap and rinse thoroughly with water. A rope washer can be used to clean or rinse the rope.

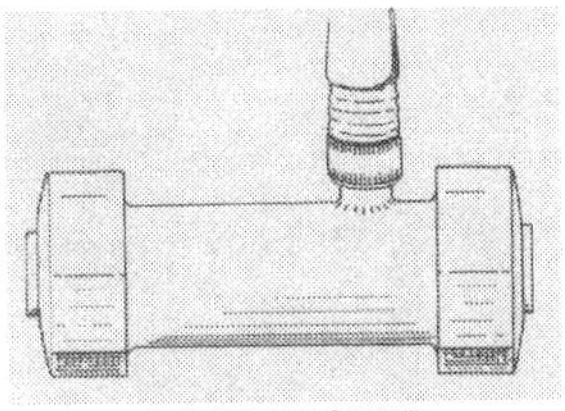

ROPE WASHER

m. When not in use, ropes should be coiled and hung on wooden pegs rather than on nails or any other metal object. They should be stored in a cool place out of the direct rays of the sun.

n. When in areas of loose rock, the rope must be inspected frequently for cuts and abrasions.

o. Always maintain an accurate Rope Log whenever using a rope.

p. Ropes 300-600 foot in length should be Mountain Coiled.

3. **INSPECTION OF ROPE**

a. All ropes have to be inspected before, during, and after all operations. Kernmantle rope is harder to inspect than a laid rope. I.e. green line. The Assault Climber must know what to look and feel for when inspecting a rope. Any of the below listed deficiencies can warrant the retirement of a rope.

(1) Excessive Fraying. Indicates broken sheath bundles or PIC breakage.

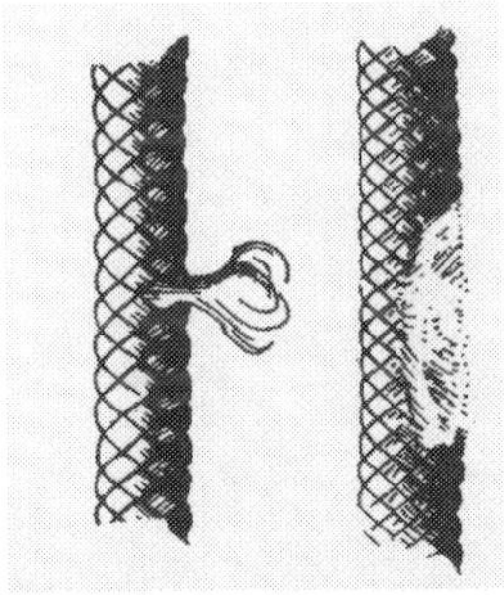

DAMAGED AND EXPOSED ROPE

(2) Exposed Core Fibers. Indicates severe sheath damage. (When you can see the inner core fibers)

(3) Glossy Marks. Signify heat fusion damage, also called a booger.

(4) Uniformity of Diameter / Size. May indicate core damage, noted by an obvious depression (hour glass) or exposure of white core fibers protruding from the sheath (puff).

(5) Discoloration. A drastic change from the ropes original color may indicate chemical change or damage.

(6) Stiffness or Soft Spots. Could signify core damage.

NOTE: Dynamic ropes measuring between 10mm and 12mm are marked at each end of its pigtails with a number "1" indicating that the rope is UIAA approved for single rope lead climbing. Dynamic ropes measuring between 8mm and 9mm are marked at each end of its pigtails with a "1/2" number indicating that two of these ropes are required to conduct a lead climb.

4. **ROPE LOG.** The purpose of the Rope Log is to maintain an accurate record for the use of each rope contained within the Marine Assault Climbers Kit (MACK). Due to the turnover of personnel, and the fact that no one person may have the same rope twice, the rope log is used to ensure the safe use, serviceability, and account of each rope.

a. Serial number. By assigning each rope a serial number, responsible units can determine information about a rope. As soon as the ropes are cut to the desired length for their intended purpose, each rope will be assigned a serial number by the responsible unit. That rope should then be labeled with that serial number in some permanent manner. The best method for this is as soon as the ends of the new rope have been whipped and fused, mark both ends of the rope with the serial number and then dip the ends of the rope, in effect, laminating the serial number to the rope ends.

(1) A rope serial number has five parts. The proper format for a rope serial number is as follows:

1. Type of rope: (S) - Static or (D) - Dynamic.

2. Last two digits of the year the rope was manufactured. Each rope has a shelf life of two years, after that it must not be used for any mountaineering purpose.

3. Four-digit number for that individual rope which is assigned by the responsible unit, and should be assigned sequentially as new ropes are issued out.

4. The length and diameter of the rope. The length may be recorded in feet, and the diameter in millimeters.

5. Responsible unit code. Example: S-96-0001-150/11mm-F2/8

   Example meaning: This rope is a 150 foot, 11mm, static rope, manufactured in 1996. It is the first rope issued by Fox Co 2/8.

b. Recording Information in a Rope Log. Once a new rope has been serialized, the rope log for that rope should be started up. At a minimum, it should contain the following information:

(1) The rope serial number.

(2) Manufacturer. Depending on the manufacturer, ropes of the same type and diameter may vary in tensile strength, stretch factor, and durability.

(3) Date of manufacture. Five years from this shelf life date, ropes are considered to have reached their normal expiration date, and should be destroyed.

(4) Date in service. This is to be recorded for tracking purposes to establish how long a rope is in service. Two years is considered maximum rope life in service.

(5) Each time a rope is used, the using unit is responsible to record how that rope was used and how much use the rope received. Additionally, before checking out a rope and prior to turning it back in, the rope must be inspected by qualified personnel, and initial in the "Inspected By" block of the rope log.

NOTE: ANY TIME A DYNAMIC ROPE IS SUBJECTED TO A FALL FACTOR 2, THAT ROPE SHOULD NOT BE USED AGAIN FOR MOUNTAINEERING.

DYNAMIC					
SERIAL #					
MANUFACTURER					
DATE OF MANUFACTURE					
DATE IN SERVICE					
INDICATE THE DATE AND NUMBER OF EACH SPECIFIC USE					
CLIMBS	**FALLS (FACTOR)**	**TOP ROPES**	**RAPPELS**	**REMARKS**	**INSPECTED BY**

ROPE LOG

5. **COILING A ROPE**. There are two types of rope coils frequently used at MWTC. The Mountain Coil and the Butterfly Coil.

a. Mountain Coil. This coil is useful for carrying the rope over a pack or over a climber's shoulder and neck. It can be used for short time storage. The mountain coil can be tied in the following manner:

(1) Sit down with your leg bent at a 90-degree angle, heel on the deck. Starting at one end, the rope is looped around the leg in a clockwise fashion, going over the knee and under the boot sole until the entire rope is coiled.

(2) If coiling a 150-foot rope, use only one leg and offset the other, when coiling a 300-foot rope, use two legs and keep them together.

(3) With the starting end of the rope, form a 12-inch bight on the top of the coils.

(4) Uncoil the last loop and along the top of the coils, wrap 4-6 times towards the closed end of the bight.

(5) The end of the rope being wrapped is then placed through the closed end of the bight.

(6) The running end of the bight is then pulled snuggly to secure the coil.

(7) To prevent the coil from unraveling, the two pigtails are tied together with a square knot.

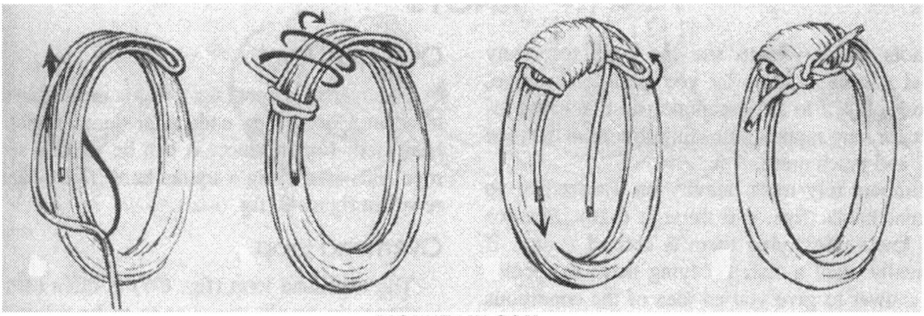

MOUNTAIN COIL

b. Butterfly Coil. This method is used for carrying a rope when the individual needs to have maximum use of his upper body, (i.e. while climbing), without the encumbrance of a large rope coil hanging across his chest.

(1) Coiling the Butterfly Coil

(a) Step 1: Find the middle of the rope, then form a three foot bight laying both ropes in the upraised palm at the two foot point.

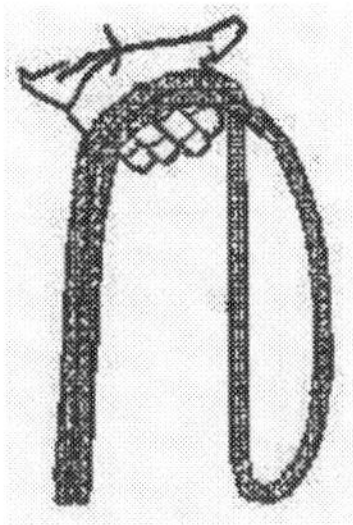

STARTING THE BUTTERLFY COIL

(b) Step 2: Form another two-foot bight with the running end. Place the rope at the two-foot bight along side on top of the original bight ensuring the running end is on the same side as the original bight.

(c) Step 3: Continue making two foot bights, laying them alternately into your palm until there is only six to eight feet remaining. At that point, begin wrapping the two pigtails horizontally four to six times at the mid way point of the ropes in a bight from bottom to top.

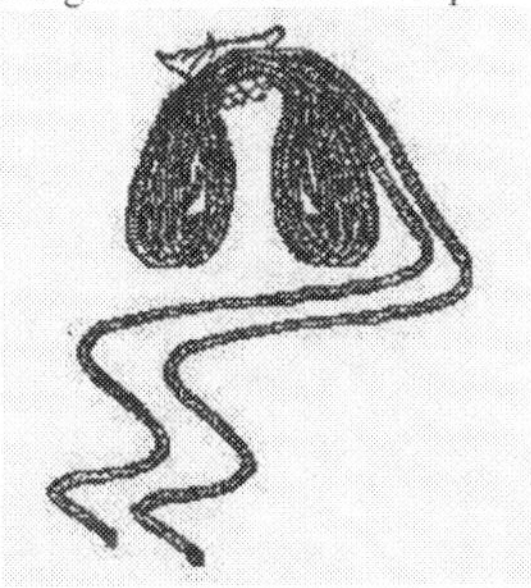

TWO FOOT BIGHTS ON BOTHS SIDES WITH 6 TO 8 FEET REMAINING

(d) Step 4: After completing your wraps, form a bight with the remaining pigtail and then thread it underneath your palm and upwards to one-foot above the coiled rope.

(e) Step 5: With the remaining pigtail, thread it through the one-foot bight in step four.

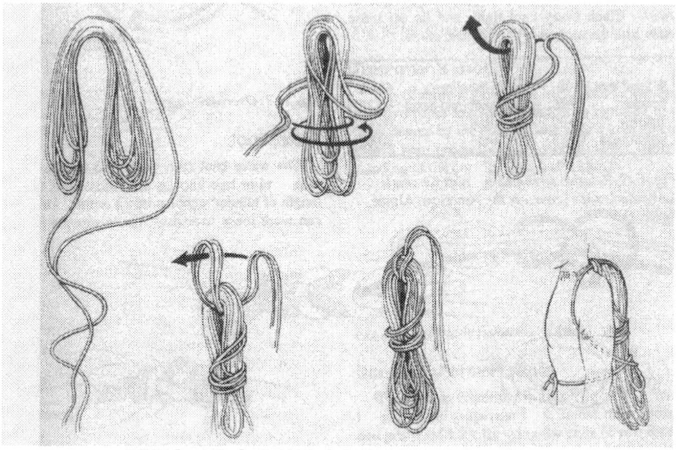

TYING AND CARRYING THE BUTTERFLY COIL

(2) Carrying the Butterfly coil. Separate the running ends, placing the coil in the center of the back of the carrier, then fun the two ends over his shoulders so as to form shoulder straps. The running ends are then brought under the arms, crossed in the back over the coil, brought around the body of the carrier and tied off with a square knot at his stomach.

6. **ROPE THROWING**. To insure that the rope will not get tangled when deployed, certain steps must be taken.

a. With a stacked rope, anchor off the standing end.

b. Take the opposite end of the rope and make 6-8 coils and place them in your strong arm. These wraps will serve as a throwing weight that you can aim.

c. 10-15 feet from the strong-arm coils create a second set of 6-8 wraps and place them in your weak arm.

d. From the edge of the cliff, sound off with the command "STAND-BY FOR ROPE". Just before you release the coils, sound off with the command "ROPE". At that time drop the weak arm coils from the cliff.

e. While taking aim, throw your strong-arm coils overhand or sidearm hard enough to hit your intended target.

f. If the throw was misdirected due to wind, tree, etc., reorganize and attempt to re-deploy the rope.

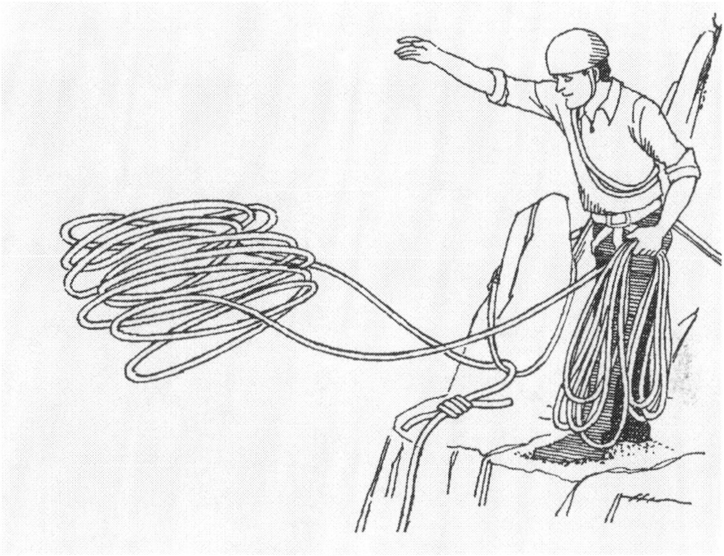

THROWING THE ROPE

6. MOUNTAINEERING KNOTS.

a. **Class I** – *End of the Rope Knots*

(1) Square Knot. Used to tie ends of two ropes of equal diameter together. It should be secured by overhand knots on both sides of the square knot.

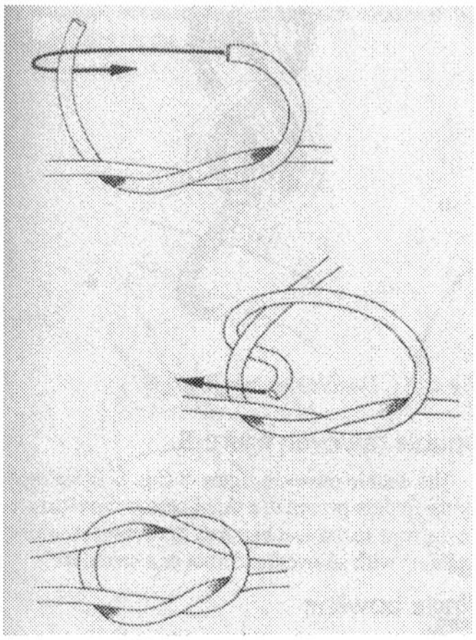

SQUARE KNOT

(2) Double Fisherman's Knot. It is a self-locking knot used for tying two ropes of equal diameter together. It can be tightened beyond untying.

DOUBLE FISHERMAN'S KNOT

(3) Water/Tape Knot. Used to secure webbing or tape runners. It is constructed by tying an overhand knot (without twists) in one end of the tape, and threading the other end of the tape through the knot from the opposite direction. After the knot is dressed down, each pigtail should be a minimum of two inches long.

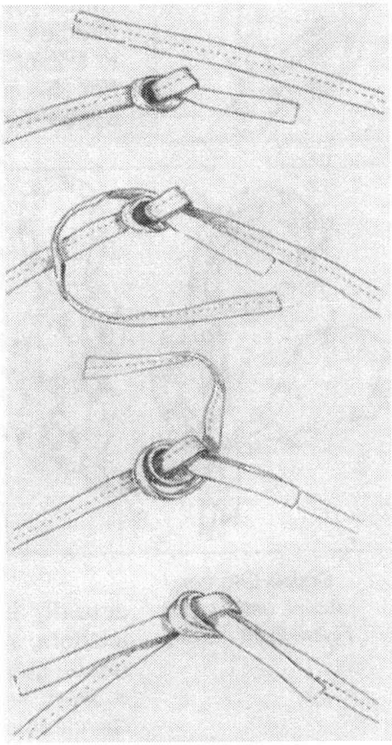

WATER TAPE KNOT

(4) Double Sheet Bend. Used to tie the ends of two or more ropes of equal or unequal diameter together.

SHEET BEND

b. **Class II** - *Anchor Knots*

(1) Bowline. Used to tie a fixed loop in the end of a rope. This knot is always tied with the pigtail on the inside and secured with an overhand knot.

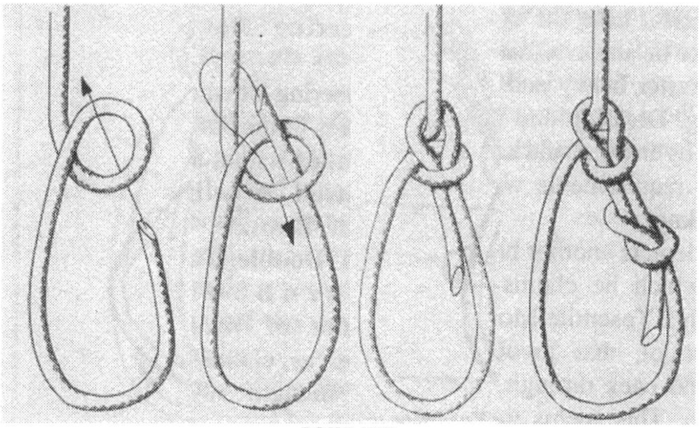

BOWLINE

(2) Round Turn with Two Half Hitches or a Bowline. A loop which runs around an object in such a manner as to provide 360 degree contact and may be used to distribute the load over a small diameter anchor. It will be secured with two half hitches or a bowline.

ROUND TURN AND TWO HALF HITCHES AND

ROUND TURN AND A BOWLINE

(3) Clove Hitch. This knot is an adjustable hitch. It could be considered a middle-of-the-rope anchor knot at the end-of-the-rope when used in conjunction with a bowline or round turn and two half hitches.

(1) Around-the-object Clove Hitch.

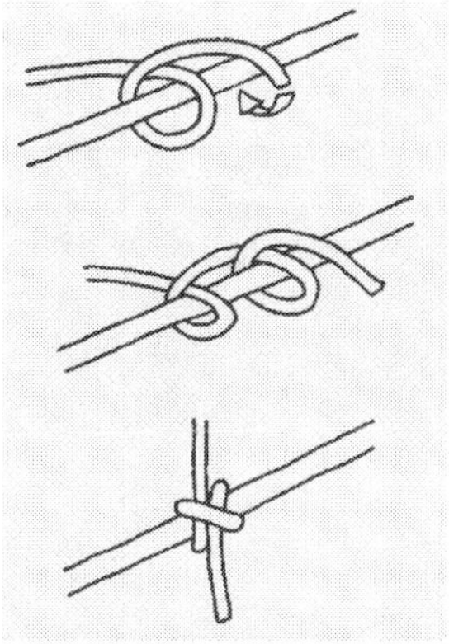

AROUND THE OBJECT CLOVE HITCH

(2) Over-the-object Clove Hitch.

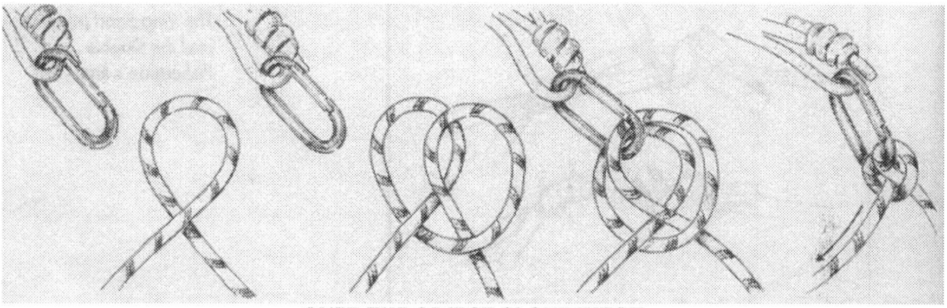

OVER THE OBJECT CLOVE HITCH

c. **Class III** - *Middle-of-the-rope*

(1) Figure-of-Eight-Loop. This is a strong knot that can be readily untied after being under load.

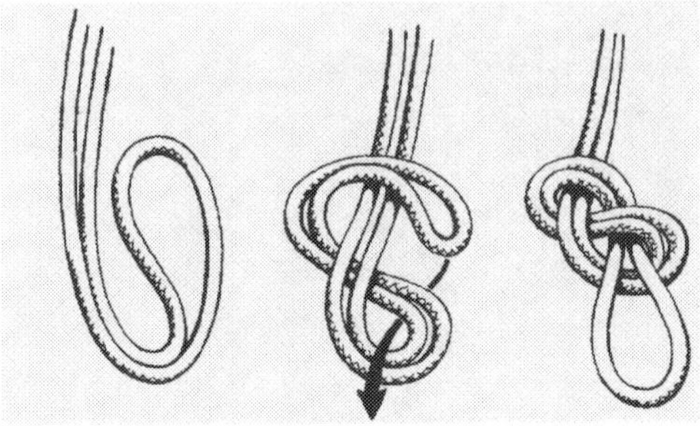

FIGURE 8 LOOP

(2) Double Figure of Eight Loop. The double Figure of Eight Loop is a strong knot and the double loop reduces the wear and strength loss from the rope bending around the carabiner by splitting the load between the two loops.

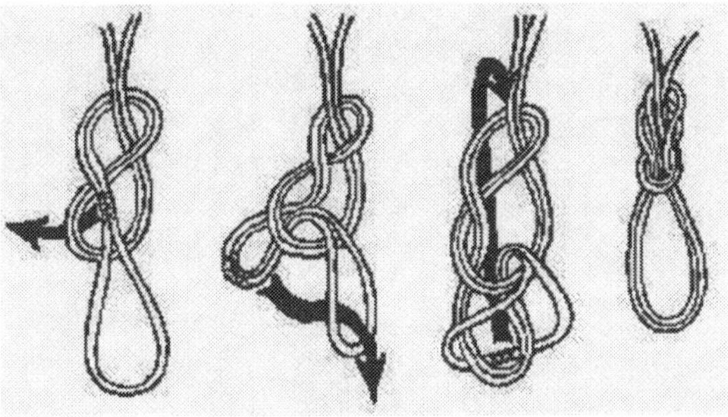

DOUBLE FIGURE 8 LOOP

(3) Two Loop Bowline. This knot can be used to construct a self-equalizing belay or to tie the middle person in on three people on a rope.

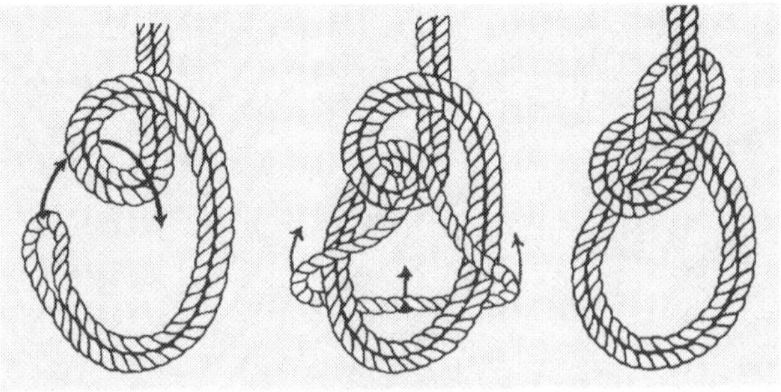

TWO LOOP BOWLINE

d. **Class IV** - *Special Knots*

(1) Prusik Knot. This knot functions by introducing friction that can be alternately set and released. For best results, tie the knot with a smaller diameter cord on a larger diameter cord. If slippage occurs, more wraps maybe used.

(a) Middle-of-the-Rope Prusik. This knot is created with an endless loop also known as a Prusik Cord. Do not tie this knot with tape due to less friction.

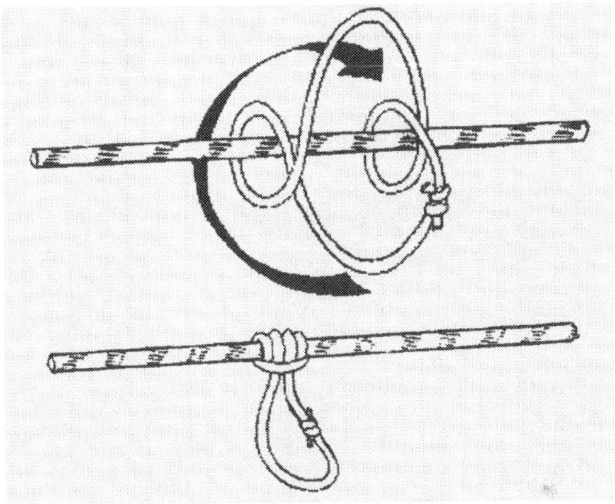

MIDDLE OF THE LINE PRUSIK

(b) End-of-the-Rope Prusik. This knot is always secured with a bowline. Do not tie this knot with tape due to less friction.

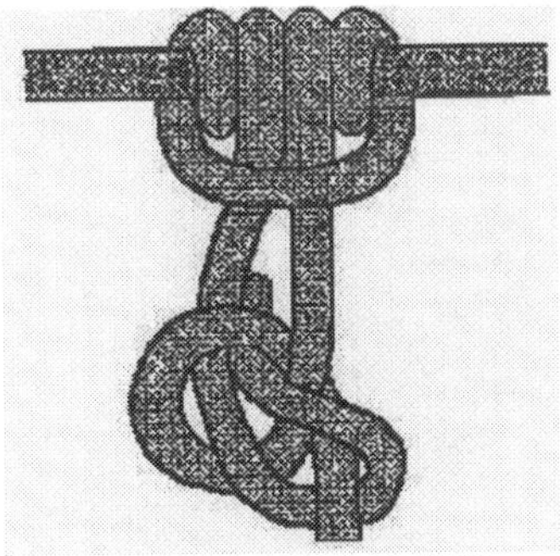

END OF THE LINE PRUSIK

(c) French Prusik. This can be constructed with a Prusik Cord or a single strand of cord with figure-of-eight loops on each end.

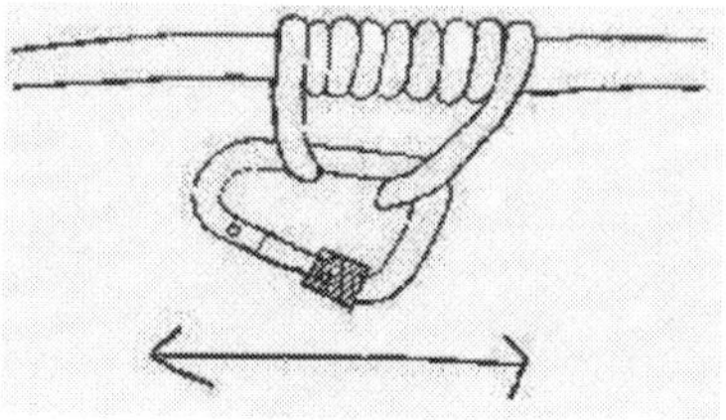

FRENCH PRUSIK

(2) Retraced Figure of Eight. Used to tie the end of the climbing rope into a harness or swammi wrap. The pigtail may or may not be secured with an overhand.

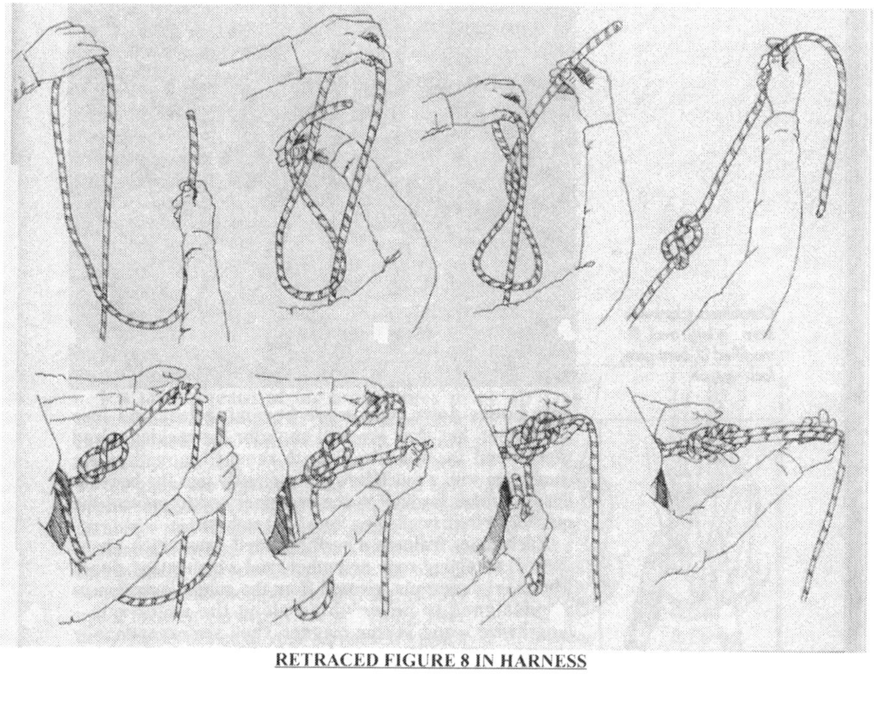

RETRACED FIGURE 8 IN HARNESS

(3) Directional Figure-Of-Eight. When tied and tensioned is applied to the standing and running ends of the rope, the knot will not pull apart. The loop will point toward the direction of pull.

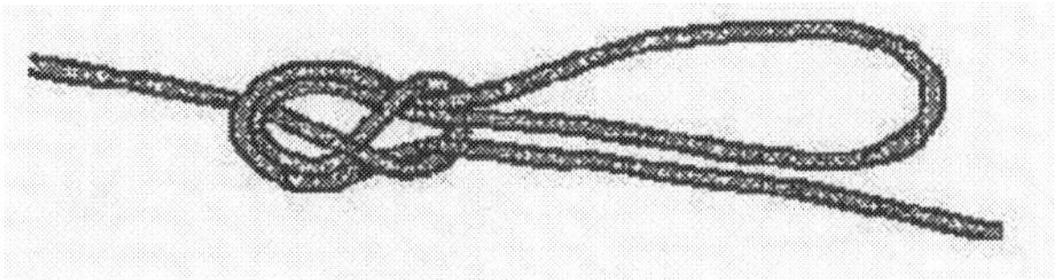

DIRECTIONAL FIGURE 8

(4) Slip Figure 8. This knot is used for retrievable anchors and fixed rope installations for the ease of untying the knot.

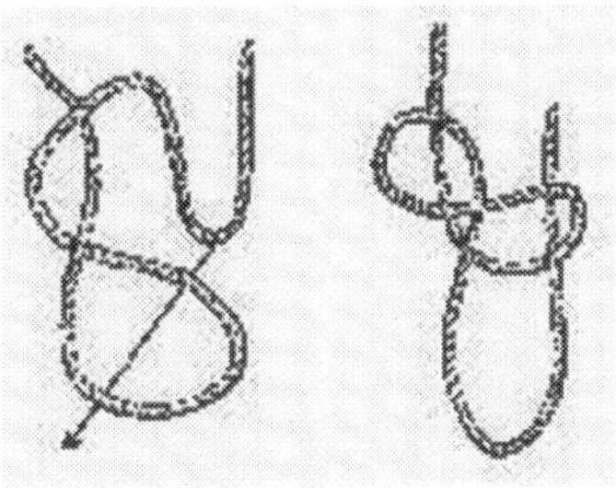

SLIP FIGURE 8

(5) Kliemheist. This is a friction knot.

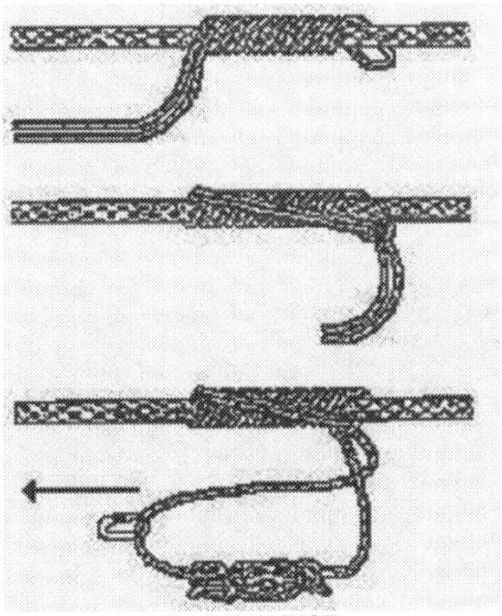

KLEIMHEIST

(6) Overhand Knot. Can be used to secure primary knots on itself.

OVERHAND

(7) M nter Hitch. This is a simple hitch in the rope that is clipped into a carabiner to put friction on the line.

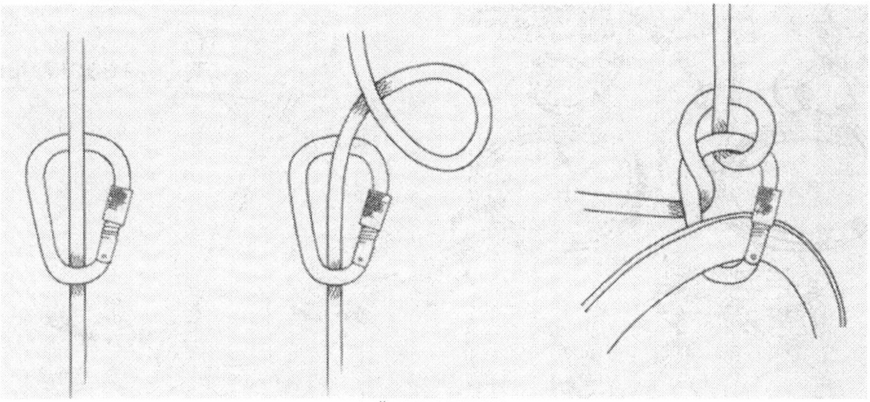

MÜNTER HITCH

(8) Timber Hitch. A Timber Hitch is used to fix a rope to a pole or equivalent for hoisting or towing purposes. It has the capability of casting off easily.

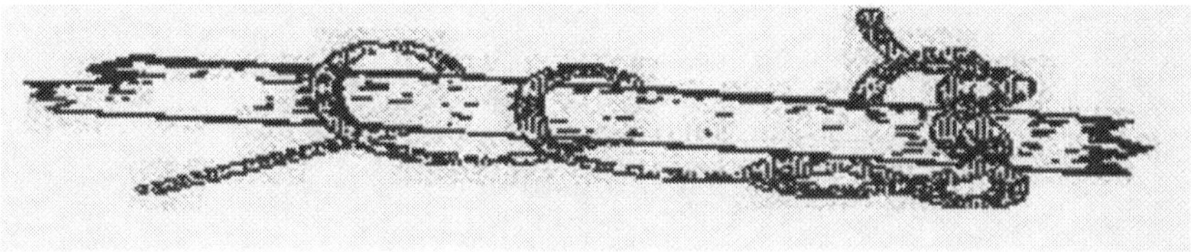

TIMBER HITCH

(9) Mariner's Hitch. This knot's advantage over the prusik is that it can be released under load.

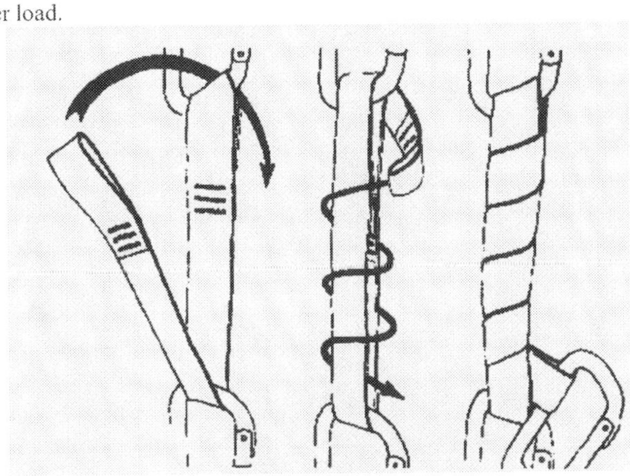

MARINER'S HITCH

(10) Krägur Knot. This friction knot is used to prevent unnecessary slippage when tied onto a rope of the same diameter.

KRÄGUR KNOT

(11) Rappel Seat. This is used as an expedient support harness for rappelling, crossing rope bridges, etc. It is constructed as follows:

(a) Center the sling rope on the left hip.

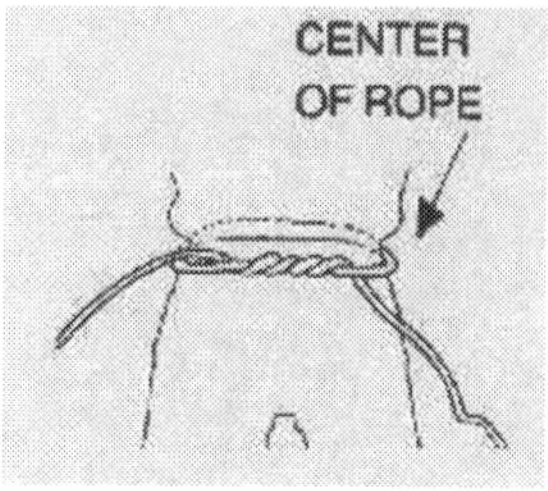

(b) Wrap the sling rope around the waist and tie at least one half hitch around itself (preferably two) in the front.

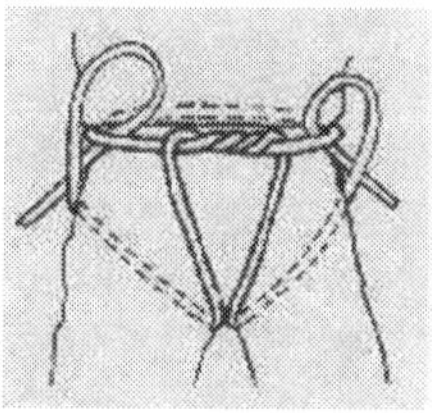

(c) Bring the running ends down through the legs, up over the buttocks, and over the original waist wrap, down between the waist wrap and the waist, and over itself, forming a bight. Cinch this up tightly.

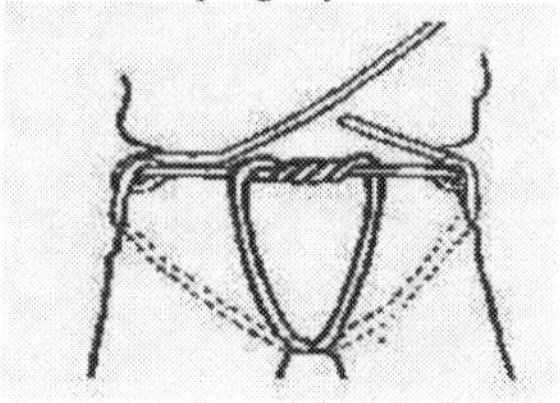

(d) Now tie a square knot with two overhand knots on the left hip ensuring that the overhand knots encompass all the wraps.

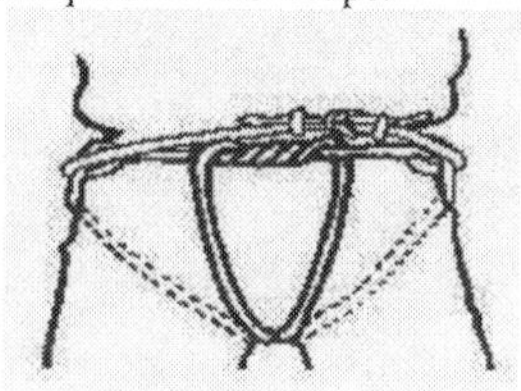

(e) Tuck any excess rope into a pocket.

(12) Bowline on a coil. Used by the first and last men on a climbing rope to tie into the rope. An overhand knot is used behind the knot. It distributes the force of a fall over a larger area of a climber's waist and is preferable to a single bowline around the waist. The bowline on a coil can also be used to take up excess rope. The bowline on a coil should have 4-6 wraps around the waist.

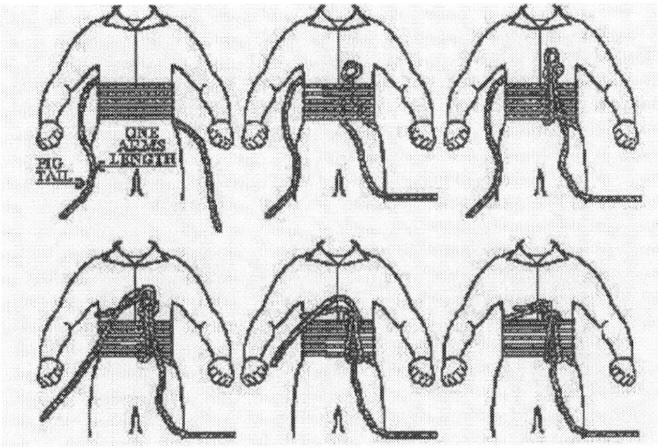

BOWLINE ON A COIL

(13) Swami wrap (Swami belt). The swami wrap (four to six wraps of rope or nylon webbing tied around the waist) popular with rock climbers who wanted to increase the working length of their climbing ropes. It also offers some improvement over the bowline on the coil in distributing the forces sustained in a fall over a larger area of the midsection of the body, but does not eliminate the problem of suffocation when hanging. The Swami wrap is constructed as follows:

(a) Take a sling rope and find the middle.

(b) Place the middle of the sling rope in the small of your back.

(c) Bring both pigtails to the front and continue to wrap around your body until both pigtails are about 15-20 inches long.

(d) Tie the pigtails on the left side with a square knot and two overhand knots.

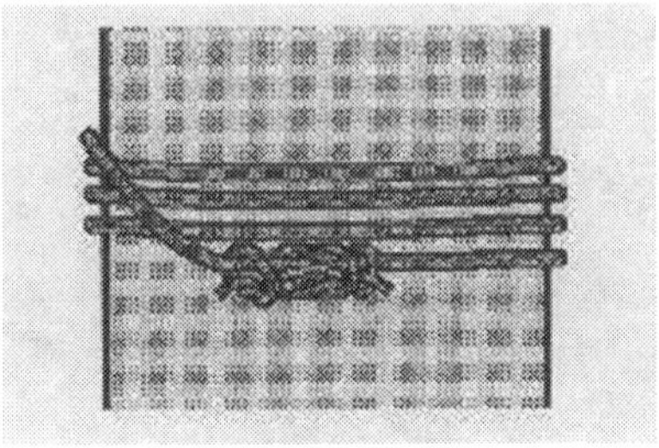

SWAMMI WRAP

(14) Three Loop Bowline. This knot will provide three bights, each of which can be adjusted against the others.

THREE LOOP BOWLINE

7 **KNOT STRENGTH.** Rope, cordage and webbing is strongest when loaded in a straight line. When bending a rope or web to create a knot, the strength of the rope is reduced. All knots should be dressed properly for maximum effective use.

KNOT	RELATIVE STRENGTH OF KNOT
No Knot	100%
Figure 8	75-80%
Bowline	70-75%
Double Bowline	70-75%

Double Fisherman	65-70%
Water Tape Knot	60-70%
Clove Hitch	60-65%
Overhand Knot	60-65%
Square Knot	45%

l. **KNOT TESTING TIME LIMIT.** The following times must be met to pass the knot tying portion of this course.

KNOT	**UNIT OPERATIONS TIME LIMIT**	**SML, ACC TIME LIMIT (BLINDFOLDED)**
Square knot	60 Seconds	30 Seconds
Double Fisherman's Knot	60 Seconds	30 Seconds
Water/Tape Knot	60 Seconds	30 Seconds
Round Turn and a Bowline	60 Seconds	30 Seconds
Round Turn and Two Half Hitches	60 Seconds	30 Seconds
Clove Hitch (around the object)	60 Seconds	30 Seconds
Munter Hitch	60 Seconds	30 Seconds
Slip Figure 8	60 Seconds	30 Seconds
Figure 8 Loop	60 Seconds	30 Seconds
Directional Figure 8	60 Seconds	30 Seconds
End of the Line Prusik	90 Seconds	45 Seconds
Retrace Figure 8	90 Seconds	45 Seconds
Rappel (Swiss) Seat	2 Minutes	60 Seconds

UNITED STATES MARINE CORPS
Mountain Warfare Training Center
Bridgeport, California 93517-5001

SMO
05/02/01

STUDENT OUTLINE

NOMENCLATURE AND CARE OF MOUNTAINEERING EQUIPMENT

LESSON PURPOSE. The purpose of this period of instruction is to introduce you to the types of equipment used here at MWTC and how to care for it so as to prevent its untimely failure. This lesson relates to all climbing and installation work that you perform here at MWTC.

OUTLINE

1. **ROPES.** All ropes used in the military must meet UIAA standards or U. S. Federal Test Standard 191A. Most ropes have a 5-year shelf life and maximum 2-year service life.

a. Static. Black in color.

(1) Construction. Kernmantle

(2) Minimum tensile strength. 7500 lbs.

(3) Maximum elongation. 1.5%

(4) Diameter. 11mm

(5) Sizes.

(a) 165 ft ± 5 ft

(b) 300 ft ± 10 ft

(6) Usage. Rescue operations and bridging where a low amount of elongation is desirable under a working load.

b. Dynamic. Olive Drab in color.

(1) Construction. Water-resistant treated Kernmantle to reduce friction.

(2) Minimum tensile strength. 6500 lbs.

(3) Maximum elongation. 6%

(4) Diameter. 10.5mm and 11mm

(5) Size. 165 ft ± 5 ft.

(6) Usage. For lead climbing/party climbing.

c. Maxim Dry Rope. Olive Drab or Multi-Colored.

(1) Construction. Water-repellent treated Kernmantle.

(2) Minimum tensile strength. 3472 lbs.

(3) Maximum elongation. 6%

(4) Diameter. 9mm

(5) Size. 150 ft ± 5 ft

(6) Usage. For glacier travel/ice climbing

d. Gold Line II.

(1) Construction. Eight strand braided nylon plymor.

(2) Minimum tensile strength. 4500 lbs.

(3) Maximum elongation. 20%

(4) Diameter. 11 mm

(5) Size. 300 ft or 600 ft spools

(6) Usage. Sling Ropes and litters only.

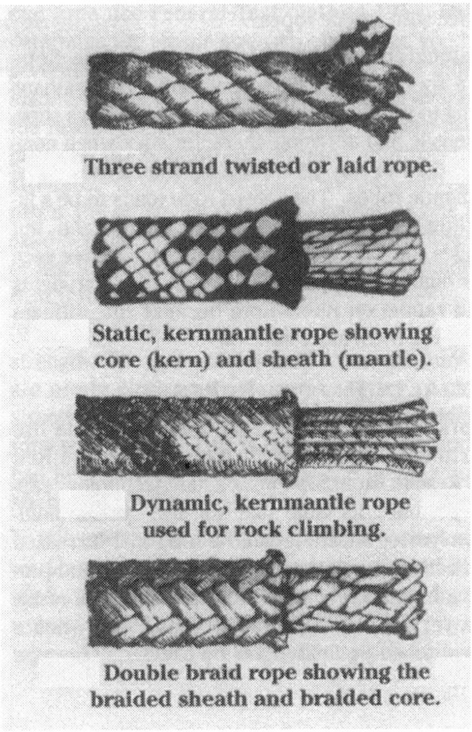

INSIDES AND OUTSIDE SHEATHS OF VARIOUS TYPES OF ROPES

NOTE: Sling ropes are made from 15 foot lengths of plymor or dynamic rope ONLY. Twenty-five foot practice coils should be constructed with static rope, but dynamic rope can be used.

2. ADVANTAGES/DISADVANTAGES

a. Advantages of Nylon Rope

(1) High strength to weight ratio.

(2) Good energy absorption in dynamic ropes.

(3) Flexible.

(4) Rot resistant, not affected by frost.

b. Disadvantages of Nylon Rope

(1) Low melting point. Nylon fuses at 400°F and melts at 480°F.

(2) Susceptible to abrasions and cuts.

(3) Affected by chemicals and light.

c. Advantages of Manila Rope

(1) Easily gripped.

(2) Hard wearing.

(3) Does not deteriorate in heat.

d. Disadvantages of Manila Rode

(1) Heavy, kinks, especially when wet. Absorbs water and swells.

(2) Burns at +300°F.

(3) Edible by rodents.

3. GENERAL INFORMATION

a. Nylon rope stretches under tension and will rupture at between 30% and 70% elongation, depending on construction.

b. Nylon rope loses as much as 30% strength when wet.

c. Temperatures as low as 250°F will damage a nylon rope.

4. NYLON WEBBING

a. The type of nylon webbing available is tubular. Tubular nylon webbing is very strong and flexible. All rules that apply to nylon rope apply to tubular nylon webbing. The size of nylon webbing used is:

(1) 1 inch tubular nylon. Tensile strength approximately 4,000 - 4,500 lbs., depending on the manufacturer.

b. Pre-sewn Spectra Runners. Tensile strength approximately 5,500 lbs.

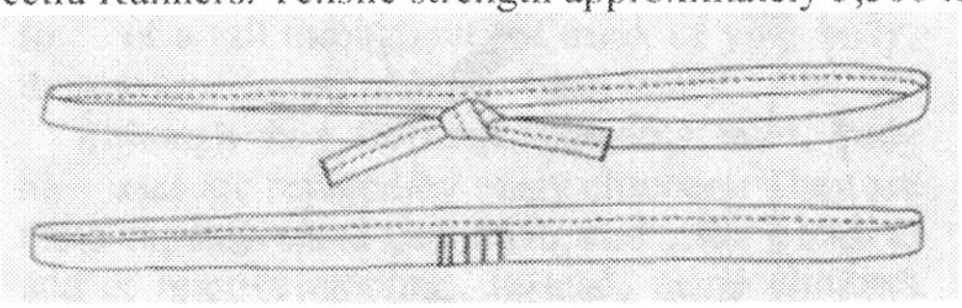

NYLON WEBBING

NOTE: These are minimum strengths. Some manufactures make even stronger webbing.

1. **CARABINERS**. Also commonly known as snap links. There are two different types of carabiners frequently used at MWTC, the steel locking carbine and the non-locking aluminum carabiners.

a. Steel Locking Carabiner. There are two different sizes of carabiners:

(1) Steel Locking Stubai 82.

(2) Large Locking "D" Stubai 85.

b. Aluminum Non-Locking Carabiner.

(1) The standard "D" non-locking carabiner with a tensile strength of 2000 lbs. or more.

2. **NOMENCLATURE OF A CARABINER**.

a. There are four parts to a non-locking carabiner. They are as follows:

(1) Gate.

(2) Gate pivot pin.

(3) Locking pin

(4) Body.

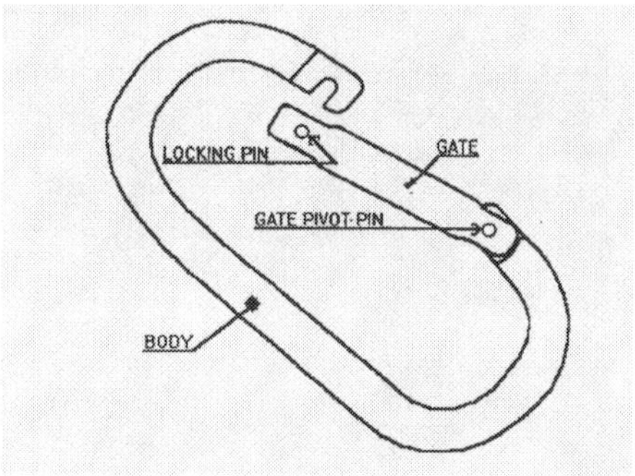

STEEL NON-LOCKING CARABINER

b. There are five parts to a locking carabiner. They are as follows:

(1) Gate

(2) Gate pivot pin

(3) Locking notch

(4) Locking nut

(5) Body

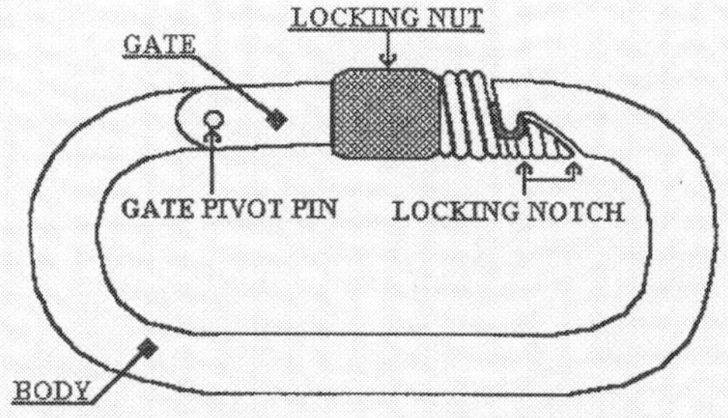

STEEL LOCKING CARABINER

3. **<u>CARE, MAINTENANCE AND SERVICEABILITY OF A CARABINER</u>**. The following inspections and maintenance should be performed on a routine basis:

a. <u>Care</u>.

(1) Avoid dropping of a carabiner onto a hard surface, hairline factures could occur.

(2) Keep the carabiners off of the bare ground because dirt and grit can get into the working parts.

(3) If the carabiner has fallen from a significant height, it should be retired.

b. <u>Maintenance</u>.

(1) Remove all dirt, moisture and grime after use.

(2) Lubricate with tri-flow graphite and wipe off with a clean cloth.

c. <u>Serviceability Checks</u>.

(1) The gate snaps shut with no friction and with no gap between the locking pin and locking notch.

(2) There is no excessive side to side movement of the gate

(3) The pivot pin is tight.

(4) That the spring-loaded gate functions properly.

(5) The locking pin is tight.

(6) The locking nut travels freely and locks securely.

(7) There are no cracks or flaws in the metal.

NOTE: The weakest part of a carabiner is the gate. If an engraver is used to mark a carabiner, it should be applied to the gate and not the load bearing side.

UNITED STATES MARINE CORPS
Mountain Warfare Training Center
Bridgeport, California 93517-5001

WMC
04/11/02

LESSON PLAN

BALANCE CLIMBING

1. LEARNING OBJECTIVES

a. TERMINAL LEARNING OBJECTIVE. In a summer mountainous environment, execute balance climbing, in accordance with the references. (WMC)

b. ENABLING LEARNING OBJECTIVES.

(1) With the aid of references, state orally the safety requirements for balance climbing, in accordance with the references. (WMC.C3a)

(2) With the aid of references, state orally the individual preparations for a balance climb, in accordance with the references. (WMC.C3b)

(3) In a summer mountainous environment, execute the duties of a spotter for a balance climb, in accordance with the references. (WMC.C3c)

(4) In a summer mountainous environment, execute the commands used between the climber and the spotter during a balance climb, in accordance with the references. (WMC.C3d)

(5) In a summer mountainous environment, execute each type of hold used in balance climbing, in accordance with the references. (WMC.C3e)

(6) Without the aid of references and given the acronym "CASHWORTH", describe orally the nine considerations for proper body position while climbing, in accordance with the references. (WMC.C3f)

OUTLINE

1. **SAFETY PRECAUTIONS**. There are two safety precautions that always apply to balance climbing. They are as follows:

C3

a. Never climb more than 10 feet above the ground. By this it is meant that the climber's feet are never more than 10 feet above the ground.

b. A spotter is required for all balance climbs.

SPOTTER WATCHING THE CLIMBER

2. **INDIVIDUAL PREPARATIONS**. Prior to beginning a balance climb there are seven things that the climber must do to prepare himself. They are as follows:

   a. Helmet with a serviceable chinstrap must be worn.

   b. Sleeves rolled down to give hand and arm freedom of movement. Blouse tucked in to your trousers. In case of fall, it may catch on a rock and cause you to flip over sideways.

   c. All watches, rings, and jewelry must be removed before climbing.

   d. Gloves will not be worn, as they can slip, and also give a false feel for the rock.

   e. Unblouse trousers, if they restrict movement.

   f. Soles of boots clean and dry.

   g. Select route where vegetation is minimal. Never use vegetation for hand or foot holds.

3. **DUTIES OF THE SPOTTER**. The spotter is the balance climber's partner who, rather than climbing himself, acts as the safety man for the climber during the climb. The five duties of the spotter are as follows:

a. Positions himself directly behind the climber before the climb starts.

b. Maintains his position facing the cliff, directly below the climber and approximately 3-4 feet away from the base of the cliff, for the duration of the climb. He will move diagonally as necessary to remain below the climber.

c. The spotter will stand with his feet shoulder width apart and arms ready to stop the climber if he falls.

d. If the climber falls, the spotter will not "catch" him; he will prevent the climber from falling further down the hill. He will do this by pushing the climber towards the base of the cliff, thereby preventing him from tumbling backwards.

e. At no time will the spotter allow anyone to come between himself and the face of the cliff while a balance climb is taking place. He will require anyone who wants to pass by his position to go behind him.

4. **SPOTTING AND CLIMBING COMMANDS**. The following are the commands used by both the spotter and climber.

COMMAND	GIVEN BY	MEANING
"Last Name, climbing."	Climber	I am ready to climb.
"Climb Climber."	Spotter	I am ready to spot you.
"Last Name, off Climb."	Climber	I am off the climb.
"ROCK!"	Climber or Spotter.	A rock has been knocked off the rock face and is falling.
"FALLING!"	Climber	I am Falling.

a. If the command "ROCK" is given, all personnel in the vicinity will take the following action:

(1) If close to the cliff face, move against the cliff face with your face against the cliff face and your hands between you and the cliff face.

(2) If not close to the cliff face, look up to locate the rock and avoid it.

5. **ACTIONS IF FALLING**. If, while making a balance climb, the climber feels himself slipping and beginning to fall, he will take the following action:

a. Sound the command "falling".

b. Push himself away from the rock face.

c. Maintain proper body position as follows:

(1) Head up.

(2) Hands out toward the rock.

(3) Body relaxed.

(4) Feet kept below the body, slightly apart.

(5) Ensure you face the cliff face as you fall.

6. **TYPES OF HOLDS**. There are five basic holds that are used in balance climbing. They are as follows:

a. Push Holds.

(1) Most effective when hands are kept low.

(2) Often used in combination with a pull hold.

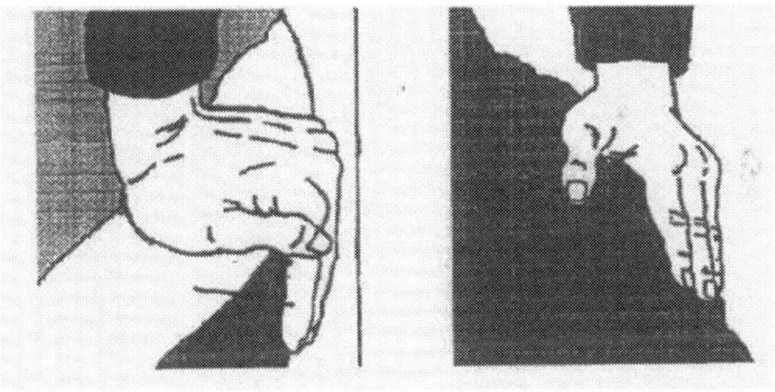

TYPES OF PUSH HOLDS

b. Pull Holds.

(1) The easiest hold to use and, consequently, often overused.

(2) Can be effective on small projections.

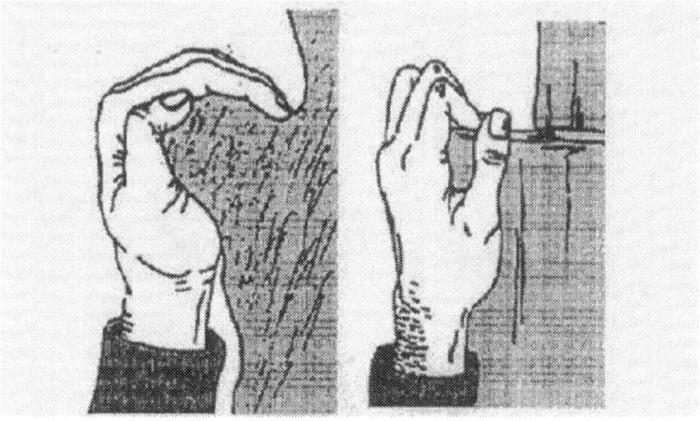

TYPES OF PULL HOLDS

c. Foot Holds

(1) Feet should be positioned with the inside of the foot to the rock.

(2) Use full sole contact as much as possible.

(3) Avoid crossing your feet. If you must cross your feet use a change step. A change step is a method of substituting one foot for the other foot on the same foothold.

(4) Making maximum use of footholds, climbing with your feet, is an effective means of conserving your body strength, since your leg muscles are stronger than your arm muscles.

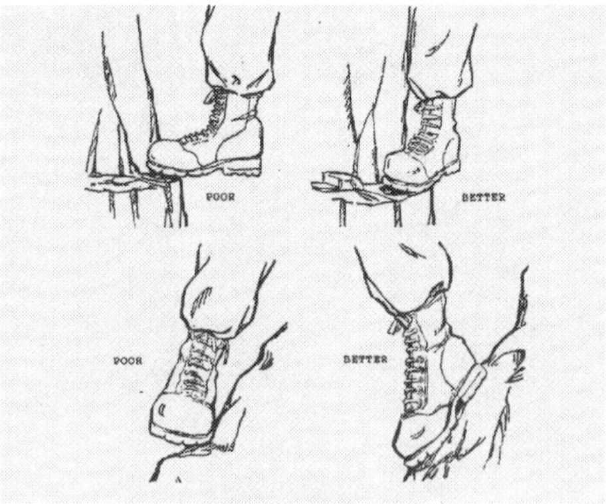

USE OF FOOT HOLDS

d. Friction Holds. A friction hold is anytime you are relying on the friction of your foot or hand against the face of the rock for traction, rather than pushing/pulling against a projection on the face of the rock.

(1) It is a type of hold that feels very insecure to an inexperienced climber.

(2) The effectiveness of this type of hold is dependent upon many things i.e., type, condition and angle of the rock face, type of boot soles, confidence, etc.

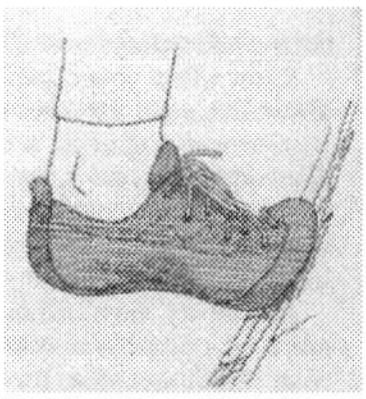

FRICTION USED IN SMEARING

e. Jam Holds. This type of hold involves jamming/wedging any part of your body or your entire body into a crack/opening in the rock.

(1) An important consideration is that you do not jam such that you cannot free that portion of your body after you complete the move. This sounds like a ridiculous statement; however, you must remember that after you complete your move, you may be withdrawing the portion of your body that you used from a different angle than you inserted it.

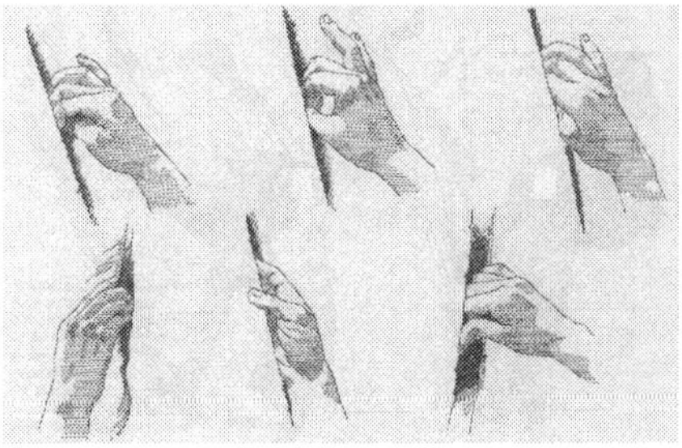

FINGER JAMS

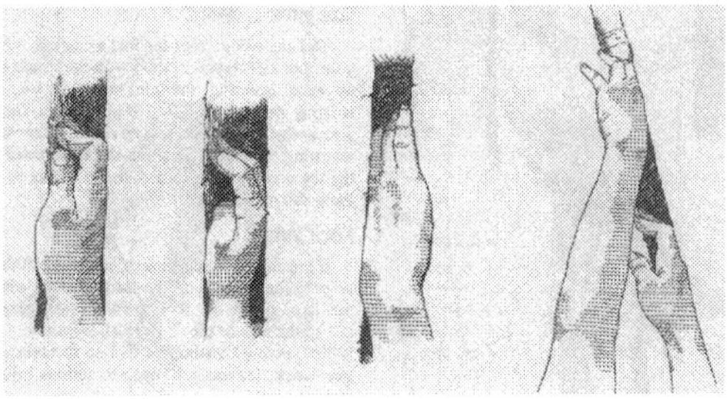

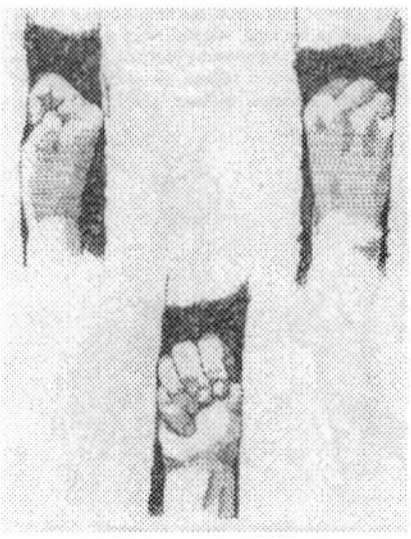

HAND JAMS **FIST JAMS**

7.) **COMBINATION HOLDS.** The five types of holds just mentioned above are not just used individually. They are most often used in combinations with each other. Some examples are:

a. Chimney Climbing. This is when you insert your entire body into a crack in the rock and by using both sides of the opening, and possibly all five types of basic holds, move up' the crack.

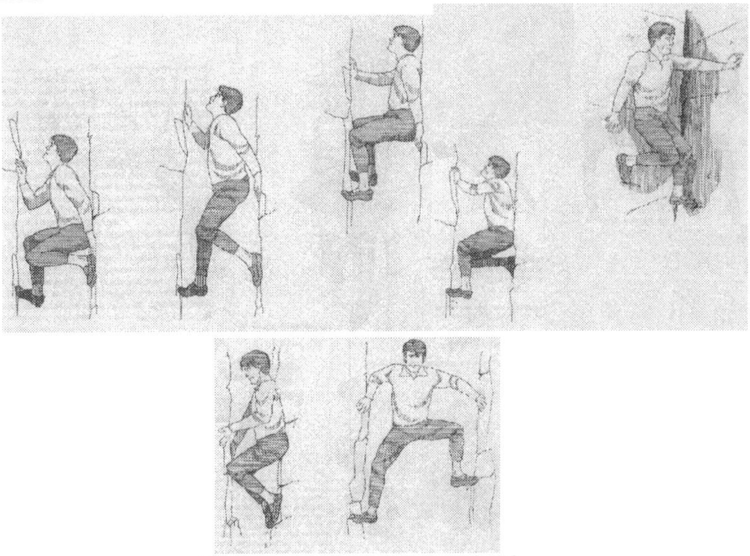

CHIMNEY CLIMBING TECHNIQUES

b. Lie-back. This is a combination of both pull holds with your hands and friction holds with your feet.

LIE BACK TECHNIQUE

c. Push-Pull. As the name implies, this is when you use a push hold and a pull hold together.

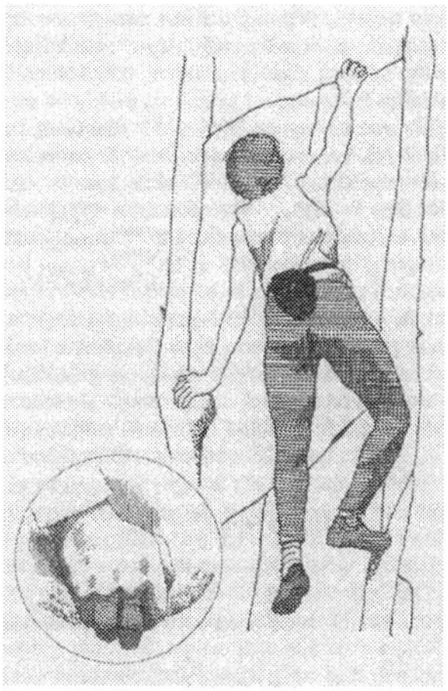

PUSH-PULL

d. Mantling. This is a technique where you continue to climb without moving your hands off a projection by pulling yourself up until your hands are at chest level and then invert your hands and push on the same projection.

MANTELING

MANTELING WITH FACE HOLD

e. Cross-Pressure in Cracks. This is a technique of putting both hands in the same crack and pulling your hands apart to hold/raise yourself.

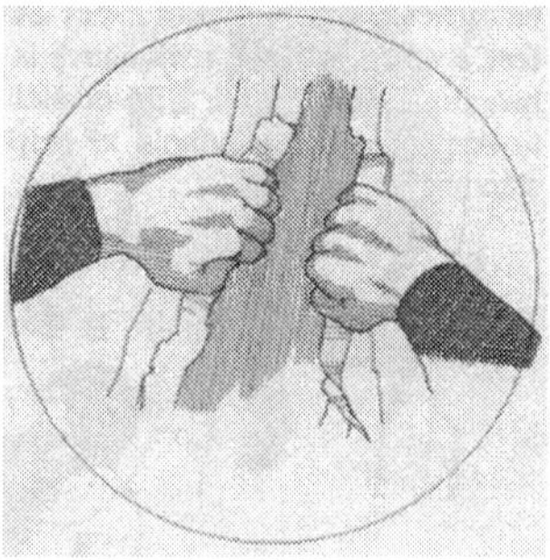

CROSS-PRESSURE METHOD

f. Inverted. Pull or push.

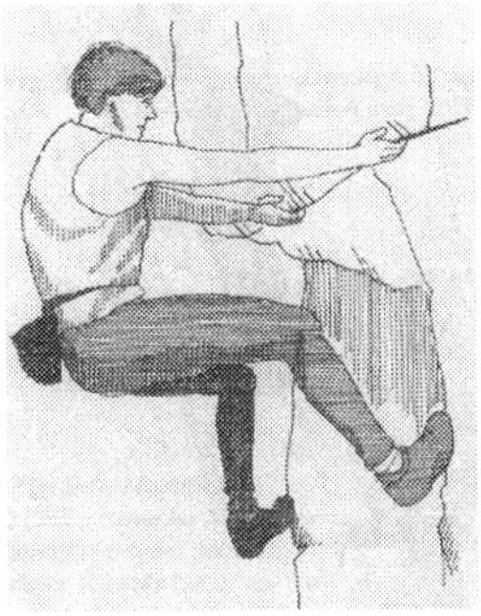

INVERTED TECHNIQUE USING COUNTERFORCE PRESSURE BETWEEN FEET AND HANDS

g. Pinch. As the name implies this is a grip used on tiny little nubbins.

PINCH

h. Stemming. The spreading of arms or legs to maintain a proper body position. (i.e. usually used in a book or chimney.)

STEMMING

8. **GENERAL USE OF HOLDS**. How you use an individual hold is dependent on your experience level, or sometimes, your imagination. Here are some general guidelines.

a. Most handholds can be used as foot holds as you move up the rock.

b. Use all holds possible in order to conserve energy.

c. Even small projections may be used as holds.

d. Do not make use of your knees or elbows due to the reason that it is skin on bone and a slip could occur if pressure is exerted on them. Knees and elbows can be used with the extension of a limb jam.

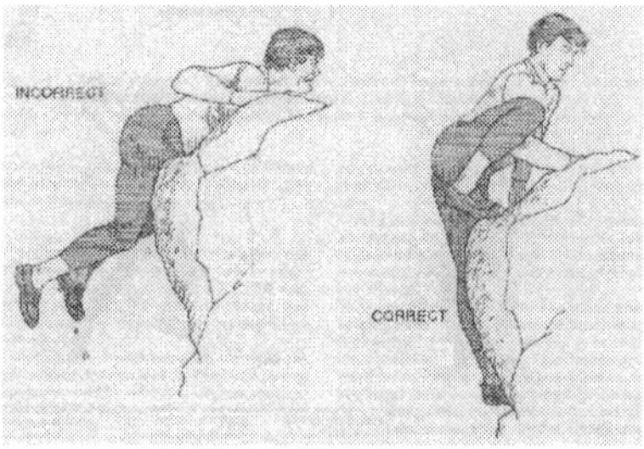

PROPER AND IMPROPER

9. **MOVEMENT ON SLAB**. Movement on slab is based on friction holds.

a. Use any and all irregularities in the slope to gain additional friction.

b. Traversing requires both hands and feet.

TRAVERSING

c. Descending steep slab may require turning inboard to face the slab and backing down.

d. The biggest mistake in slab climbing is leaning into the rock. Maintain maximum friction by keeping weight centered.

SLAB CLIMBING

10. BODY POSITION

a. The climber should climb with his body in balance by keeping his weight centered over and between his feet. Don't hug the rock. Don't over extend and become "spread-eagled". While climbing, keep in mind the acronym "CASHWORTH" for proper body position and movement. (WMC.C3F)

C - Conserve energy.

A - Always test holds.

S - Stand upright, on flexed joints.

H - Hands kept low; handholds should be waist to shoulder high.

W - Watch your feet.

O - On three points of contact.

R - Rhythmic movement.

T - Think ahead.

H - Heels kept low, lower than the toes.

UNITED STATES MARINE CORPS
Mountain Warfare Training Center
Bridgeport, California 93517-5001

WMC.C4
04/11/02

STUDENT OUTLINE

NATURAL AND ARTIFICIAL ANCHORS

TERMINAL LEARNING OBJECTIVE In a summer mountainous environment, establish an anchor system in accordance with the references. (WMC.C4)

ENABLING LEARNING OBJECTIVES.

(1) Without the aid of references and from a given list, choose the two types of anchors, in accordance with the references. (WMC.C4a)

(2) In a summer mountainous environment, establish a natural anchor point, in accordance with the references. (WMC.C4b)

OUTLINE

1. **TYPES OF ANCHORS.** (WMC.C4a)There are two types of anchors that we use. These two types are:

1) NATURAL
2) ARTIFICIAL

C4

2. **GENERAL CONSIDERATIONS FOR ANCHORS**. Anytime you employ natural or artificial anchors; there are special considerations that you must apply. These considerations apply both to the anchor itself and to the material being used to build the anchor. Some examples of these considerations are:

a. Whether using natural or artificial anchors, the installing unit must insure that the anchor is suitable for the load.

b. The anchor position must be relative to the direction of pull on the anchor.

c. The angle between the anchor points should not exceed 90 degrees. This is to ensure that no added stress is put upon the anchors, as well as the equipment being used to construct the anchor.

(1) To decrease the angle between anchor points, materials (i.e. sling ropes, web runners, Prusik cord, etc..) could be used to extend the anchor which will decrease the angle between anchor points.

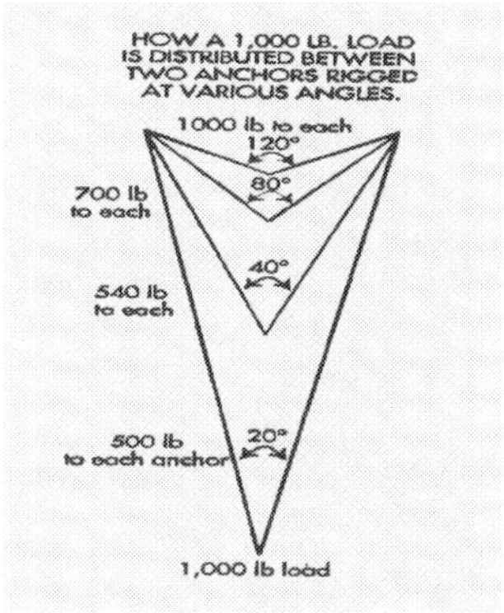

3. **NATURAL ANCHORS.** A natural feature is the preferred type of anchor point. Some examples and considerations are as followed:

a. Types of natural anchors.

(1) Trees.

(a) Select a tree that has not been chopped, burned or is rotten.

(b) The tree should be at least 6" in diameter and strong enough to support the intended load.

(c) Trees growing on rocky terrain should be treated with suspicion, since the roots normally are shallow and spread out along a relatively flat surface.

(2) Shrubs and Bushes.

(a) Select a shrub or bush that is alive and is not brittle, charred or loose.

(b) To avoid leverage, locate the central root and construct the anchor as near to the base as possible.

(3) Rocks and Boulders.

(a) Stability is of prime importance when considering a rock or boulder for an anchor. It must be strong enough and secure enough for the intended load.

(b) All surfaces of the rock or boulder should be inspected for any rough or sharp points. These areas must be padded to protect the rope from being abraded or cut.

(4) Spikes and Flakes.

(a) To check stability of a spike or flake, thump it with the heel of your hand. Anything that sounds hallow is suspicious.

(b) They should be checked for cracks or other signs of weathering that may impair their firmness.

(c) Sharp edges must be padded to protect the rope against cuts and abrasion.

(5) Threads and Chockstones.

(a) A thread is when the rock weathers or cracks to form a hole in the main wall. A chockstone is a rock wedged in a crack.

(b) Check a thread by thumping it with the heel of your hand. Make a visual inspection for cracks and weather. Common sense will prevail for choosing this type of anchor.

(c) When choosing a chockstone, ensure that it has substantial contact with the crack and that the stone's symmetry corresponds with the intended direction of pull.

b. Types of Natural High Tension Anchors. The following are types of Natural High Tension Anchor Systems used to construct installations. These can be tied on any suitable natural anchor point:

(1) Tree Wrap Anchor. The tree wrap is an anchor used to relieve tension on the actual knot itself. This system requires more rope when tied around a large anchor point.

(a) Tie a Figure 8 Loop on the standing end of the rope and wrap the 4-6 times around an anchor point.

(b) Attach a locking carabiner through the knot's loop and clip it onto the running end of the rope.

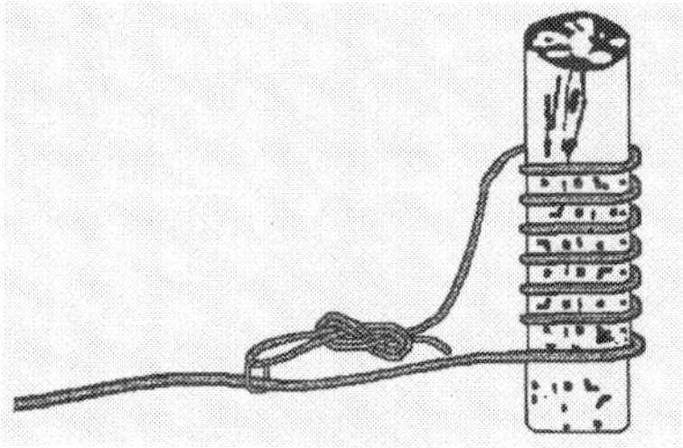

NOTE: If the anchor point to be wrapped is larger then 18' around, then three wraps will suffice.

(2) Swami Wrap Anchor. When tying the Swami Wrap, ensure that the joining knot is kept on the side of the anchor point. The Swami Wrap can be loosened under load if need be. This system can be substituted for a tree wrap when the installation's rope length is an issue.

(a) Select a suitable anchor point and tie a Swami Wrap around it.

(b) Clip a locking carabiner into as many wraps as possible. This will serve as the attachment point for the anchor system.

(3) Figure of 8 Anchor. The Figure of 8 Anchor is a quick and efficient system. This system cannot be loosened under load. It is tied in the following manner:

(a) Tie a Figure of 8 Loop on the standing end of the rope and wrap the knot around a suitable anchor point. Attach a locking carabiner through the knot's loop and clip it onto the running end of the rope.

(b) Before tensioning the anchor system, adjust the rope so the running end of the rope runs smoothly through the carabiner towards the direction of pull. This will prevent any lateral tension.

(b) If two ropes are used, the upper rope's carabiner has its gate upwards, and the lower rope's carabiner has it's gate downward.

c. Primary and Secondary Natural Anchor Point System. The two anchor points should be in line with the direction of pull. The primary anchor is the point nearest the running end while the secondary anchor point is directly behind the primary. It is constructed in the following manner:

(1) The primary anchor knot should be an around the object clove hitch. This is chosen for ease of untying the system after tension has been placed on the rope.

(2) The secondary knot is tied around a suitable anchor point ensuring that the rope is taut between the two anchor points.

4. **ARTIFICIAL ANCHOR SYSTEMS**. Any time we use anything other then a natural feature, we are using an artificial anchor point. Artificial anchors can be constructed in the ground, or on the rock itself. The following are artificial anchor systems:

a. Single Timber Deadman. This system is constructed in the ground and it requires considerable time and effort. The steps of it's construction are as follows:

(1) Dig a trench 6 feet long and 3 feet deep and wide enough to work in at a 90 degree angle to the direction of the pull.

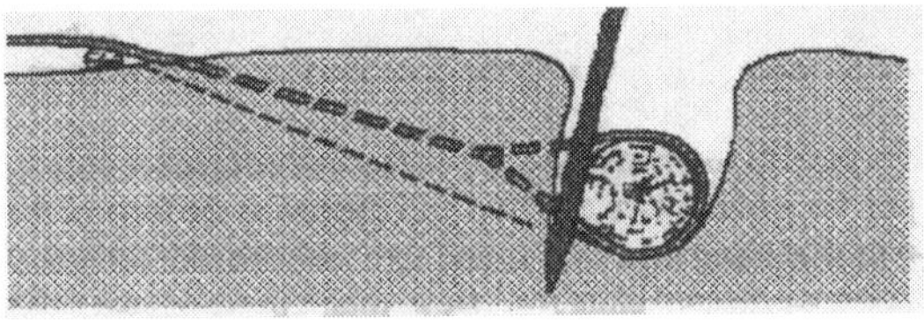

(2) Dig another trench about 12 inches wide. This trench is dug so that it intersects the main trench at a right angle in the middle. The bottom of this trench should be parallel to the direction of pull and should join the bottom of the main trench.

(3) Take an anchoring device (i.e. log, engineer stakes, bundled up branches etc.) that is strong enough to support the intended load. The anchor is placed into the main trench and covered with dirt with the exception of that part of the anchor that joins the second trench. Stakes approximately 3 feet long should be driven approximately 1-1/2 feet into the ground between the dead-man and the slanted side of the trench to assist in holding the dead-man in place if the soil is soft.

b. Picket Hold Fast. The picket hold fast is an easier anchor to construct than the Deadman, and can be used almost anywhere. The strength of the picket system depends on the pickets and the soil or snow conditions. The picket system can be used for both high and non-tension systems. Construction is as follows:

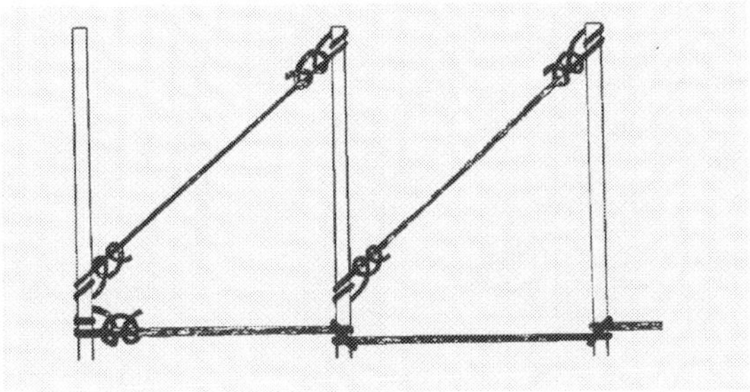

(1) Three stakes (i.e. logs, engineer stakes, etc.) are driven into the ground at a 30 degree angle against the direction of pull.

(2) The line of pickets should be driven in line with the direction of pull. The distance between the pickets can be anywhere from 3-12 feet apart depending on the terrain and soil conditions.

(3) Before tying any rope to the picket anchors, the base of the pickets must be buffed or padded.

(4) To tie a rope to the pickets, go to the furthest picket away from the cliff edge. Tie a round turn and two half hitches at the base of the picket. Take the running end of the rope and tie an over the object clove hitch to the base of the middle picket. Then with the running end of the rope tie an over the object clove hitch to the picket closest to the edge of the cliff face. Ensure that there is tension on the rope in between each picket.

(5) To tie off the pickets to themselves, go to the furthest picket from the cliff edge. And with a sling rope, tie a round turn and two half hitches at the base of that picket. With the running end of that sling rope tie a round turn and two half hitches at the top of the middle picket. Using a second sling rope tie a round turn and two half hitches to the base of the middle picket. Then with the running end of that sling rope tie a round turn and two half hitches to the top of the picket closest to the cliff edge.

c. Equalized Anchor. This system is built with a minimum of three pieces of protection. It can be tied with the standing end of a rope or by utilizing a practice coil.

(1) Construction with the Standing End of the Rope:

(a) Tie a Figure of 8 Loop in the standing end of the rope.

(b) Place a carabiner in each artificial anchor point and attach the knot in either of the outside carabiners.

(c) Clip the rope into the remaining carabiners.

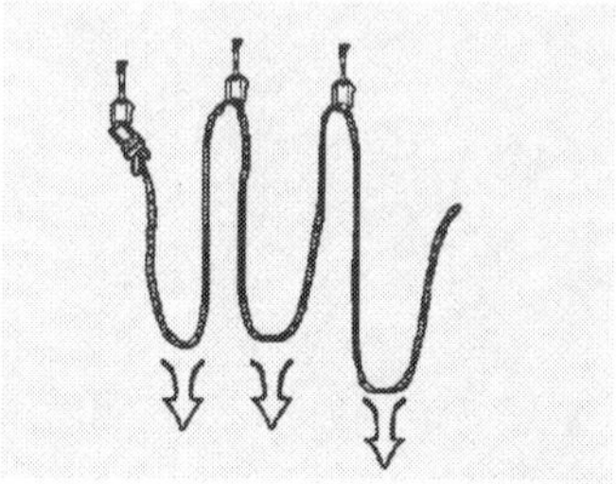

(d) A bight of rope is then pulled down after each carabiner into the anticipated direction of pull. With all three bights, tie an overhand knot around itself to include the running end of the rope.

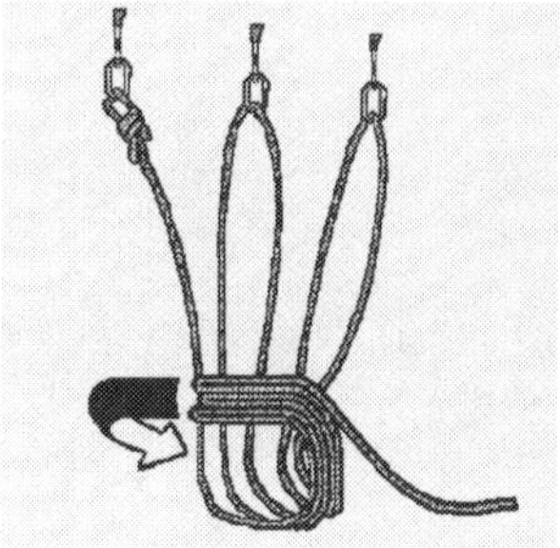

(e) With the running end of the rope coming from the overhand knot, tie a Figure of 8 Loop and attach it to the overhand knot's loop with a locking carabiner.

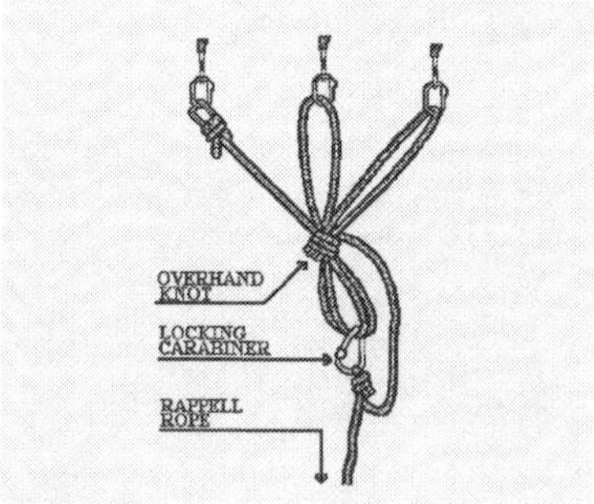

(f) Ensure that there is sufficient slack in the dead rope to prevent disfiguring the knot thus ruining the structural integrity of the knot.

(2) Construction with the Cordalette Method:

(a) Create an endless loop with a practice coil and clip it into each of the artificial anchor points.

(b) A bight of rope is then pulled down between each carabiner into the intended direction of pull.

(c) With all three bights, tie an overhand knot on itself.

(d) Place a locking carabiner into the overhand knot's loop. This will serve as the attachment point for the anchor system.

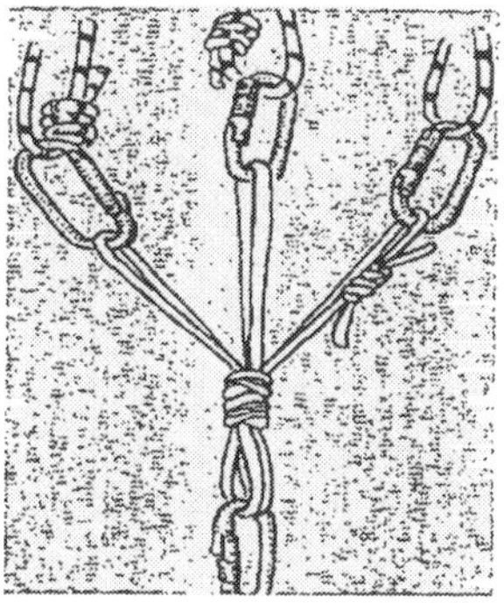

NOTE: If constructing a two-rope high-tension installation, a minimum of three artificial anchors per rope will be used.

UNITED STATES MARINE CORPS
Mountain Warfare Training Center
Bridgeport, California 93517-5001

SML
ACC
02/11/02

LESSON PLAN

SIT HARNESS

LEARNING OBJECTIVES

TERMINAL LEARNING OBJECTIVE: Without the aid of references, properly wear and maintain the MWTC sit harness, in accordance with the references.

ENABLING LEARNING OBJECTIVES

(1) Without the aid of references and given a diagram, name in writing the parts of a sit harness, in accordance with the references.

(2) Without the aid of references, wear a sit harness, in accordance with the references.

BODY

1. (5 Min) **NOMENCLATURE**

   a. Waist belt.

   b. Leg loops (adjustable).

   c. Buttocks straps (adjustable).

   d. Fastex buckle - 2.

   e. Doughnut.

   f. Equipment loops.

   g. D-ring (older models).

   h. Waist belt tie-in point.

   i. Crotch strap.

WEARING OF THE SIT

C5

HARNESS

a. First, disconnect the fastex buckle at the rear of the harness.

b. Hold the harness in front of you, put your feet through the leg loops ensuring that the buckles on the leg loops are outboard, on your thighs.

c. Fasten the waist belt into the buckle, ensuring that it is a tight but comfortable fit. **You must ensure that the waist belt is threaded back through the buckle, this action locks the waist belt to the buckle. Failure to do this will cause the waist belt to slip through the buckle when it is under load.**

d. Adjust the leg loops so that they are high on your thigh, once you have them adjusted, get your buddy to clip the fastex buckle male to the most comfortable female fastex. Then adjust the buttock straps so as the leg loops are held up.

e. If all the above are done correctly, the harness should now be a comfortable but snug fit, after you have fitted it for the first time there is no need to go through the same procedure each time you put it on. Simply hold the harness in front of you, step into it and attach the waist belt to the buckle in the approved manner.

2. TYING IN THE ROPE (END OF ROPE)

a. Pass the end of the climbing rope up through the crotch strap, then through the doughnut and through the waist belt tie-in point.

b. Tie the rope using a retraceable figure eight, adjusting the knot to get it as close as possible to the body.

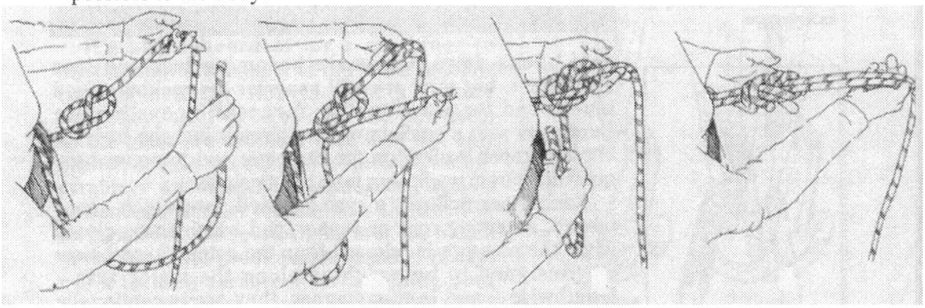

RETRACE FIGURE 8 IN SIT HARNESS

3. TYING IN THE ROPE (MIDDLE OF THE ROPE)

a. Take up a bight of rope and tie a figure eight loop, then take a steel locking carabiner and attach the carabiner to the harness by securing the crotch strap, doughnut, and the waist belt tie-in loop. Now attach the figure eight loop to the carabiner.

4. (5 Min) **CARE AND MAINTENANCE**

a. Avoid contact with chemicals, as this will damage the nylon.

b. Regularly inspect for signs of abrasions and normal wear. Pay particular attention to wear points such as the tie-in loops, buckles, and sewn joints.

c. Keep away from heat such as open flames, cigarettes, etc.

d. If soiled by grit and seawater, wash in lukewarm water with pure soap and allow to dry in a warm room away from direct heat.

e. Two to three years of life can be expected during normal climbing use.

f. It is recommended that a harness that has experienced a serious fall should be discarded.

g. Under no circumstances will you ever tie into the equipment rack of any sit harness as a belay, anchor point, etc.

UNITED STATES MARINE CORPS
Mountain Warfare Training Center
Bridgeport, California 93517-5001

FMST.07.33
05/02/01

STUDENT OUTLINE

ESTABLISHMENT OF RAPPEL POINTS AND RAPPELLING

TERMINAL LEARNING OBJECTIVE Given a mountainous environment, mountaineering equipment, helmet, sit harness, rappel gloves, 165 ft static rope, rescue 8, prusic cord, and one or more assistants, conduct rappelling operations to support mission requirements per the reference. (FMST.07.33)

ENABLING LEARNING OBJECTIVES.

(1) Without the aid of references, select from a given list the three criteria involved in site selection for a rappel site, in accordance with the references. (FMST.07.33a)

(2) In a summer mountainous environment, execute the commands used between a rappeller and his belay man when conducting a rappel, in accordance with the references. (FMST.07.33b)

(3) Without the aid of references, select from a given list three of the nine duties of the rappel point NCOIC, in accordance with the references. (FMST.07.33c)

(4) Without the aid of references, select from a given list three of the six duties of the first man down a rappel rope, in accordance with the references. (FMST.07.33d)

(5) Without the aid of references, select from a given list the three types of rappels, in accordance with the references. (FMST.07.33e)

(6) In a summer mountainous environment, rappel down a cliff face using a seat-hip rappel, in accordance with the references. (FMST.07.33g)

C6

OUTLINE

1. **INHERENT DANGER OF RAPPELLING**. Rappelling is inherently dangerous because rappellers rely totally on the equipment.

   a. To ensure a safe training evolution, two ropes will be utilized at MWTC.

   b. All the rappels taught at MWTC can be completed with a single rope if the situation arises.

   c. If utilizing a one-rope system with the carabiner wrap, the rope will be attached to the carabiner with two wraps vice one.

2. **SITE SELECTION**. (FMST.07.33a) when selecting a rappel site consider these three factors.

   a. There must be a good anchor. As previously taught in *NATURAL* and *ARTIFICIAL ANCHORS*, natural anchors are preferred.

   b. The rappel route down should be as free of obstacles (i.e., vegetation, debris) as possible.

   c. There must be suitable loading and unloading platforms.

NOTE: The evaluation of a site for the above factors should be made by the rappel point NCOIC, who should be the most experienced rappeller in the unit.

3. **DUTIES OF THE RAPPEL POINT NCOIC**. (FMST.07.33c)Once a rappelling site has been selected, one person will be appointed to each rappel lane as a rappel point NCOIC. These individuals should have experience as rappellers. The rappel point NCOIC has ten duties and responsibilities, which are:

   a. Ensures that the anchor points are sound and that the knots are properly tied.

   b. Ensures that loose rock and debris is cleared from the loading platform.

   c. Allows only one man on the loading platform at a time and ensures that the rappel point is run in an orderly manner.

   d. Ensures that each man is properly prepared for the particular rappel; i.e. gloves on, sleeves down, helmet secured, rappel seat tied correctly and secured properly.

   e. Attaches the rappeller to the rope and ensures the rappeller knows the proper braking position for that particular rappel.

   f. Ensures that the proper commands or signals are used.

   g. Dispatches each man down the rope.

h. The rappel point NCOIC will be the last man down the rope.

i. The rappel point NCOIC will ensure that the ropes are inspected after every 50 rappels.

j. The rappel point NCOIC will maintain a rope log.

4. **TYPES/USE**. (FMST.07.33e) The three types of rappels and when they are preferred to be used are:

a. Hasty Rappel: It is used when carrying loads down moderate slopes.

b. Seat-Shoulder Rappel: It is used for heavily laden troops over vertical to near vertical cliff faces.

c. Seat-Hip Rappel: It is used when carrying loads over vertical to near vertical faces.

5. **HASTY RAPPEL**. The hasty rappel is the easiest type of rappel to prepare for. It requires no equipment other than a rope and gloves.

a. Conduct. A hasty rappel is conducted in the following manner:

(1) Sleeves will be rolled down and gloves will be put on.

(2) Face slightly sideways.

(3) Place the rappel rope across your back, grasping it with both hands, palms forward, and arms extended.

(4) The hand nearest the anchor is the guide hand. The hand farthest from the anchor is the brake hand.

(5) Lean out at a moderate angle to the slope.

(6) Descend down the hill facing half-sideways, taking small steps and continually looking downhill while leading with the brake hand.

(7) Feet should not cross and the downhill foot should lead at all times.

b. Braking. The steps for braking during a hasty rappel are as follows:

(1) Bring the lower (brake) hand across the front of the chest to brake.

(2) At the same time, turn to face up toward the anchor point.

6. **SEAT-SHOULDER RAPPEL**. The seat-shoulder rappel relies on friction as the main effort of controlling the descent. It is very efficient for men with heavy packs because it provides support for heavy loads on the back.

a. Conduct. A seat-shoulder rappel is conducted in the following manner:

(1) Put on your rappel-seat, roll down your sleeves and put on your gloves.

NOTE: To avoid causing a possible injury, it is advised that the rappel seat be constructed at the rear of the rappel site loading platform. If the rappel seat is worn for too long, the rappel seat could loosen up enough to cause you to slip out of your seat while rappelling.

(2) The steel locking carabiner is placed on the rappel-seat so that the gate opens down and away, to prevent the gate from opening once the wraps are placed into the carabiner.

(3) Step-up to the rope with you're left shoulder facing the anchor.

(4) The rappel rope is attached to the rappeller's hard point carabiner as follows:

(a) Snap the rope into the locking carabiner.

(b) Taking slack from the standing (anchor) end of the rope, make one wrap with the rope around the body of the carabiner and back through the gate.

(c) Ensure that the locking nut of the carabiner is fastened to lock the carabiner closed.

NOTE: If you are using only one rope to rappel with due to the tactical situation or equipment availability, the procedures are the same, EXCEPT the individual will make two wraps around the body of the carabiner instead of one.

(5) Take the rope across your chest, over your left shoulder, diagonally across your pack and down to the right (brake) hand.

(6) Descend by walking down the cliff using the braking procedure to control the rate of descent. Look under your brake arm for possible obstacles to avoid.

b. Braking. The steps for braking during a seat-shoulder rappel are as follows:

(1) Lean back.

(2) Face directly uphill while bringing the brake hand across the chest.

7. **SEAT-HIP RAPPEL**. The seat-hip rappel is the most commonly used rappel.

a. Conduct. A seat-hip rappel is conducted in the following manner:

(1) Construct the rappel seat; roll down sleeves and put gloves on.

(2) The steel locking carabiner is placed on the rappel-seat so that the gate opens up and away.

(3) Step up to the rope with your left shoulder facing the anchor.

(4) The rappel rope is snapped into the carabiner as follows:

(a) Snap the rope into the locking carabiner.

(b) Taking slack from the standing (anchor) end of the rope, make one wrap with the rope around the body of the carabiner and through the gate again.

(c) Ensure that the locking nut of the carabiner is fastened to lock the carabiner closed.

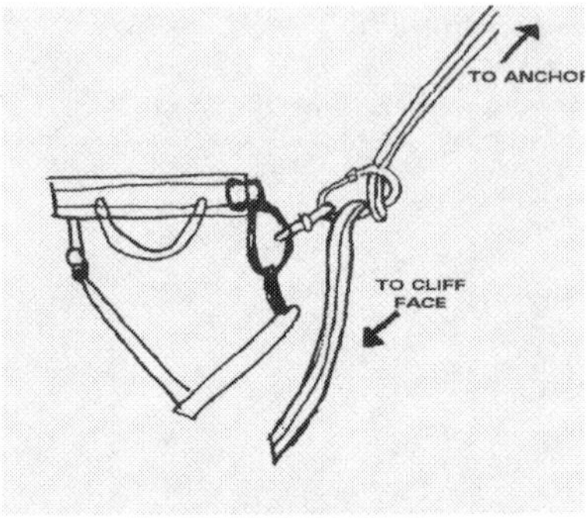

SEAT-HIP RAPPEL

NOTE: If you are using only one rope to rappel with due to the tactical situation or equipment availability, the procedures are the same, EXCEPT the individual will make two wraps around the body of the carabiner instead of one.

(5) The rappeller will grasp the running end of the rope with the brake (right) hand, palm down and turned slightly inboard, near the hip.

b. Braking. The steps in braking for a seat-hip rappel are as follows:

(1) Grasp the rope tightly with the brake hand.

(2) Take the brake hand and place it in the small of the back. This will create enough friction to stop all momentum.

NOTE: At no time will you bound or jump while you are descending. You "walk down" the cliff face using the proper braking procedure to control your rate of descent.

8. **SAFETY OF THE FIRST MAN DOWN**. To ensure the safety of the first man down, the rappeller will:

a. Before deploying the ropes, tie the two ends of the ropes together with an overhand knot.

b. Tie a friction knot on the standing end of the rappel rope with a Prusik cord.

c. Attach the rappel ropes to the harness's hard point with a locking carabiner.

d. Attach the Prusik cord to the harness's hard point with a locking carabiner. This cord is also referred to as a safety Prusik.

NOTE: If the Safety Prusik is too short, it can be extended with a web runner.

9. **DUTIES OF THE FIRST MAN DOWN**. (FMST.07.33d) The first man down the rope has specific duties. They are as follows:

a. Selects a smooth route down for the ropes.

b. Clears the route of loose rocks and debris.

c. The first man down will untie the overhand knot and straighten the ropes out once he reaches the bottom.

d. The first man down belays the next man down the rope.

(1) To belay correctly, the belay man will stand facing the cliff face, arms up with palms facing upward and over lapping. The rope will pass through the opening of the over lapping hand's thumb and index fingers.

(2) To stop a fallen rappeller, the belay man will grab the rope with closed fists and pull straight down bringing the forearms parallel to the deck.

e. Take charge of personnel as they arrive at the bottom to include appointing a belay man.

10. **RAPPELLING COMMANDS.** (FMST.07.33b) In order to conduct rappelling operations safely, it is essential that everyone understands the sequence of events. The following voice commands or rope tugs will be utilized:

VOICE COMMANDS	GIVEN BY	MEANING
"Lane # on Rappel"	Rappeller	I am ready to begin rappelling
"Lane # on Belay"	Belayer	I am on belay and you may begin to rappel
"Lane # off Rappel"	Rappeller	I have completed the rappel and am off the rope
"Lane # off Belay"	Belayer	I have completed the belaying of the rappeller

ROPE TUGS	GIVEN BY	MEANING
Three Tugs	Rappel Point NCOIC	The rappeller is ready to begin rappelling
Three Tugs	Belayer	I am on belay and the rappeller may begin to rappel
Three Tugs	Belayer	The rappeller is off the rappel rope

11. **TYING-OFF**. Occasionally, it may be necessary to stop during a rappel before reaching the bottom of a cliff. The following sequence is used:

a. The commands used in tying-off are as listed below:

COMMAND	GIVEN BY	MEANING
"Lane # Tying-off"	Rappeller	I am ready to tie-off, give me some slack.
"Lane # Tying-off"	Belayer	I have given you enough slack and you may tie-off.
"Lane # On rappel"	Rappeller	I have completed tying-off and I am ready to resume rappelling.
"Lane # On rappel"	Belayer	I am on belay, you may rappel.

b. The procedure used when tying-off is as follows:

(1) The rappeller gives the command, "Lane #, tying-off."

(2) The Belayer gives the rappeller slack and gives the command "Lane # Tying-off". He continues to hold the rope in the belay position, remaining alert and watching the rappeller.

(3) The rappeller quickly brings his brake hand to the twelve o'clock position, so that the running end and standing end are parallel.

(4) The rappeller grasps all ropes with his guide hand as close to the carabiner as possible.

(5) The rappeller steps over the rope so that the running end is going between his legs.

(6) The rappeller releases the rope with his brake hand then reaches under the running end, over the standing end. He then takes up a bight from the running end about two feet long and pulls it over the standing end and under the running end, forming a half hitch.

(7) He pulls the half hitch tightly against the guide hand.

(8) He works the half hitch down snugly against the carabiner while maintaining contact with the guide hand as long as possible.

NOTE*:* A loose half hitch could bind into the carabiner causing difficulties in clearing the knot out. Make sure that the first half hitch is dressed down tightly before moving it against the carabiner.

(9) Place another half hitch above the one already tied.

(10) To untie, reverse the steps. Remove the safety half hitch, then shrink the first half hitch down to a small loop by grabbing the running end of both ropes and pulling them straight out to the left. Place your right hand in the middle of the two bights until they are snug on all four fingers. Then remove one finger and make the bights snug on three fingers. Repeat this process until it's down to one finger and both the bights are equal.

(11) Now, grasp the running end with both hands and smartly jerk the running end of the ropes upward to pop the small loop out from the first half hitch. From this position, keep the guide hand around both of the ropes next to the carabiner; step back over the rope so that the running end is to your right side. Grab just the running end with your brake hand and quickly set the brake behind the small of your back then, readjust you guide hand onto the standing end. You now should be in the seat-hip rappel position.

NOTE: The wraps may bind up some after untying, where no further movement down is possible. To alleviate this, keep the brake on and force your body weight down to pop the wraps in the carabiner back to their intended position and then continue on with the rappel.

(12) The rappeller gives the command "Lane #, on rappel."

(13) The Belayer takes up the slack and gives the command "Lane #, on belay."

(14) The rappeller resumes to rappel down.

12. **RESCUE TECHNIQUES**. When conducting rappelling operations, the possibility of a rappeller getting caught on the rope due to either his clothing or equipment can occur. There are two types of rescues that can be performed to free the rappeller from this situation.

a. The Self-Rescue Technique.

(1) After realizing that you are caught up on the rope, check with your belay man to ensure that he has a solid brake set.

(2) Using a safety Prusik, tie a friction knot onto the rappel rope approximately an arms length above the area that is fouled. Anchor the other end of your safety Prusik to your rappel seats hard point. This is the same technique used as discussed earlier with the first man down the rope.

(3) Work the friction knot up until there is no tension on the rappel device. This will give you the necessary slack to free the malfunction.

(4) Once you have corrected the problem, continue to rappel down using the first man down method.

b. The Buddy Rescue Technique. Himself uses this method when the rappeller is unable to correct the problem.

(1) A rescuer will rappel down on another rope to the disabled rappeller and tie-off.

(2) The rescuer will then establish a safety Prusik onto the victim as taught in the self-rescue technique.

(3) After clearing the malfunction, continue with rappelling operations.

c. Rappelling Casualty Rescue. There are three different methods in which to get an injured rappeller to the bottom.

(1) The first method is to allow the belay man to lower the casualty by slowly releasing the tension on the rappel rope.

(2) The second method is used for critical injuries or when the belay man cannot properly control the casualty's descent.

(a) The rescuer will rappel down using another rope. Once the rescuer gets to the casualty, the rescuer will have his belay man brake him off, enabling the rescuer to use both hands.

(b) If necessary, the rescuer will perform the basic first aid needed. Once the casualty is ready to be lowered, the rescuer will call down to both of the belay men to simultaneously lower both men. The rescuer will hold onto the casualty the entire way down so that he doesn't bounce off the rock face.

(3) The third method is called a tandem rappel. Its used to rescue a casualty who has sustained an injury serious enough that he cannot operate the rappel device himself and requires the assistance of a second rappeller. This type of rescue can be conducted from the top of the cliff face or while on rappel, depending on where the injury occurred.

(a) If the injury occurs while on top of the cliff face, the following steps should be taken:

1. Take either a long sling rope or Prusik cord and tie a figure-eight knot offset so that one length of the cord is longer then the other.

2. Tie a figure-eight knot at each end.

3. Take a rappel device and attach it to the offset figure-eight loop with a carabiner and attach it to the rappel rope.

4. Next, attach the casualty's hard point to the short end of the sling rope/Prusik cord and the rescuer to the other end.

5. The rescuer will then utilize a safety Prusik in the same method as the first man down technique, except that the friction knot will be tied below the rappelling device and controlled with the brake hand.

6. The rescuer will then maneuver himself under the casualty to provide assistance and support.

7. At this point you are ready to rappel both rescuer and casualty at the same time.

NOTE: Your rappel device should be far enough in front of you so that the casualty will not be able to reach it.

(b) If the injury occurs during the rappel, the following steps are taken:

1. The rescuer will preset the same system on a separate rappel rope and rappel down to the casualty.

2. The rescuer will then take the short end of the sling rope and attach it to the casualty's hard point.

3. With a knife, the rescuer will cut the casualty's rappel rope away.

4. Rappel down to the bottom of the cliff face.

NOTE: THIS IS THE ONLY TIME THAT A KNIFE WILL BE USED DURING A RAPPEL. EXTREME CAUTION SHOULD BE USED. IT IS NOT TO BE DONE IN TRAINING, BUT ONLY DURING AN ACTUAL RESCUE AND AS A LAST RESORT.

13. **RETRIEVABLE RAPPELS**. Once a unit has rappelled down a vertical obstacle, it may be necessary to retrieve the rope(s). Depending on the height of the obstacle either one or two ropes will used to construct the rappel lane.

a. One-Rope Retrievable Rappel:

(1) Find the middle of the rope and place it directly behind a suitable anchor point.

(2) Join the pigtails of the rope with an overhand knot and deploy the rope down the obstacle.

(3) On one side of the rope in front of the anchor point, tie an over-the-object clove hitch onto a locking carabiner.

(4) On the other side of the rope, tie another over-the-object clove hitch and attach it to the same carabiner.

(5) The first man down will utilize a safety Prusik and untie the overhand knot after he reaches the bottom.

(6) All others conduct proper rappel procedures.

(7) The last man down will disconnect the carabiner from the rope, and ensuring that the middle of the rope is directly behind the anchor point, rappel down the rope.

(8) Pulling on either end of the rope then retrieves the rope.

b. Two-Rope Retrievable Rappel. This type of system is identical to the one rope retrievable rappel except for a few considerations.

(1) The reason for using two ropes vice one rope is that the height of the rappel is greater.

(2) When two ropes are used they should be joined together using a square knot. This knot will be placed directly behind the anchor point.

(3) All else remains the same as the one rope retrievable rappel except:

(a) The last man down will disconnect the carabiner from the rope and will move the joining knot as close to the vertical obstacle's edge as possible. This will prevent the knot from possibly getting caught up while retrieving.

(b) The last man down will then place a carabiner on the rope below the knot. This will enable him to know which line to pull for retrieval.

14. **EQUIPMENT**. Equipment should be worn in accordance with unit SOP. The unit has the responsibility to determine which methods it feels are most beneficial to the mission. Weapons should be worn across shoulder; muzzle down and away from the brake hand with a tight sling securely attached to the weapon. Weapons should also be dummy corded to the individuals.

UNITED STATES MARINE CORPS
Mountain Warfare Training Center
Bridgeport California, 93517-5001

WMC
04/11/02

STUDENT OUTLINE

TOP ROPING

LEARNING OBJECTIVES.

TERMINAL LEARNING OBJECTIVE. In a summer mountainous environment, and given a rock face and a belay man, conduct top roping with cartridge belt and rifle, in accordance with the references. (WMC)

ENABLING LEARNING OBJECTIVE

(1) In a summer mountainous environment, establish a belay point from the top of a cliff for use in a top rope climb, in accordance with the references. (WMC.a)

(2) In a summer mountainous environment, demonstrate the use of climbing commands while conducting a top rope, in accordance with the references. (WMC.b)

OUTLINE

1. **ESTABLISHING A BELAY STANCE FROM THE TOP**.

   a. The Belayer will establish a sitting belay stance on the cliff head by:

      (1) Constructing a suitable anchor with the standing end of the rope.

      (2) With the direction of pull away from the anchor, tie a directional figure 8 loop near enough to the cliff's edge so that the climber can be observed, if possible.

      (3) Tying a swami Wrap around oneself, clip a locking carabiner through all the rear wraps and clip a large locking carabiner through all the front wraps.

      (4) Secure the rear locking carabiner into the directional figure of 8 loop.

(5) Secure the running end of the rope to the large locking carabiner utilizing a suitable friction belay (i.e., Munter hitch, stitch plate, etc.).

NOTE: Gloves **will not** be worn while belaying a climber during a top rope.

2. **SECURING THE CLIMBER TO THE ROPE**.

a. The climber will tie into the end of the top rope by:

(1) Constructing a Swami wrap around oneself.

(2) Tying a retrace figure 8 loop through all of the Swami's wraps or;

(3) A figure of 8 loop clipped into a locking carabiner secured through all the Swami's wraps.

b. Alternative methods.

(1) A bowline on a coil as discussed in ROPE MANAGEMENT or;

(2) Tie into the hard point of the harness as discussed in SIT HARNESS.

3. **ESTABLISHING A BELAY STANCE FROM THE BOTTOM**.

a. Belaying from the bottom is commonly referred to as a "yo-yo" or "Sling Shot" belay. This system can be constructed for either a direct or indirect belay.

(1) Construct a suitable top anchor utilizing either a pulley or one steel locking or two non-locking carabiners as the attachment point for the anchor.

(2) The rope will travel up from the bottom and pass through the attachment point then return to the bottom.

(3) The climber will be secured to one end of the rope and the Belayer from the other end.

4. **CLIMBER'S RESPONSIBILITIES**.

a. The climber ensures that the Belayer is anchored and on belay, by use of commands or prearranged signals prior to beginning the climb.

b. The climber will not out climb the belayer; this will cause slack in the rope between the belayer and the climber.

c. Avoid placing excess pressure/weight on belay man.

d. Weapons will be worn across shoulder, muzzle down and to the left, with a tight sling securely attached to the weapon.

5. **CLIMBING COMMANDS AND SIGNALS.** In order to conduct top rope operations safely, it is essential that everyone understands the sequence of events. The following voice commands or rope tugs will be utilized:

VOICE COMMANDS	GIVEN BY	MEANING
"Lane #, Up Rope"	Climber	Belayer needs to take in the slack.
"That's Me"	Climber	Excess slack has been taken up between us.
"Lane #, On Belay"	Belayer	I am on belay.
"Lane #, On Climb"	Climber	I am ready to climb.
"Climb Climber"	Belayer	You may begin to climb.

ROPE TUGS	GIVEN BY	MEANING
Three Tugs	Climber	Belayer needs to take in the slack.
Three Tugs	Belayer	I have taken up the slack and you may begin to climb.
Three Tugs	Climber	Climbing.

UNITED STATES MARINE CORPS
Mountain Warfare Training Center
Bridgeport, California 93517-5001

FMST.07.40
03/02/02

STUDENT OUTLINE

MOUNTAIN CASUALTY EVACUATIONS

TERMINAL LEARNING OBJECTIVES. In a summer mountainous environment and given a simulated/actual casualty, evacuate casualties, in accordance with the references. (FMST.07.40)

ENABLING LEARNING OBJECTIVES

(1) With the aid of references, state orally the general considerations for medevac procedures, in accordance with the references. (FMST.07.40a)

(2) In a summer mountainous environment, and with the aid of references, construct an expedient litter, in accordance with the references. (FMST.07.40b)

(3) In a summer mountainous environment, and with the aid of references, secure a simulated/actual casualty in a SKEDCO litter, in accordance with the references. (FMST.07.40c)

(4) In a summer mountainous environment, and with the aid of references, attach a belay line to a SKEDCO litter for movement up/down moderate slopes, in accordance with the references. (FMST.07.40d)

(5) In a summer mountainous environment, with the aid of references, and given a simulated/actual casualty, conduct a medevac using the barrow-boy method, in accordance with the references. (FMST.07.40e)

C8

OUTLINE

1. **GENERAL CONSIDERATIONS** (FMST.07.40a) the general considerations are a set of guidelines that can be used no matter how serious the casualty is. They are remembered by a simple acronym, **APASSNGG.**

   a. Apply Essential First Aid. (i.e. splints, pressure bandage, etc.)

   b. Protect the Patient from the Elements. Provide the casualty with proper insulation from the ground. Ensure that he is warm and dry. If there are any natural hazards (i.e. rock fall, lighting, etc.) either move the casualty as quickly as possible or ensure that he is well protected.

   c. Avoid Unnecessary Handling of Patient.

   d. Select Easiest Route. Send scouts ahead if possible, to break trails.

   e. Set Up Relay Points and Warming Stations. If the route is long and arduous, set up relay points and warming stations with minimum amount of medical personnel at warming stations to:

      (1) Permit emergency treatment. Treat for shock, hemorrhage, or other conditions that may arise.

      (2) Reevaluate the patient constantly. If patient develops increased signs of shock or other symptoms during the evacuation, he may be retained at an emergency station until stable.

   f. Normal Litter Teams Must Be Augmented in Arduous Terrain.

   g. Give Litter Teams Specific Goals. This litter teams job is extremely tiring, both physically and mentally. The litter teams must be given realistic goals to work towards.

   h. Gear. Ensure all of the patient's gear is kept with him throughout the evacuation.

2. **ONE -MAN CARRIES**

   a. Sling Rope Carry. The sling rope carry requires two men and a 15-foot sling rope. One as the bearer, and an assistant to help in securing the casualty to the bearer. Conscious or unconscious casualties my be transported in this manner:

(1) Bearer kneels on all fours and the assistant places casualty face down on bearer's back ensuring the casualty's armpits are even with the bearer's shoulders.

(2) He then finds the middle of the sling rope and places it between the casualty's shoulders and the ends of the sling rope are run under the casualties armpits, crossed and over the bearer's shoulders and under his arms.

(3) Then the ropes are run between the casualty's legs, around his thighs, and tied with a square knot with two overhands just above the bearer's belt buckle.

(4) Ensure the rope is tight. Padding, when available, should be placed where the rope passes over the bearer's shoulders and under the casualties thighs.

b. <u>The rope coil carry.</u> This requires a bearer and a rope coil. It can be used to carry a conscious or unconscious casualty.

(1) Position the casualty on his back.

(2) Separate the loops of the mountain coil into two approximately equal groups.

(3) Slip ½ of the coil over the casualty's left leg and ½ over his right leg so that the wraps holding the coil are in the casualty's crotch, the loops extending upward the armpits.

(4) The bearer lies on his back between the casualty's leg and slips his arms the loops. He then moves forward until the coil is extended. When using the rope coil the bearer ties the coil to himself vice slipping his arms through the loops.

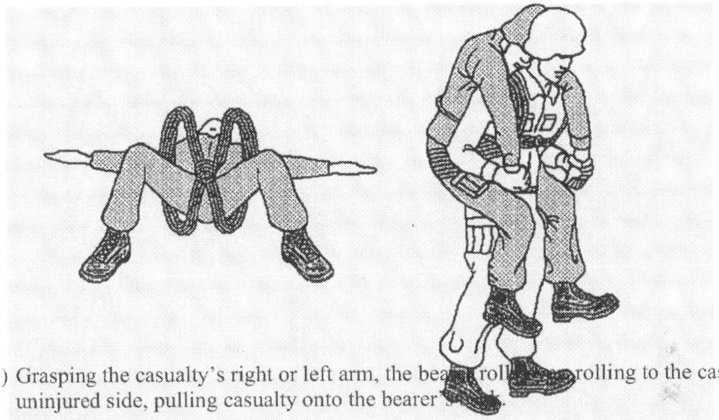

(5) Grasping the casualty's right or left arm, the bearer rolls over, rolling to the casualty's uninjured side, pulling casualty onto the bearer's back.

(6) Holding the casualty's wrists, the bearer carefully stands, using his legs to lift up and keeping his back as straight as possible.

NOTE: The length of the coils on the rope coil and the height of the bearer are to be considered. If the coils are too long and the bearer happens to be a shorter person, it will require the coils to be uncoiled and shortened. If this is not done, then the casualty will hang too low on the bearer's back and make it a very cumbersome evacuation. A sling rope harness can be used around the victim's back and the bearer's chest, which will free the bearer's hands.

3. **TWO-MAN CARRY**

a. Pole Carry. The pole carry method is a field expedient method and should be considered as a last resort only, when narrow ledges must be traversed; vegetation limits the bearers to a narrow trail. This method is difficult for the bearers and uncomfortable for the casualty. Two bearers, four sling ropes and a 12 foot pole - 3 inches in diameter, are required for this carry.

(1) The casualty is placed on his back in a sleeping bag or wrapped in a poncho or blanket, then placed on an insulated pad.

(2) One sling rope is placed under the casualty below the armpits and tied with a square knot across the casualty's chest.

(3) The second sling rope is tied in the same manner at the casualty's waist.

(4) The third sling rope is placed at the casualty's legs below the knee.

(5) The fourth sling rope is tied around the ankles.

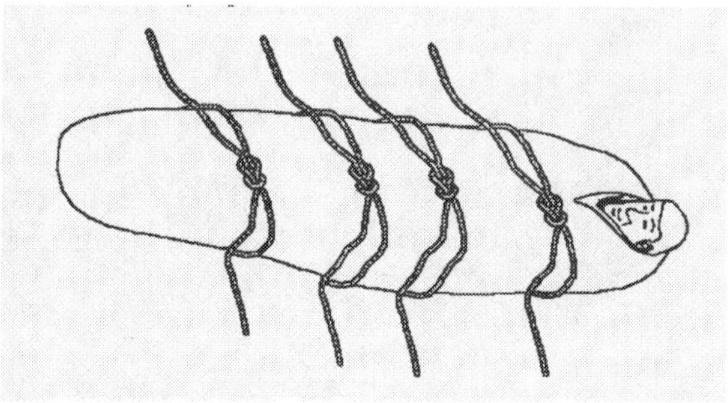

(6) The pole is placed along the casualty's length and secured using square knots with two overhands with the ends of the sling ropes. The square knots should be so tight that the overhands are tied onto themselves

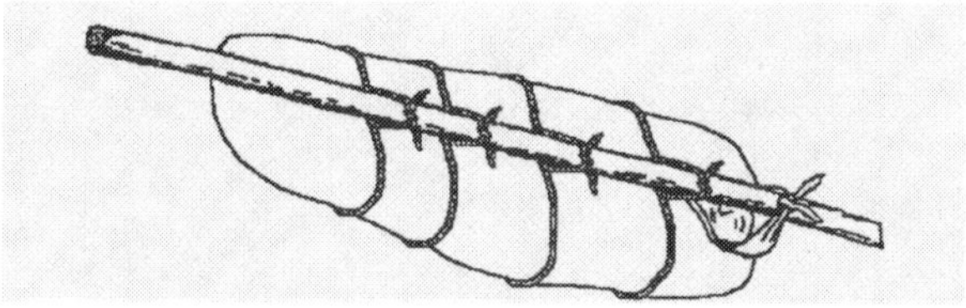

(7) The casualty should hang below the pole, as close to the pole as possible, to prevent swinging during movement.

(8) The casualty's head may be supported using a triangular bandage or a cartridge belt passed around the pole.

(9) For additional support and of movement, two additional bearers may be required, as well as a mountain coil.

(a) Mountain coil is split into two equal coils.

(b) Place knot of mountain coil under casualty's lower back.

(c) Additional bearers slip into each half of the hasty coil one on each side of casualty, aiding in support and movement of the casualty.

4. **EXPEDIENT LITTERS.** (FMST.07.40b) The two types of expedient litters that we will talk about are the alpine basket and the poncho litter.

a. Alpine Basket. To belay the alpine basket, the pre-rigs are attached to the bights formed coming through the loops.

(1) If barrow boy is to be used, the procedure previously discussed will be adhered to.

(2) If barrow boy is not used, then a tag line from the bottom must be implemented to keep the casualty away from the cliff face on the decent.

(3) Construction of the Alpine Basket:

(a) Start by making the same amount of bights as in the rope litter, but start from one end and tie a figure-of-eight loop to run the first bight through.

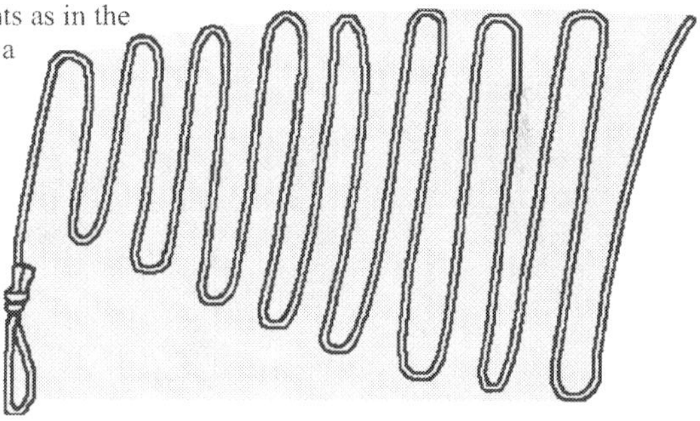

(b) Place padding i.e., isopor mat, on top of the bights and then lay the casualty on the padding and bights.

(c) Start at the casualty's feet and pull the first bight up around the casualty's ankles and through the figure-of-eight loop tied into the starting end of the rope.

(d) Go to the opposite side of the casualty and pull up the second bight and pull it through the loop formed by the bight that was pulled through the figure-of-eight.

(e) Continue until you get to the casualty's armpits, bring the second to the last bight up over the casualty's shoulder and into the bight and then bring the last bight up over the casualty's other shoulder and into the last bight formed.

(f) Secure the end of the bight with an appropriate anchor knot.

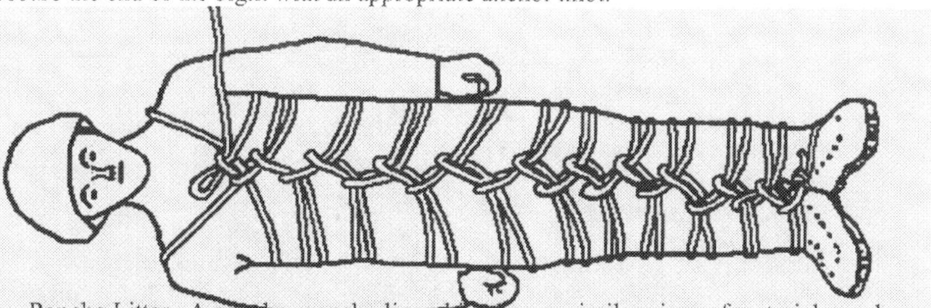

b. Poncho Litter. A poncho, poncho liner, bivy bag or similar piece of material may be used. In addition you will need six individuals with sling ropes.

(1) Lay poncho litter flat on the ground.

(2) Select six rocks about the size of a golf ball. Place one rock in each of the corners and one in the middle on each side in the middle of the litter. The rocks are placed on the under side of the poncho or like material. If a bivy bag is used the casualty should be zipped inside the bivy bag. The rocks should then be arranged in the same manner only on the inside below the zipper.

(3) Tie the sling rope together with an overhand knot. Take the middle of the rope and secure it around the rock with a clove hitch.

(4) An isomat may be laid on the poncho to help make the litter firmer.

(5) The casualty is then placed in the litter. The sling ropes are adjusted by feeding the pigtails of the over hand knot through itself to adjust for length. The loop is then put over the inboard shoulder of the carriers. Insure that the casualty is carried level.

5. **LITTERS.** There are two kinds of litters that are used for casualty evacuations in moderate to vertical terrain. The SKEDCO and the Stokes litter. Each has a variation of procedures for securing a casualty and rigging the litters for either raising or lowering. Let's first look at the SKEDCO litter.

(1) Securing a Casualty to the SKEDCO (FMST.07.40c)

(a) First unroll the litter. The litter must be re-rolled the opposite way to allow the litter to lay flat. Then lay the litter next to the casualty.

(b) If the casualty has any possibility of a spinal injury the Oregon Spine Splint must be used. Secure the splint to the casualty by use of the color-coded buckles. Experienced medical personal are recommended if spinal immobilization is necessary.

(c) Once the casualty is on the SKEDCO use the four body straps to secure the casualty to the litter. Unless injuries prevent, the casualty's arms should be at his sides to prevent further injuries to himself or the rescuers.

(d) Once the casualty is secured with the body straps the feet straps must be secured. The feet straps are secured last to ensure the casualty is in the proper position on the SKED. The feet can be positioned in three ways. The first position is feet together with the straps running on the outside of the feet. The second position is feet apart with the straps running on the inside of the feet. The last position is the feet stacked. This is the most uncomfortable position and not recommended for casualties with possible spinal cord injuries. Placing the heel of one foot on top of the toes of the other forms this position. This position will only be used for a casualty in confined spaces. Once the feet are positioned the feet straps must be secured. Start by bending the feet end of the SKED to form a plat form for the feet. Then loop the feet straps through the second grommets on each side.

(e) The last thing to do is to form the head end to protect the casualty's head. If possible the casualty should wear a helmet. Form the head end tying the pull strap up and secure it to the first body straps.

(2) Medevacing a Casualty. There are many ways to move a casualty once in the SKEDCO. However the medevac team must keep the general considerations in mind. The two methods that we will talk about next are the simplest and require the least amount of additional rigging

(a) The first way is to drag the casualty by the drag strap located at the head end of the SKED. You can also use the SKEDCO's carrying bag as a harness in conjunction with the pull strap and towing harness. If additional people are required, cordage can be added to the pull strap or the front carrying handles.

(b) The second way is to carry the casualty using the carrying handles. By using the set of four removable webbing the litter team can be augmented. To do this each pieces of webbing is tied to make an endless loop. Then pass a bight of the loop through one of the grommets to create additional handles.

(3) Rigging the SKED for Vertical Terrain and Helicopter Lift. In this type of terrain special requirements must be taken to raise or lower a casualty in the SKEDCO.

(a) Rigging the SKED for a Vertical Employment. A vertical raise or lower is used when moving a casualty on steep earth to avoid any further injury to the casualty. On vertical terrain the vertical raise or lower can be used if the terrain is not uniformed or there is a chance of rock fall. Ensure that the casualty's head is always above his feet.

(1) Identify the 30-foot piece of cordage that comes with the SKEDCO. Then tie a figure eight in the middle of the rope. If the rope is worn or missing, the same process can be done with two sling ropes.

(2) Next pass each end of the rope through the grommets at the head of the SKEDCO. Leaving approximately 1-2 feet of rope between the stretcher and the knot.

(3) Continue to feed each end through the grommets and the carrying handles towards the foot end of the SKEDCO. Pass the ends of the rope through the last grommets at the foot end and secure the two ends with a square knot without over hands.

(4) Bring the pigtails up and over the casualty's feet and pass the ends through the
a carrying handles towards the middle of the casualty. Then tie a square knot
b with two over hands.

c

(4) Rigging the SKEDCO for a Horizontal Employment. A horizontal raise or lower is preferred on uniformed vertical terrain. The horizontal employment allows the rescuer to assist the casualty on either a raise or lower. It also allows the rescuer the ability to monitor the casualty's condition and can easily treat the casualty if the need arises.

(a) Identify the two 4 inch nylon straps. They should be two lengths; one four inches shorter than the other. The shorter strap should be marked HEAD STRAPS.

(b) Insert one end of the head strap into a slot near the head of the litter. Then wrap the rest of the straps under the SKEDCO and pass the other end through the opposing slot. Do the same at the foot end of the SKEDCO with the other strap. Ensure that the strap runs smoothly under the SKEDCO.

(c) Connect the strap ends with the large locking caribiner that comes with the SKEDCO. If the caribiner is worn or missing, opposing Stubai 85 locking carabiners will suffice.

6. **STOKES LITTER**. The stokes litter is a litter that is constructed of metal tubing with a plastic covering. The litter is formed in a rectangular basket shape with mesh attached to the frame. Using the stokes for an evacuation (as with any evacuation) it should be padded for the casualty.

a. Securing the casualty to the stokes litter. In the event that the "seat belts" are missing from the stokes litter, sling ropes can be used to the lash the casualty. The steps involved are:

(1) Tie two sling ropes together using square knots and two overhands.

(2) Tie a stirrup hitch around ankles and feet, feed the two pigtails through the right angles of the stokes. Do not cross the ropes at the ankles.

(3) Lace the sling rope towards the casualty's head by passing the rope through the right angles (not over the top of the rails of the stokes).

(4) Secure the ends of the sling ropes by tying a clove hitch with two half hitches on the thick vertical bar located by the victim's shoulder.

7. **YOSEMITE PRE-RIG.** The title of this method, pre-rig implies the meaning ready to use, but in this case the rig must be constructed. Its purpose is for attaching a belay line to a litter and to make the litter easily adjusted. This is the one method that is used to secure a litter to a belay line.

a. Construction: This method normally requires four sling ropes. The steps are: (1) using one sling rope, tie a figure-of-eight loop with one tail.

(1) Take the remaining tail and run it through the window of the stokes litter or in the stirrups of the collapsible litter.

(2) Tie a Kragur knot onto the same sling rope.

(3) Repeat steps (1), (2), and (3) with the three remaining sling ropes.

(4) Suspend the litter to ensure that the comers are balanced.

8. **ASCENTS OR DESCENT OVER STEEP TO MODERATE SLOPES.** When the litter team is ascending or descending a slope they must consider the potential for further injury to the casualty or to themselves. If the risk of injury is high a belay line may be used to prevent injury to the casualty and the rescuers.

a. Preparing Casualty for Ascents or Decent over Steep to Moderate Terrain. This procedure will be depending on several things. Initially, site selection should contain the following features.

(1) Suitable anchor points.
(2) Clearance for casualty along the route
(3) Loading and unloading points.

b. Additional considerations.

(1) The casualty will always be rigged for vertical employment when on steep to moderate terrain.
(2) The smoothest possible route must be selected.
(3) Ensure that the casualty's head is above his feet.

c. Rescuers positions. There are two methods that can be used for the rescuers for moving a casualty in steep to moderate terrain.

(1) Two to four men will position themselves on each side of the litter. They can then carry the litter by the carrying handles. In steep terrain a second belay line may be used to assist the rescuers. We will discuses the belay line later in the chapter.

(2) The Caterpillar method will require as many personnel as possible. The personnel will split in half and position themselves on each side of the litter forming a tunnel. As the litter is raised or lowered each member will hand the litter to the next member in the tunnel. As the litter passes each person in the tunnel he will peel off and assume the lead either at the top or bottom of the tunnel. This will continue until the litter reaches its desired destination.

d. Belay Line. (FMST.07.40d) For belaying of a casualty, one rope will be used and from the top using one of two methods depending on the application.

(1) Body Belay. This method should only be used over moderate terrain. The belay man will establish a sitting position behind a suitable anchor (i.e., rock, tree, etc.) and pass the standing end of the rope behind his back. The running end of the rope will feed out from the belay man's right side. A figure of eight loop is tied to the end of the running end of the rope. It is then attached to the litter's figure eight loop with a locking carabiner. The belay man will then remove all of the slack between himself and the litter. The standing end of the rope should be stacked on the belay man's left side and run through his left side. As the casualty is lowered, the belay man will feed the rope from behind his back allowing it to run through his right hand. If the belay man needs to stop the casualty, he will clench the rope in his left hand, and

bring the rope to the center of his chest.

(2) Direct Belay. This method is the safest for either raising or lowering a casualty in either moderate to steep terrain.

(a) To Lower a Casualty. First a swami wrap will be tied around a suitable anchor point. Two locking carabiners will be clipped into all of the wraps of the swami wraps, gates up. A figure of eight loop is tied into the end of the static rope and attached to the litter with a locking carabiner. After all the slack has been taken up between the litter and the anchor, the rope must be tied through an appropriate belay device. The belay device is attached to the anchor through one of the two locking carabiners on the anchor. A safety (French) prussic will be tied to the running end of the rope and clipped into the second locking carabiner on the anchor. While the casualty is being lowered, one person will control the rope running through the belay device. The safety prussic will be controlled by a second person. Should the primary belay man lose control, the person operating the safety prussic simply lets go and the prussic will bind onto the rope, stopping the casualty.

(b) To Raise a Casualty. The anchors will be established in the same manner as discussed in lowering the casualty with one minor change. The one change is that instead of running the rope through a belay device, the rope will only run through a locking carabiner. The load will be raised by the use of a mule team. The mule must consist of as many people as possible. The mule team will raise the load in as straight a line from the anchor as possible. If the space does not permit a ninety-degree angle away from the anchor is also an option. The mule team will walk backward until the last man reaches his limit of advance. Once he reaches that limit he will peel off the end and return to the front of the mule team. This process is continued until the casualty reaches the top. If the load becomes unmanageable, the safety prussic will be allowed to bind on the rope while the mule team repositions themselves. If the person operating the safety prussic can not see the casualty a Point NCO will be in charge of communicating with the mule team.

(c) To Belay the Rescuers. If the route is too steep or the footing is poor the rescuers may need some assistance either on the raising or lowering of a casualty. If this is the case a separate belay line will be established for the rescuers. The anchor and the belay line are established in the same manner. The same anchor can be used if it is suitable for the load. The rescuers will then tie either around the chest bowlines or swami wraps. A figure eight loop will be tied into the end of the static rope and connected to the bottom rescuer with a locking carabiner. The other rescuers will connect themselves to the same rope with middle of the line prussic. They will be connected in this manner so that they can adjust their position to the casualty.

9. **BARROW BOY.** (FMST.07.40e) A barrow boy is no more than an assistant to the litter on vertical to near vertical cliff faces. The barrow boy can be used for either the Stokes or the

SKED litters. For this situation the Stokes litter should only be used in the horizontal position. However the SKED can be employed in either the horizontal or the vertical positions.

NOTE: Static load, lowered slowly, little friction, so don't need 2 ropes.

a. Rigging the Barrow Boy.

(1) First the rescuer must ensure that a suitable anchor has been established, a proper belay has been constructed, and that a safety prussic has been constructed.
(2) Then the rescuer must ensure that if an A-frame is used that it has been constructed and anchored properly.

(3) Next the rescuer will tie a rappel seat on. (a sit harness can also be used) Then he will ensure that an around the body bowline is tied onto the casualty. A figure eight will be tied on to the end to act as the casualty's safety.

(4) Then after running the running end of the rope though the carabiners or pulley of the A-frame he will tie a figure eight loop at the end of the static rope. He will then attach the figure eight loop to his hard point with a locking carabiner. Then he will tie a middle of the line prussic above the figure eight and attach it to the same locking carabiner in his hard point. This is called the adjustment prussic. It is used to adjust the position of the Barrow Boy in relation to the litter. Next he will take six to eight feet of slack from the end of the line figure eight and tie a directional figure of eight with direction of pull down. (Note: the prussic should be between the end of the line figure of eight and the directional figure of eight.) The directional figure of eight is the attaching point for the litter and the casualty's safety.

(5) Once the rescuer and the litter are secured, the belay man must take all the slack out of the system. The rescuers will the maneuver the litter through the apex of the A-frame with the help of the point NCO.

(6) Once onto the cliff face the rescuer will then position him self with his adjustment prussic so that he can be of most assistance to the litter on the raise or lower. The rescuer will pull the litter out away from the cliff face so that the casualty rides smoothly up or down the cliff face.

(7) The Point NCO will be in charge of the belay men or the mule team. He will also communicate with the rescuer about the rate of speed, if the rescuer needs to be stopped along the route, and when he reaches the top or bottom of the cliff.

10. **<u>TANDEM LOWERING.</u>** The tandem lowering system can be used for the walking wounded, POW's, or more serious casualties when situation would not permit using the barrow boy.

a. The assistant to the casualty should first tie a rappel seat on himself and then assist the casualty with his.

b. The assistant will take only one belay line and tie a end of the line figure 8 loop, and clip this into his rappel seat.

c. A directional figure 8 will be tied approximately 12 inches up the rope from the figure 8 loop (with the loop pointed down) and this will be clipped to the casualty's rappel seat.

d. If needed, safety prussic cords should be tied above the casualty's directional figure 8. These safety cords should be attached in the same manner as the first man down a rappel.

e. The casualty and the assistant will lower as one, with the assistant helping on the way down the cliff.

11. **OTHER CONSIDERATIONS.** All of the techniques we have discussed for the evacuation of a casualty from top to bottom can also be used on a suspension traverse or rope bridge, with a slight variation in the belay line. Two belay lines may be used for rope bridges and the suspension traverse, if they are available. No matter what type of litter is used, the individuals involved in the evacuation must ensure that the head is always uphill or not lower than the feet.

UNITED STATES MARINE CORPS
Mountain Warfare Training Center
Bridgeport, California 93517-5001

FMST.0
7.34
04/12/02

STUDENT HANDOUT

ONE-ROPE BRIDGE

LEARNING OBJECTIVES

TERMINAL LEARNING OBJECTIVES. In a summer mountainous environment, conduct bridging, in accordance with the references. (FMST.07.34)

ENABLING LEARNING OBJECTIVES

(1) Without the aid of references, list in writing the criteria for site selection for a one-rope bridge, in accordance with the reference. (FMST.07.34a)

(2) In a summer mountainous environment with an assault load , cross a one-rope bridge using the rappel seat method, in accordance with the reference. (FMST.07.34b)

(3) Without the aid of references, list in writing the rescue techniques, in accordance with the reference. (FMST.07.34c)

(4) In a summer mountainous environment transport a SKED stretcher across a one rope bridge, in accordance with the reference. (FMST.07.34d)

BODY

1. **SITE SELECTION**

a. The two criteria for site selection for a one-rope bridge are: (FMST.07.34a)

(1) There must be suitable anchors on both sides of the stream.

(2) The anchors must offer good loading and unloading platforms.

b. Other considerations involved are:

(1) The site chosen for the initial crossing does not have to be at the location for the construction of the bridge, just as long as the rope can be taken to the selected site the

C9

crossing.

2) The site chosen for the lead swimmer to cross should be as free as possible from obstacles in the water, such as large boulders, stumps or logs as discussed in *STREAM CROSSING*.

3) The anchors must be close enough for the 150 foot coil to reach both near side and far side anchors. Keep in mind that it will take approximately 1/3 of the 150 foot rope for tightening and anchoring of the bridge.

2. **DISTANCE ESTIMATION**. The follow methods can be used to determine the distance between anchor points:

a. Azimuth Method. Shoot an azimuth to a point on the far side of the intended obstacle to cross. Then move LEFT or RIGHT (perpendicular to the azimuth) until you get a 15 degree offset of your previous azimuth. Next, measure the distance that you paced in feet from your first azimuth to your second azimuth, and multiply that distance by three. This total will give you the approximate distance across the obstacle in feet.

b. Unit Average Method. Get three Marines to judge (best guess) the distance across the intended obstacle. Add up the total accumulated distance, then divide by three. This will give an estimation of the distance across the obstacle.

Example:	Marine 1	240 ft
	Marine 2	230 ft
	Marine 3	250 ft
		720 ft

Divided by 3 = 240 ft

3. **ORGANIZATION**. The organization of construction is broken down into four groups. They are as follow:

a. The Bridging Team - consisting of the bridge NCO and another Marine.

b. The Safety Line Team - consisting of the lead swimmer and his belay man.

c. The Mule Team - consisting 3 Marines for a 9:1, and 6 Marines for a 3:1.

d. The Security Team – consisting of the rest of the party.

4. **CONSTRUCTION**. This bridge is ideal for squad and platoon sized units, for its quick construction and minimal amount of equipment required.

a. Once the site for the bridge has been designated by the bridge NCO, the Safety Line Team will move up stream from this site to enter the water. Their first step is to flake out the rope. Once this is accomplished, the lead swimmer's will take a bight of the rope and tie a figure 8 loop at the end that will be going to the far side, ensuring the knot has a 18"-24" loop. The lead swimmer's upstream arm will go in it, with the belay man tending his rope the lead swimmer will cross the never using the flow to assist him. (Ferry angle as taught in STREAM CROSSING)

b. At the same time as the lead swimmer is crossing, the bridging team will be preparing the near side anchor. The bridging team will tie a swammi wrap around the anchor, using a sling rope or practice coil, and ensuring the square knot is behind or on the side of the anchor. Once the anchor is secure, flake out the bridging line in order to send it across.

c. Once the lead swimmer is across he will move to the site where the far side anchor will be established and secure his rope. The belay man will then move down to the near side anchor, and as close to the waters edge as possible. He will attach the bridging line to the safety by tying a middle of the rope in a figure 8 knot to the safety line and an end of the rope figure 8 knot to the bridging line connecting the two ropes with an 85 carabiner. The lead swimmer will then pull the rope across. The belay man with the help of another Marine will belay the bridging line across keeping tension on the rope to keep the rope out of the water.

d. After the bridging line is across the lead swimmer will detach the bridging line from the safety line and secure it to the far side anchor using a tree wrap, wrapping from right to left, ensuring there are 3-5 wraps.

e. Once the bridging line is secure. The lead swimmer and the belay man will secure the safety line on both banks, ensuring the rope is creating a ferry angle (minimum 45 degrees). This will be the safety line for the Marines crossing the bridge.

NOTE: If a fall from a rope bridge would result in death or serious injury, then utilize a safety line that will be anchored parallel with the bridge and hand tightened. The Marine crossing will clip into both ropes prior crossing.

f. On the near side bank:

1. The bridging team will need the following gear to build the mechanical advantage:

   (a) Four steel locking carabiners & 3 -Three foot prusiks

2. Once the bridging line is secured on the far side, the bridging team will take one 85 and clip it into the near side anchor (swammi wrap). Then they will take the bridging line and clip into that 85, then using one three foot long prusik (16" loop) the bridging line will be secured to the 85 by tying a French Prusik, which will act as a braking knot. The bridging team will now pull the rope taut, and begin construction of the mechanical advantage.

3. The first step is to take one three foot prusik (16" loop) and tie a French prusik on the bridging line as far away from the anchor as possible. Then take one 85 and clip it to the loops of the French prusik.

4. Taking a bight from the running end of the rope clip it into the 85 hanging from the tails of the French prusik.

5. Then bring a bight back from the clipped in rope to the anchor, and taking one 85 clip it into the 85 created in step (2), take the bight of rope and clip into this 85.

6. Take one three foot prusik and tie a French prusik onto the bottom end of the bight clipped into the 85 created in step (3), take one 85 and clip it into the tail loops of the French prusik.

7. From the anchor, take a bight of rope from the 85 created in step (6) and clip it into the 85 created in step (6).

8. You have made approximately a 9:1 mechanical advantage.

NOTE: If the manpower is available the squad/platoon has the option of building a 3:1 mechanical advantage vise a 9: 1, the mule team will just have to add another body having no more then 3 bodies on a 9:1 and 6 bodies on a 3:1

g. Tensioning of the Bridge

1. Now that the bridge is built, the bridge NCO can call up the mule team to tension the bridge.

2. This procedure is facilitated by the mechanical advantage system that was just put into place. The braking knot (French prusik) is used to hold tension on the bridging line while the bridge is being tensioned.

3. The mule team will begin to pull on the running end of the rope coming out of the mechanical advantage. They will pull the rope straight back, trying to keep it in line with the bridge as best as possible.

4. The mule team will tension the rope as much as possible, the bridge NCO will monitor the system. Once the mule team cannot pull anymore tension, the bridge NCO will then have them hold the rope in place, and will reset the braking knot. Once the braking knot is set the bridge NCO will ask for slack from the mule team and cycle the system out. This process will continue until the bridge is tight.

5. On the last cycle the bridge NCO will set the brake knot and with the help of another Marine they will make a bight out of the running end and bring it around the tree while keeping tension.

6. Take a bight and make a complete round turn on the body of the 85 created in step (2). Last, the bridge NCO will tie two half hitches encompassing all the ropes just behind the anchor.

7. At this stage the bridge is tight and secured, the bridge NCO will now call up the remainder of the squad/platoon to cross. The bridge NCO will monitor the crossing of all Marines, and will be the last one to cross.

TRANSITION: Now that we have discussed the construction of the bridge, are there any questions over the material? Next we will discuss the crossing of the bridge.

5. **CROSSING.** (FMST.07.34b) The method used to cross is known as a horizontal traverse. This traverse can be accomplished in the following ways.

a. Rappel Seat Method

1. The Marine ties himself into a rappel seat and inserts a large steel locking carabiner with the gate facing down and away.

2. The Marine faces the bridge with his right hand towards the near anchor to snap into the bridge, once the 85 is locked onto the bridge rope, a helper flips the 85 over so that the locking nut screws down. A carrier rope with a girth hitch may be added to assist the Marine in mounting the rope.

3. The Marine hangs below the bridge from his rappel seat with his head pointing in the direction of the far anchor and allows his legs to hang free.

4) Pulling with his arms makes progress.

5) This method is the safest and therefore the preferred method.

6) If the Marine must take a pack across, he may wear it with the waistband secured: However, the preferred method is to have another carabiner (aluminum non-locking is sufficient) attached to the pack frame at the top and attach this to the bridge behind the Marine putting his legs through the shoulder straps and pulling the pack across with him.

7) One man at a time will cross, although one can load and another can unload concurrently.

8) Weapons will be worn across the shoulder; muzzle down with a tight sling securely attached to the weapon.

b. Pulley Method. This method is used when the one rope bridge is long, uphill, or speed is vital and the Marines crossing it have a lot of heavy personal equipment; M24OG's, radios, etc.

1) Equipment required. One pulley, four steel locking carabiners, and a hauling line twice the length of the obstacle.

2) Setting up the system. Construction of the first suspension point is done by attaching a pulley to the one rope bridge, one 85 carabiner is attached to the pulley, gate down, and a second 85 is attached to the first 85 with the gate facing the near side bank. Into this second 85, attach the hauling rope with a figure-8 loop. The figure-8 is placed halfway along the hauling rope and attached to the 85, which is then locked. Four feet down the line another figure-8 on a bight is placed into the second suspension point.

3) Connection. The Marine clips his 85 into the lower 85, his equipment is clipped into the second 85. The mule team starts to pull.

c. Helmets will be worn for all methods of crossing the bridge. Gloves are optional.

NOTE: There are additional methods of crossing a one rope bridge such as the Commando crawl, Monkey crawl and hand over hand techniques, however, they are not taught at MWTC due to safety reasons.

6. **RESCUE TECHNIQUES** If an individual is unable to complete the crossing on the one rope bridge using the rappel seat method of crossing (i.e. from a fall or exhaustion), a rescue will have to be made in the following manner: (FMST.07.34c)

a. Reach. First try to reach the victim by using an object such as a pole, your hand, if the victim is close enough, etc ...

b. Throw. If reaching the victim is impossible, try throwing a rope to the victim and have him attach the rope to himself, preferably his seat, and pull him back to the desired side of the installation.

c. Tow. If the victim is unable to catch or reach a rope being thrown to him or the victim is unconscious, tie a figure-of-eight loop into the middle of the safety line and then connect the rope to the rope bridge with a steel locking carabiner. The mule team will pull the carabiner up against the victim's seat and begin towing the victim to the desired side of the installation. The rope used to tow the victim should be twice the length of the span of the rope bridge (If necessary, ropes can be tied together to accommodate the span). This will allow towing from either side without having to throw or have a Marine carry the safety line back across the bridge each time a towing rescue is performed.

d. Go. If all else fails, the last option will be to go after the victim. The rescuer will move out onto the rope bridge with a safety line attached to himself at approximately eight feet from the end of the safety line. In the end of the safety line tie a figure-of-eight loop and insert a steel locking carabiner in the end of it. Once the rescuer has made contact with the victim. He will attach the steel locking carabiner to either the victim's seat (preferably the victim's carabiner that is attached to the rope bridge), or to the bridge itself. To ensuring that the carabiner is placed so that it pulls against the victim's carabiner. The mule team now starts to pull both the victim and the rescuer to the desired side of the installation.

e. Cut. If while crossing the rope over water the individual goes underwater and no other rescue technique can be employed, the rope will be cut.

7. RETRIEVING THE ONE ROPE BRIDGE

a. Before the bridge NCO sends the last Marine to cross, they must make the bridge retrievable.

b. The first step is to break down the mechanical advantage, ensuring the braking knot is set before doing so. The system will be broken down until the bridging line is attached only to the brake knot.

c. They will now take the rope around to the back of the anchor, and tie a slip figure 8, with the loop of the slip 8, take it and attach it to the bridging line just in back of the braking knot with a 85.

d. Now pull the running end coming out of the slip 8 until it tightens, and bites on the rope. Once the rope bites, tie a thumb knot behind the slip 8 as close to the slip figure 8 as possible.

e. The bridge NCO, with the help of another Marine, will now untie the swammi wrap, until all of the tension is on the bridging line.

f. The safety line will now be brought up, and attached to the end of the bridging line with an overhand knot.

g. The bridge NCO and the other. Marine will now cross the bridge.

NOTE: It is possible that the bridge may lose tension after breaking the bridge down. The packs of the last two Marines across can be sent over with other Marines, prior to the slip figure being tied.

h. After both Marines have crossed, the safely line will be brought up from down stream.

i. The tree wrap will now be taken off. Once the tree wrap is off, all knots and carabiners will be taken out.

j. The safety line will now be pulled, the bridge is retrieved.

UNITED STATES MARINE CORPS
Mountain Warfare Training Center
Bridgeport, California 93517-5001

WMC
04/12/02

STUDENT OUTLINE

SUMMER MOUNTAIN WARFIGHTING LOAD REQUIREMENTS

LESSON PURPOSE. The purpose of this period of instruction is to introduce the student to the different levels of gear requirements in accordance with MWTC's uniform codes. This lesson relates to mountain walking and tactical evolutions

OUTLINE

1. **BASIC UNIFORM REQUIREMENTS.**

a. Marines operating in mountainous terrain will generally wear the standard issued utility uniform. However, due to the wide range of temperatures, and sudden changes in weather in the mountains, Marines will find the need to be continually adding and removing layers. The amount of clothing worn may vary depending upon the severity of the weather, the activity level of the Marine, and the individual metabolism of the Marine. Every person in that unit should still maintain the same outer camouflage layer.

b. As part of the basic uniform, each man will be required to have in his possession, at all times, seven required pocket items. These seven items should be carried in the pockets of your utility uniform:

(1) Pocketknife

(2) Whistle

(3) Pressure Bandage

(4) Chapstick and sunscreen

(5) Sunglasses

(6) Survival Kit and rations (fire starting material, signaling material, food gathering material, water procuring material, 1st aid, shelter material, and some high energy, lightweight snacks.)

(7) Notebook with pen/pencil

c. Some additional items that should be carried in your pockets at all times:

(1) Contact gloves

(2) Paracord (10 meters)

(3) Flashlight with tactical lens and spare batteries

(4) Chemlights

2. **ASSAULT LOAD.**

a. The Assault Load is equipment in addition to the basic uniform requirements. It is carried in the load-bearing vest (LBV), butt pack and the pack system. This is the equipment carried for short duration missions such as security patrols or during the assault. It is also carried at all times when away from the bivouac site.

(1) An extra insulating layer (Field jacket liner, wooly pully, etc.)

(2) Protective layer (ECWCS parka and trousers)

(3) LBV with 2 quarts of water and first aid kit

(4) Helmet

(5) Rations for the time away from your bivouac site

(6) Cold weather hat or balaclava

(7) Individual mountaineering gear

a. Sling rope

b. Carabiners

c. Rappelling Gloves

(8) Specialized mountaineering equipment

(9) Mission essential gear

3. **COMBAT LOAD**. The Combat Load is the equipment carried for longer duration missions.

The following items are in addition to the items already being carried in the Assault Load:

a. Sleeping bag with bivy bag

b. Isopor mat

c. Stove

d. Fuel bottle

e. Poncho (for expedient shelters or medevac purposes)

4. **EXISTENCE LOAD**. The Existence Load is to be packed for longer duration combat missions, where it is necessary to replace worn out items. These items could be uniforms, boots, personal hygiene gear, etc. The Existence Load should be brought to the forward elements once the situation allows.

5. **PACKING CONSIDERATIONS.** Because most Marines are familiar with how to pack a pack, these are general guidelines only.

a. When not wearing your protective layer or insulating layers, keep it handy at the top of the pack. When taking a break during a movement, you will be able to quickly don a layer to prevent getting chilled.

b. Keep your stove and fuel bottle in the outside pockets of your pack. They may leak and soak your equipment with fuel if stored inside your pack.

c. Keep the climbing rope in the rope bag. This will reduce the rope's exposure to petroleum products, ultra-violet light, cuts, and abrasions.

d. Ice ax is placed through the ice ax loops with spike up, adze outboard, and secured in place.

e. Crampons are attached to the outside of the pack with the tips covered and facing outboard.

Table of Contents

Chapter 3. Weapons Handling

Chapter 4. Fundamentals of Marksmanship

Chapter 5. Rifle Firing Positions

Chapter 6. Use of Cover and Concealment

Chapter 7. Rifle Presentation

Chapter 8. Effects of Weather

Chapter 9. Zeroing

Chapter 10. Engagement Techniques

Appendices

Chapter 1. Introduction to Rifle Marksmanship

All Marines share a common warfighting belief: "Every Marine a rifleman." This simple credo reinforces the belief that all Marines are forged from a common experience, share a common set of values, and are trained as members of an expeditionary force in readiness. As such, there are no "rear area" Marines, and no one is very far from the fighting during expeditionary operations. The Marine rifleman of the next conflict will be as in past conflicts: among the first to confront the enemy and the last to hang his weapon in the rack after the conflict is won.

1001. ROLE OF THE MARINE RIFLEMAN

Marine Corps forces are employed across the entire range of military operations. At one end is war, which is characterized by large-scale, sustained combat operations. At the other end of the scale are those actions referred to as military operations other than war (MOOTW). MOOTW focuses on deterring aggression, resolving conflict, promoting peace, and supporting civil authorities. These operations can occur before, during, and after combat operations. Training and preparation for MOOTW should not detract from the Corps' primary mission of training Marines to fight and win in combat. MOOTW normally does not involve combat. However, Marines always need to be prepared to protect themselves and respond to changing threats and unexpected situations. Whenever the situation warrants the application of deadly force, the Marine rifleman must be able to deliver well aimed shots to eliminate the threat. Sometimes the need for a well aimed shot may even be heightened by the presence of noncombatants in close proximity to the target. The proficient rifleman handles this challenge without unnecessarily escalating the level of violence or causing unnecessary collateral damage. The Marine rifleman must have the versatility, flexibility, and skills to deal with a situation at any level of intensity across the entire range of military operations.

To be combat ready, the Marine must be skilled in the techniques and procedures of rifle marksmanship and take proper care of his rifle. Even when equipped with the best rifle in the world, a unit with poorly trained riflemen cannot be depended upon to accomplish their mission. Usually, poorly trained riflemen either fail to fire their weapon or they waste ammunition by firing ineffectively. To send Marines into harm's way without thorough training in the use of their individual weapons carries undue risks for every Marine in the unit. On the other hand, well trained riflemen can deliver accurate fire against the enemy under the most adverse conditions. A well trained rifleman is not only confident that he can help his unit accomplish it's mission, he is confident that he can protect his fellow Marines and himself.

1002. CONDITIONS AFFECTING MARKSMANSHIP IN COMBAT

Many factors affect the application of marksmanship in combat; among them are—

- Most targets are linear in nature and will consist of a number of men or objects irregularly spaced along covered or concealed areas.
- Most targets can be detected by smoke, flash, dust, noise or movement, but will only be visible for a brief moment before taking cover.
- The nature of the target, irregularities of terrain and vegetation will often require a rifleman to use a position other than prone in order to fire effectively on the target.
- The time in which a target can be engaged is often fleeting.

1003. COMBAT MINDSET

In a combat environment, the Marine must be constantly prepared for possible target engagement. When a target presents itself in combat, there may be very little time to take action. A Marine must be able to engage the target quickly and accurately. The unique demands of combat (i.e., stress, uncertainty) dictate that the Marine be both physically and mentally prepared to engage enemy targets. It will not be enough to simply know marksmanship techniques and procedures. The Marine must develop the mental discipline to prepare for enemy contact. In the confusion, noise, and stress of the combat environment, the Marine must have the ability to eliminate any hesitation, fear or uncertainty of action and to focus on the actions required to fire well-aimed shots. This is accomplished through establishment of a combat mindset. The key factors in the development of a combat mindset include both physical and mental preparation.

Physical Preparation

In combat, targets can present themselves without warning. Therefore, it is essential for the Marine to maintain proper balance and control of his weapon at all times so he can quickly assume a firing position, present the weapon, and accurately engage the target. However, speed alone does not equate to effective target engagement. The Marine should fire only as fast as he can fire accurately, never exceeding his physical ability to apply the fundamentals of marksmanship. To be effective in combat, the Marine must train to perfect the physical skills of shooting so those skills become second nature. Mastery of physical skills allow the Marine to concentrate on the mental aspects of target engagement; e.g., scanning for targets, detection of targets, selection and use of cover. The more physical skills that a Marine can perform automatically, the more concentration he can give to the mental side of target engagement.

Mental Preparation

While combat is unpredictable and constantly changing, the Marine can prepare himself mentally for confrontation with the enemy. The stress of battle, coupled with the often limited time available to engage targets, requires concentration on the mental aspects of target engagement; e.g., scanning for targets, detection of targets, and the selection and use of cover.

Knowledge of the Combat Environment

The Marine must be constantly aware of the surroundings to include the terrain, available cover, possible areas of enemy contact, backdrop of the target, etc. This awareness will enable the Marine to select and assume a firing position and to quickly and accurately engage targets.

Plan of Action

In combat, the situation will dictate the action to be taken. The Marine must understand the situation, identify and evaluate possible courses of action, and then develop a plan for target engagement that accomplishes the mission.

Confidence

The Marine must believe in his ability to engage targets accurately in any combat situation. A Marine's level of confidence is rooted in the belief that future challenges will be overcome—particularly the challenge of firing well aimed shots in a combat environment where the enemy may be returning fire. A key factor in a Marine's level of confidence is the degree to which he has mastered the techniques and procedures of the rifle marksmanship. Mastery of rifle marksmanship can only be obtained by quality instruction. Quality instruction is the foundation for practical application of the marksmanship fundamentals during range and field firing.

Chapter 2. Introduction to the M16A2 Service Rifle

Note

+The procedures in this manual are written for right-handed Marines; left-handed Marines should reverse instructions as necessary.

2001. DESCRIPTION

The M16A2 service rifle is a lightweight, 5.56 millimeter (mm), magazine-fed, gas-operated, air-cooled, shoulder-fired rifle. (Fig. 2-1 shows a right-side view and fig. 2-2, on page 2-2, shows the left-side view.) The rifle fires in either semiautomatic (single-shot) mode or a three-round burst through the use of a selector lever. The M16A2 service rifle has a maximum effective range of 550 meters for individual or point targets. The bore and chamber are chrome-plated to reduce wear and fouling. The handguards are aluminum-lined and are vented to permit air to circulate around the barrel for cooling purposes and to protect the gas tube. An aluminum receiver helps reduce the overall weight of the rifle. The trigger guard is equipped with a spring-loaded retaining pin that, when depressed, allows the trigger guard to be rotated out of the way for access to the trigger while wearing heavy gloves. An ejection port cover prevents dirt and sand from getting into the rifle through the ejection port. The ejection port cover should be closed when the rifle is not being fired. It is automatically opened by the action of the bolt carrier. The muzzle compensator serves as a flash suppressor and assists in reducing muzzle climb.

2002. OPERATIONAL CONTROLS

Selector Lever

The selector lever has three settings—safe, semi, and burst. The setting selected depends on the firing situation. See figure 2-3 on page 2-2.

Safe

The selector lever in the safe position prevents the rifle from firing.

Semi

The selector lever in the semi position allows one shot to be fired with each pull of the trigger.

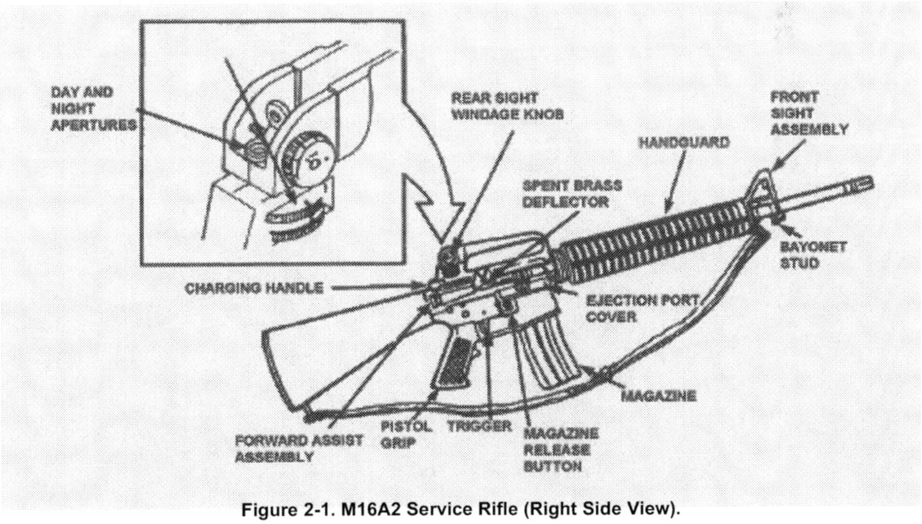

Figure 2-1. M16A2 Service Rifle (Right Side View).

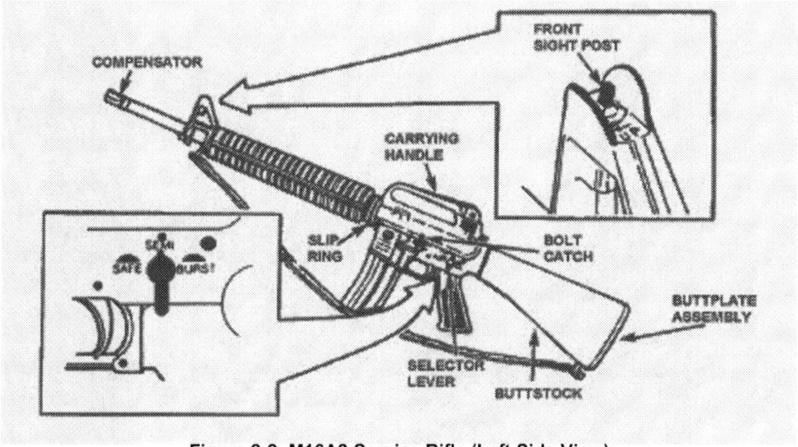

Figure 2-2. M16A2 Service Rifle (Left Side View).

Burst

The selector lever in the burst position allows the rifle to continue its cycle of operation until interrupted by the burst cam. With each pull of the trigger, the burst cam limits the maximum number of rounds fired to three. The burst cam is not "self-indexing." If burst is selected, the burst cam does not automatically reset to the first shot position of the three-round burst. One, two or three shots may be fired on the first pull of the trigger. Each subsequent pull of the trigger results in a complete three-round burst unless the trigger is released before the cycle is complete. If the trigger is released during the burst and the three-round cycle is interrupted, the next pull of the trigger fires the rounds remaining in the interrupted three-round cycle.

Magazine Release Button

The magazine release button releases the magazine from the magazine well. See figure 2-4.

Charging Handle

When the charging handle is pulled to the rear, the bolt unlocks from the barrel extension locking lugs and the bolt carrier group moves to the rear of the receiver. See figure 2-5.

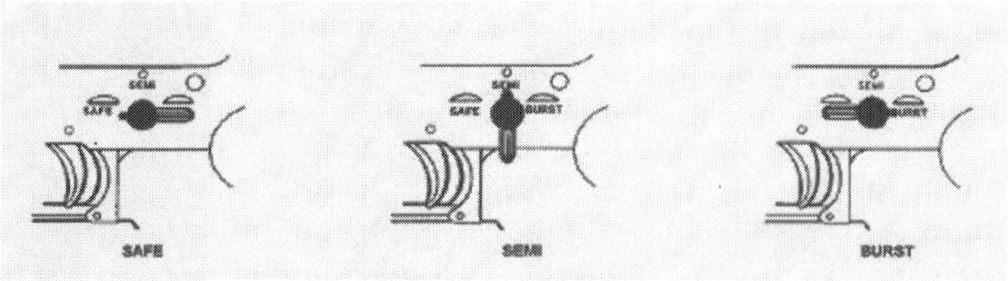

Figure 2-3. Selector Lever.

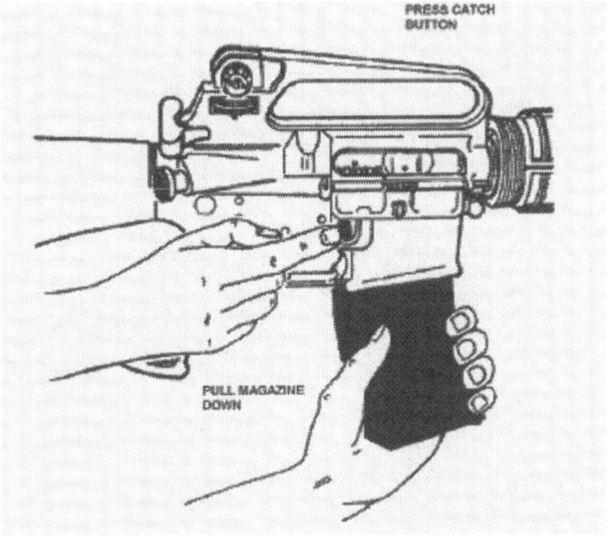

Figure 2-4. Magazine Release Button.

Bolt Catch

If the charging handle is pulled to the rear when the lower portion of the bolt catch is depressed, the bolt carrier group will lock to the rear. When the bolt carrier group is locked to the rear and the upper portion of the bolt catch is depressed, the bolt carrier group will slide forward, driven by the buffer assembly and action spring, into the firing position.

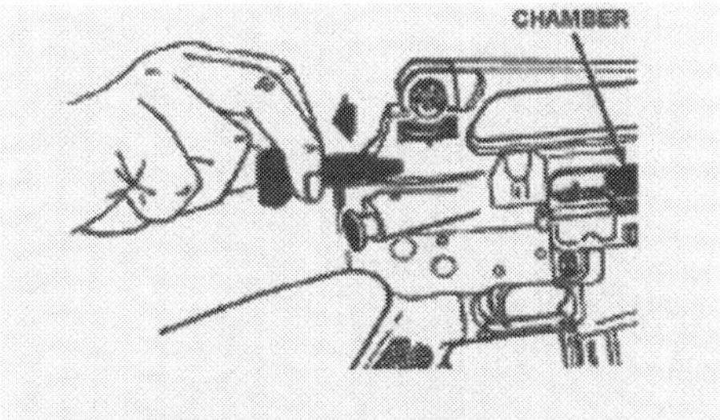

Figure 2-5. Charging Handle.

2003. CYCLE OF OPERATION

Firing

The hammer releases and strikes the head of the firing pin, driving the firing pin into the round's primer. The primer ignites the powder in the cartridge. Gas generated by the rapid burning of powder propels the projectile through the barrel. After the projectile passes the gas port, a portion of the expanding gas enters the gas port and gas tube. The gas tube directs the gas rearward into the bolt carrier key and causes the bolt carrier to move rearward. See figure 2-6 on page 2-4.

Unlocking

Figure 2-7 on page 2-4 illustrates unlocking of the bolt. As the bolt carrier moves to the rear, the bolt cam pin follows the path of the cam track located in the bolt carrier. This causes the bolt assembly to rotate until the bolt-locking lugs are no longer aligned behind the barrel extension locking lugs.

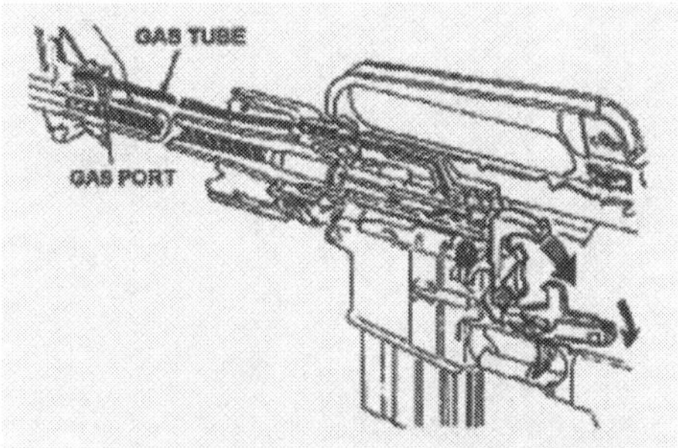

Figure 2-6. Firing.

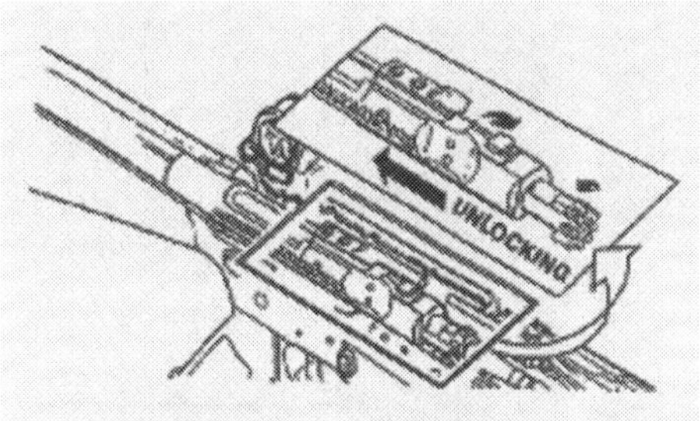

Figure 2-7. Unlocking.

Extracting

As the bolt carrier group continues to move to the rear, the extractor claw withdraws the cartridge case from the chamber. See figure 2-8.

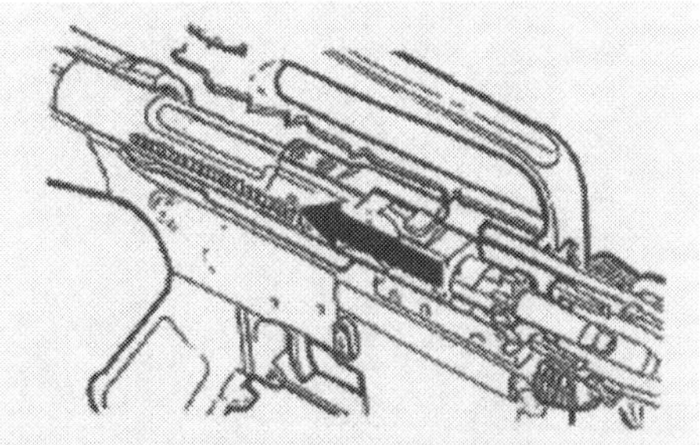

Figure 2-8. Extracting.

Ejecting

The ejector, located in the bolt face, is compressed into the bolt body by the base of the cartridge case. The rearward movement of the bolt carrier group allows the nose of the cartridge case to clear the front of the ejection port. The cartridge case is thrown out by the action of the ejector and spring. See figure 2-9.

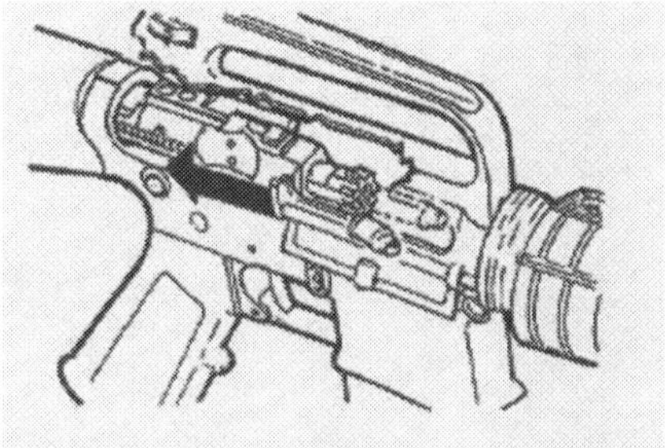

Figure 2-9. Ejecting.

Cocking

Continuing its rearward travel, the bolt carrier overrides the hammer, forces it down into the receiver, compresses the hammer spring, and causes the disconnector to engage the lower hammer hook. See figure 2-10.

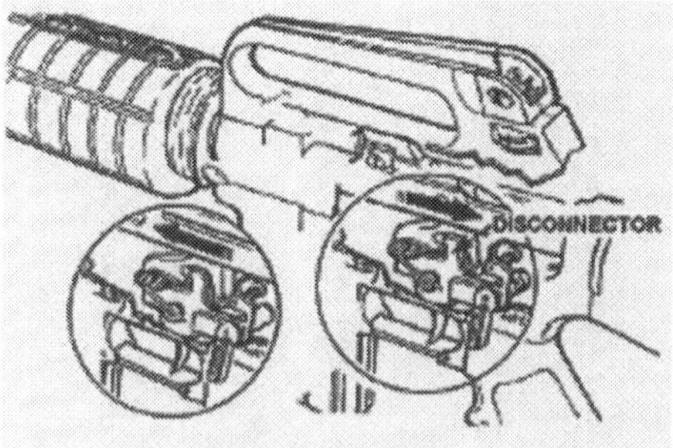

Figure 2-10. Cocking.

Feeding

Once rearward motion causes the bolt carrier group to clear the top of the magazine, the expansion of the magazine spring forces a round into the path of the bolt. After the action spring overcomes and absorbs the rearward motion of the bolt carrier group, it expands and sends the buffer assembly and bolt carrier group forward with enough force to strip a round from the magazine. See figure 2-11.

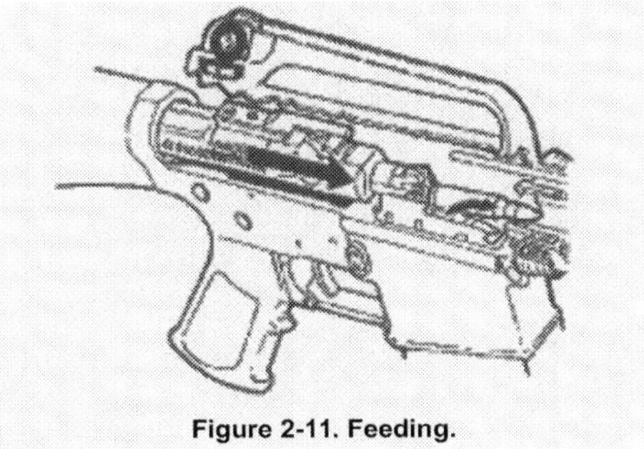

Figure 2-11. Feeding.

Chambering

As the bolt carrier group continues to move forward, pushing a fresh round in front of it, the face of the bolt thrusts the new round into the chamber. The extractor claw grips the rim of the cartridge case. The ejector is forced into its hole, compressing the ejector spring. See figure 2-12.

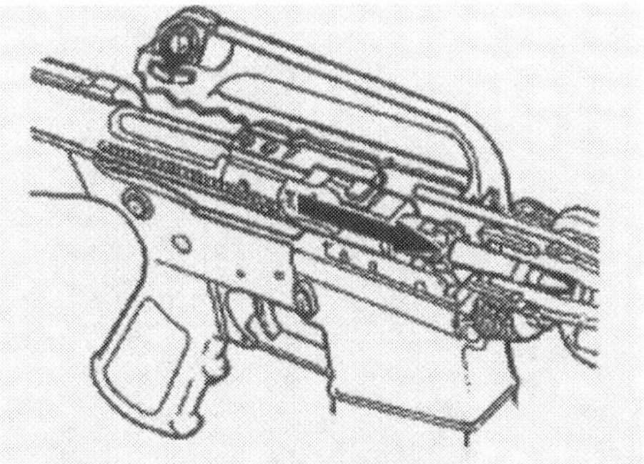

Figure 2-12. Chambering.

Locking

As the bolt carrier group continues to move forward, the bolt-locking lugs are forced against the barrel extension and the bolt cam pin is forced along the cam track. The bolt rotates and aligns the bolt locking lugs behind the barrel extension locking lugs. The weapon is ready to fire. See figure 2-13.

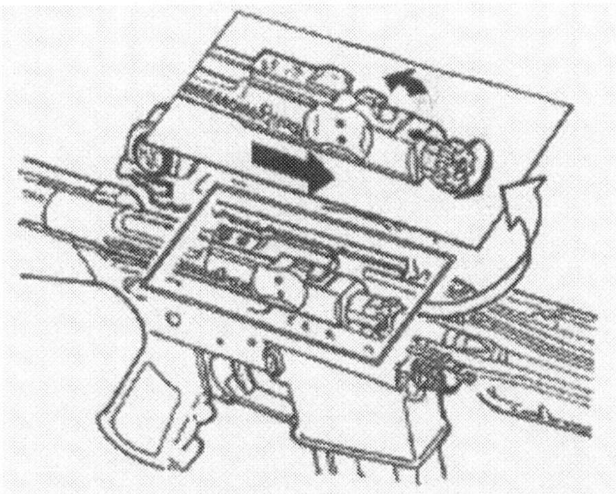

Figure 2-13. Locking.

2004. Ammunition

Four types of ammunition are authorized for use with the M16A2 service rifle: ball (M193 and M855), tracer (M196 and M856), dummy (M199), and blank (M200) (see fig. 2-14).

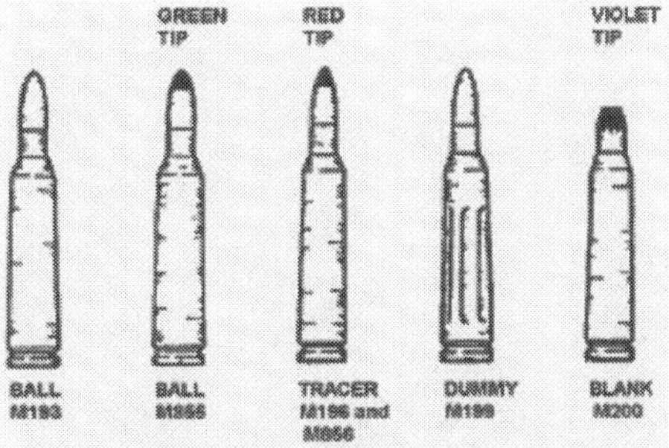

Figure 2-14. Authorized Ammunition.

M193 Ball

This ammunition is a 5.56mm centerfire cartridge with a 55-grain gilded-metal jacket, lead alloy core bullet. The primer and case are waterproofed. The M193 ball ammunition has no identifying marks.

M855 Ball

This ammunition is the primary ammunition for the M16A2 rifle. Identified by a green tip, its 5.56mm centerfire cartridge has better penetration than the M193. It has a 62-grain gilded-metal jacket bullet. The rear two-thirds of the core of the projectile is lead alloy and the front one-third is a solid steel penetrator. The primer and case are waterproofed.

M196 and M856 Tracer

This ammunition has the same basic characteristics as ball ammunition. Identified by a bright red tip, its primary uses include observation firing, incendiary effect, and signaling. Tracer ammunition should be intermixed with ball ammunition in a ratio no greater than 1:1. The preferred ratio is one tracer to four balls (1:4) to prevent metal fouling in the bore.

M199 Dummy

This ammunition has six grooves along the side of the case. It contains no propellants or primer. The primer well is open to prevent damage to the firing pin. The dummy cartridge is used during dry fire and other training purposes.

M200 Blank

This ammunition has the case mouth closed with a seven-petal rosette crimp. It contains no projectile. Blank ammunition, identified by its violet tip, is used for training purposes.

2005. Preventive Maintenance

Normal care and cleaning of the rifle will result in proper functioning of its all parts. Only issue-type cleaning materials maybe used. Improper maintenance can cause stoppages, reducing combat readiness and effectiveness.

Main Group Disassembly

The M16A2 service rifle is disassembled into three main groups (see fig. 2-15). Before disassembling the M16A2 service rifle—

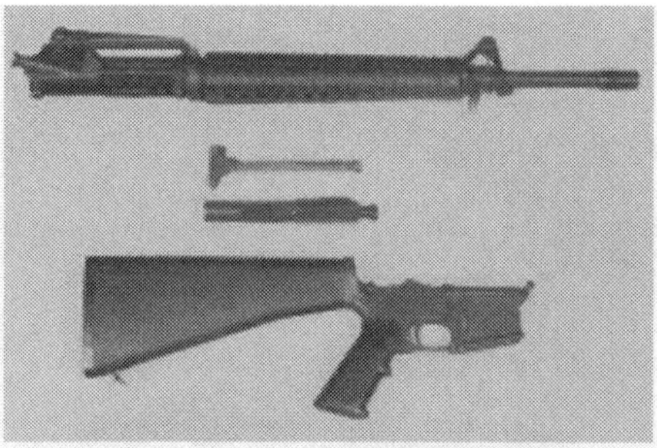

Figure 2-15. Three Main Groups.

- Ensure the rifle is in Condition 4 (see para. 3002).
- Remove the sling.

Note

To facilitate control and ease of disassembly, the handguards may be removed using the "buddy system" as described in Technical Manual (TM) 05538C-10/1A, *Operator's Manual w/Components List Rifle, 5.56-mm, M16A2E W/E (1005-01-128-9936).*

Upper Receiver

To disassemble the upper receiver—

- Place the rifle on the butt-stock. Press down on the slip ring with both hands. Pull the handguards free (see fig. 2-16).

  Use caution when the handguards are off the rifle. Handguards provide protection for the gas tube. Damage to the gas tube adversely affects the functioning of the rifle.

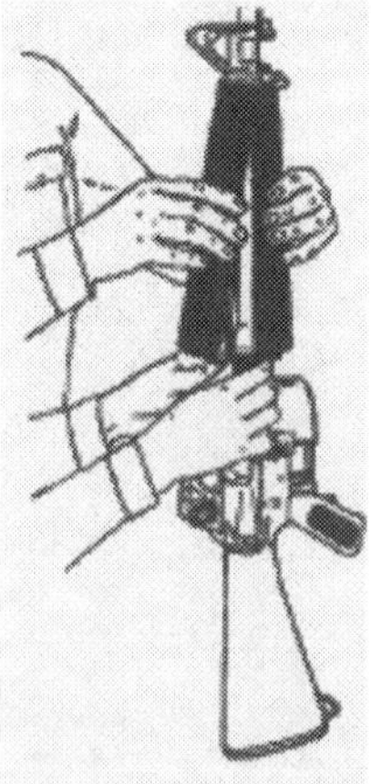

Figure 2-16. Removing the Handguards.

- Move the take down pin from the left to the right as far as it will go to allow the lower receiver to pivot down from the upper receiver.
- Move the receiver pivot pin from left to right as far as it will go and separate the upper and lower receivers.
- Pull back the charging handle and bolt carrier about 3 inches and remove the bolt carrier group.
- Remove the charging handle by sliding it back and down, out of the upper receiver.

No further disassembly is conducted on the upper receiver group.

Bolt Carrier

To disassemble the bolt carrier—

- Remove the firing pin retaining pin.
- Push the bolt back into the bolt carrier to the locked position.
- Tap the base of the bolt carrier against the palm of your hand so the firing pin will drop out.
- Rotate the bolt cam pin one-quarter turn and lift the bolt cam pin out.
- Withdraw the bolt assembly from the carrier.
- Press on the extractor's rear and use the firing pin to push out the extractor-retaining pin. Remove the extractor and spring (the spring is permanently attached to the extractor). (See fig. 2-17.)

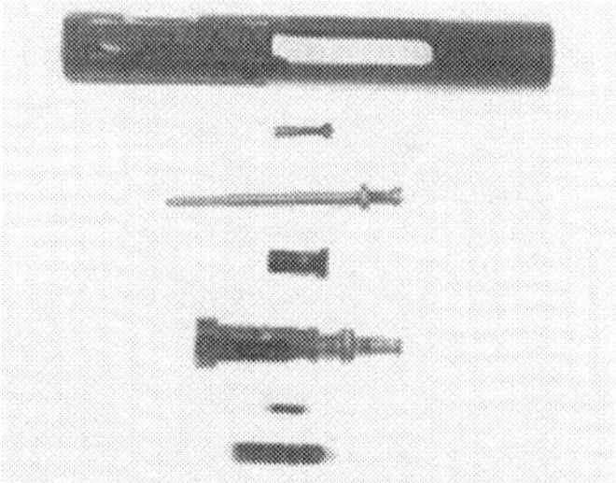

Figure 2-17. Bolt Carrier Disassembled.

CAUTION

Be careful not to damage the tip of the firing pin while pushing out the extractor-retaining pin.

Note

The extractor assembly has a rubber insert within the spring. Do not attempt to remove it. If the spring comes loose, put the large end of the spring in the extractor and seat it. Push in the extractor pin.

Lower Receiver

To disassemble the lower receiver—

- Press in the buffer and depress the buffer retainer.

Note

It may be necessary to use the edge of the charging handle to depress the buffer retainer.

- Press the hammer downward and ease the buffer and action spring forward and out of the receiver.
- Separate the parts. See figure 2-18.

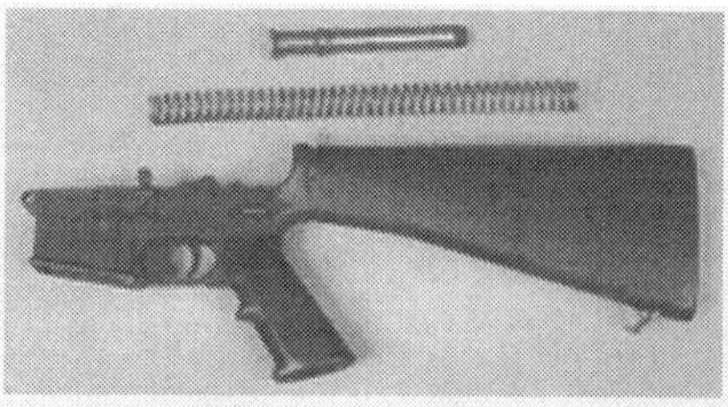

Figure 2-18. Lower Receiver Disassembled.

No further disassembly of the lower receiver is performed.

Note

In combat situations, the rifle may be partially disassembled in any sequence. However, combat situations are the exception, not the rule. Under normal operating circumstances, disassemble the rifle in the sequence just performed. Any further disassembly of the rifle is to be performed by a qualified armorer.

Magazine Disassembly

The magazine should be disassembled regularly for cleaning to avoid the possibility of malfunction or stoppage of the rifle caused by dirty or damaged magazines. To disassemble the magazine—

- Pry up and push base plate out from the magazine.
- Jiggle the spring and follower to remove. Do not remove the follower from the spring. See figure 2-19.

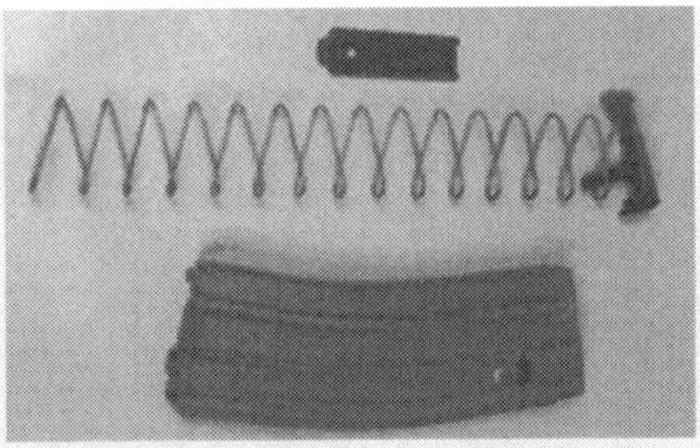

Figure 2-19. Magazine Disassembled.

Cleaning

Cleaning Materials

The following cleaning materials are used in preventive maintenance (see fig. 2-20):

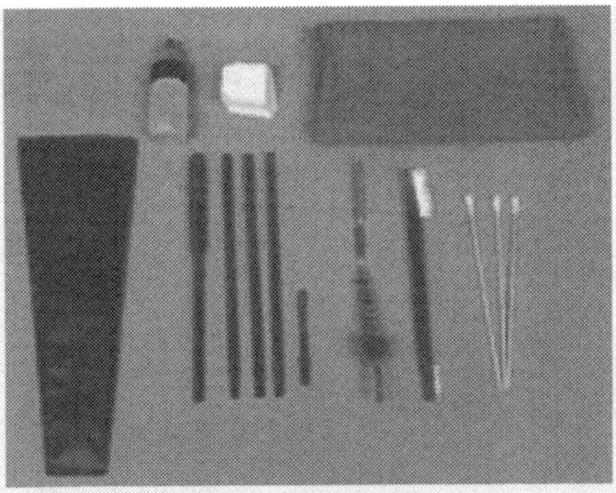

Figure 2-20. Cleaning Materials.

- Cleaner, lubricant, and preservative (CLP). Always shake the bottle well before use.
- Rod in three sections and a handle assembly.
- Patch holder section, swabs, patches, pipe cleaners, and clean rags.
- Brushes: bore, chamber, and general purpose.

Cleaning the Upper Receiver

Basic cleaning of the upper receiver group should include the following:

- Attach the three rod sections together but leave each one about two turns short of being tight.
- Attach the patch holder onto the rod.
- Point the muzzle down and insert the non-patch end of the rod into the chamber. Attach the handle to the cleaning rod section and pull a CLP-moistened 5.56mm patch through the bore.
- Attach the bore brush to the rod but leave it two turns short of being tight. Put a few drops of CLP on the bore brush. Insert the rod into the barrel from the chamber end, attach the handle, and pull the brush through the bore. Repeat 3 times. Remove bore brush and attach the patch holder to the rod with a CLP moistened patch insert the rod into the barrel from the chamber end, attach the handle, and pull the patch through the bore.
- Inspect the bore for cleanliness by holding the muzzle to your eye and looking into the bore.
- Repeat the above steps until the patches come out of the bore clean.
- Attach the chamber brush and one section of the cleaning rod to the handle. Moisten it well with CLP and insert it into the chamber.
- Scrub the chamber and bolt lugs using a combination of a plunging and clockwise rotating action.

Note

Do not reverse direction of the brush while it is in the chamber.

- Clean the interior portion of the upper receiver with the general-purpose brush and CLP.
- Dry the bore, chamber, and the interior of the receiver with rifle patches, swabs, and clean rags until they come out clean. Then moisten all interior surfaces with CLP.
- Wipe the barrel, gas tube, and handguards clean with a rag.

Cleaning the Bolt Carrier Group

- Clean the outer and inner surfaces of the bolt carrier with a general-purpose brush.
- Clean the bolt carrier key with a pipe cleaner.
- Clean the locking lugs, gas rings, and exterior of the bolt with the general-purpose brush.
- Insert a swab into the rear of the bolt and swab out the firing pin recess and gas ports.

- Clean the extractor with the general-purpose brush, ensuring all the carbon is removed from underneath the extractor lip.
- Clean extractor pin, firing pin, and firing pin retaining pin using the general-purpose brush and CLP.
- Clean charging handle assembly with the general-purpose brush and patches.

Cleaning the Lower Receiver

- Wipe dirt from the firing mechanism using a general-purpose brush, clean patch, pipe cleaners, and swabs.
- Clean the outside of the receiver with the general-purpose brush and CLP. Clean the buttplate and rear sling swivel, ensuring drain hole is clear of dirt.
- Wipe the inside of the buffer tube, buffer, and action spring.
- Wipe the inside of the magazine well with a rag.
- Wipe out the inside of the pistol grip and ensure that it is clean.

Cleaning the Magazine

- Clean the inside of the magazine with the general-purpose brush and CLP.
- Wipe dry.
- Keep the spring lightly oiled.

Inspection

While cleaning the rifle, and during each succeeding step in the preventive maintenance process, inspect each part for cracks and chips and ensure parts are not bent or badly worn. Report any damaged part to the armorer. Inspection is a critical step to ensure the combat readiness of your rifle. It is performed normally during rifle cleaning (prior to lubrication), however, it can be performed throughout the preventive maintenance process.

Lubrication

Lubrication is performed as part of the detailed procedure for preventive maintenance. Lubrication procedures are also performed in preparation for firing.

Lubricant

In all but the coldest arctic conditions, CLP is the lubricant for the rifle. Remember to remove excess CLP from the bore and chamber before firing.

- Lightly lube means that a film of CLP barely visible to the eye should be applied.
- Generously lube means the CLP should be applied heavily enough that it can be spread with the finger.

Upper Receiver

- Lightly lube the inside of the upper receiver, bore, chamber, outer surfaces of the barrel, and surfaces under the handguard.
- Depress the front sight detent and apply two or three drops of CLP to the front sight detent. Depress several times to work lubrication into the spring.
- Lubricate the moving parts and elevation screw shaft of the rear sight.

Bolt Carrier Group

- Generously lube the outside of the cam pin area, the bolt rings, and outside the bolt body.
- Lightly lube the charging handle and the inner and outer surfaces of the bolt carrier.

Lower Receiver

- Lightly lube the inside of the lower receiver extension.
- Generously lube the moving parts inside the lower receiver and their pins.

Reassembly

Reassembling the Rifle

- Return all cleaning gear into the buttstock of the rifle and close the buttplate.
- Connect the buffer and action spring and insert them into the buffer tube/stock.
- Place the extractor and spring back on the bolt. Depress the extractor to align the holes and reinsert the extractor pin.
- Insert the bolt into the carrier. Do not switch bolts between rifles.
- Hold the bolt carrier with the bolt carrier key at 12 o'clock. Insert the bolt into the bolt carrier with the extractor at 12 o'clock.

- Rotate the bolt counterclockwise until the cam pinhole aligns to the cam pin slot in the bolt carrier.

WARNING WARNING WARNING WARNING

WARNING Ensure the cam pin is installed in the bolt group or the rifle may explode while firing.

- Insert the bolt cam pin through the bolt carrier and into the bolt. Rotate the cam pin 1/4 turn right or left. Pull the bolt forward until it stops.
- Drop in the firing pin from the rear of the bolt carrier and seat it.
- Replace the firing pin retaining pin. Ensure the head of the firing pin retaining pin is recessed inside the bolt carrier. The firing pin should not fall out when the bolt carrier group is turned upside down.
- Place the charging handle in the upper receiver by lining it up with the grooves in the receiver. Push the charging handle partially in.
- With the bolt in the unlocked position, place bolt carrier key into the groove of the charging handle.
- Push the charging handle and bolt carrier group into the upper receiver until the charging handle locks.
- Join the upper and lower receivers and engage the receiver pivot pin.
- Ensure the selector lever is on safe before closing the upper receiver. Close the upper and lower receiver groups. Push in the takedown pin.
- Install the handguards.
- Attach the sling.

Reassembling the Magazine

To reassemble the magazine—

- Insert the follower and jiggle the spring to install.
- Slide the base under all four tabs until the base catches. Make sure the printing is on the outside.

2006. Function Check

A function check is performed to ensure the rifle operates properly. To perform a function check:

- Place the weapon in Condition 4 (see para. 3002).
- Pull the charging handle to the rear and release. Ensure the selector lever is on safe and pull the trigger. The hammer should not fall.
- Place the selector lever on semi. Pull the trigger and hold it to the rear. The hammer should fall. Pull the charging handle to the rear and release. Release the trigger and pull again. The hammer should fall.
- Pull the charging handle to the rear and release. Place the selector lever on burst. Pull the trigger and hold it to the rear. The hammer should fall. Pull the charging handle to the rear three times and release. Release the trigger and pull again. The hammer should fall.
- Pull the charging handle to the rear and release. Place the selector lever on safe.

2007. User Serviceability Inspection

Individual Marines must perform user serviceability inspections on their weapons before firing them. This inspection ensures the weapon is in an acceptable operating condition.

- Place rifle in Condition 4 (see para. 3002).
- Conduct a function check.
- Check the rifle to ensure the following:
- Compensator is tight.
- Barrel is tight.
- Front sight post is straight.
- Front sight post is adjustable.
- Handguards are serviceable.
- Rear sight elevation and windage knobs are adjustable and have distinct clicks.
- Stock is tight on the lower receiver.
- Weapon is properly lubricated for operational conditions.
- Barrel is clear of obstructions.
- Gas rings are serviceable.
- Ensure magazines are serviceable.
- Load the rifle with an empty magazine. Ensure that the magazine can be seated.
- Without depressing the bolt catch, pull the charging handle to the rear. Ensure that the bolt locks to the rear.
- Depress the upper portion of the bolt catch and observe bolt moving forward on an empty chamber.

Ensure the bolt moves completely forward and locks in the chamber.

- Repeat this procedure with all magazines.

2008. Field Maintenance

Preventive maintenance in the field is performed when detailed disassembly and cleaning is not practical due to operational tempo or the level of threat. To perform limited field preventive maintenance—

- Place the rifle in Condition 4 (see paragraph 3002).
- Break the rifle down by removing the rear take down pin and rotating the upper receiver and barrel forward.
- Remove the bolt carrier group.
- Do not disassemble the bolt carrier group further.
- Clean the bolt carrier group.
- Clean the upper and lower receiver groups (without further disassembly).
- Clean the bore and chamber.
- Lubricate the rifle.
- Reassemble the rifle and perform a user serviceability inspection.

2009. Cleaning the Rifle in Various Conditions

The climatic conditions in various locations require special knowledge about cleaning and maintaining the rifle. The conditions that will affect the rifle the most are: hot, wet tropical; hot, dry desert; arctic or low temperature; and heavy rain and fording.

Hot, Wet Tropical

- Perform normal maintenance.
- Clean and lubricate your rifle more often. Inspect hidden surfaces for corrosion. Pay particular attention to spring-loaded detentes.
- Use lubricant more liberally.
- Unload and check the inside of the magazine more frequently. Wipe dry and check for corrosion.
- When practical, keep the rifle covered.

Hot, Dry Desert

Hot dry climates are usually areas that contain blowing sand and fine dust. Dust and sand will get into the rifle and magazines, causing stoppages. It is imperative to pay particular attention to the cleaning and lubrication of the rifle in this type of climate.

Corrosion is less likely to form in these environments, and lubrication will attract more dirt. For this reason, use lubrication more sparingly.

Whenever practical, keep the rifle covered.

Arctic or Low Temperature

- Clean and lubricate the rifle in a warm room, with the rifle at room temperature, if possible. Lubricating oil, arctic weapons (LAW) can be used below a temperature of 0 degrees Fahrenheit and must be used below -35 degrees Fahrenheit.
- Keep the rifle covered when moving from a warm to a cold environment to allow gradual cooling of the rifle. This prevents the condensation of moisture and freezing. Condensation will form on the rifle when it is moved from outdoors to indoors. If possible, leave the rifle in a protected but cold area outdoors. When bringing the rifle inside to a warm place, it should be disassembled and wiped down several times as it warms.
- Always try to keep the rifle dry.
- Unload and perform a function check every 30 minutes, if possible, to help prevent freezing of functional parts.
- Do not lay a warm rifle in snow or ice.
- Keep the inside of the magazine and ammunition wiped dry. Moisture will freeze and cause stoppages.

Heavy Rain and Fording

- Keep the rifle dry and covered when practical.
- Keep water out of the barrel if possible. If water does get in, drain and (if possible) dry with a patch. If water is in the barrel, point the muzzle down and break the seal by doing a chamber check so the water will drain. If water is in the stock of the weapon, ensure the drain hole in the stock is clear so the water can run out.
- Perform normal maintenance.

CHAPTER 3. WEAPONS HANDLING

Weapons handling procedures provide a consistent and standardized way for a Marine to handle, operate, and employ the rifle safely and effectively. Proper weapons handling procedures ensure the safety of Marines by eliminating negligent discharges and reinforcing positive identification of targets before engagement. Weapons handling procedures apply at all levels of training and during combat operations.

Note

+The procedures in this manual are written for right-handed Marines; left-handed Marines should reverse instructions as necessary.

3001. Safety Rules

Safe handling of the rifle is critical. If proper weapons handling procedures are not used, a Marine risks his safety and the safety of his fellow Marines. During combat, a Marine must react quickly, safely, and be mentally prepared to engage targets. To ensure that only the intended target is engaged, a Marine must apply the following safety rules at all times:

Rule 1—Treat every weapon as if it were loaded. When a Marine takes charge of a rifle in any situation, he must treat the weapon as if it were loaded, determine its condition (see para. 3003), and continue applying the other safety rules.

Rule 2—Never point a weapon at anything you do not intend to shoot. A Marine must maintain muzzle awareness at all times.

Rule 3—Keep your finger straight and off the trigger until you are ready to fire. A target must be identified before taking the weapon off safe and moving the finger to the trigger.

Rule 4—Keep the weapon on safe until you intend to fire. A target must be identified before taking the weapon off safe. This rule is intended to eliminate the chance of the weapon discharging by accident (e.g., brush snagging the trigger).

3002. Weapons Conditions

A weapon's readiness is described by one of four conditions. The steps in the loading and unloading process take the rifle through four specific conditions of readiness for live fire.

Condition 1. Safety on, magazine inserted, round in chamber, bolt forward, ejection port cover closed.

Condition 2. Not applicable to the M16A2 rifle.

Condition 3. Safety on, magazine inserted, chamber empty, bolt forward, ejection port cover closed.

Condition 4. Safety on, magazine removed, chamber empty, bolt forward, ejection port cover closed.

3003. Determining a Weapon's Condition (Chamber Check)

A Marine must know the condition of his weapon at all times. When a Marine takes charge of a weapon in any situation, he must determine its condition. Situations include coming across an unmanned rifle in combat, taking charge of any weapon after it has been unmanned (e.g., out of a rifle rack, stored in a vehicle), or taking charge of another Marine's weapon. To determine the condition of the weapon in any of these situations, the Marine should:

- Determine if a magazine is present.
- Ensure the rifle is on safe.

- Conduct a chamber check.
- Bring the left hand back against the magazine well.
- Extend the fingers of the left hand and cover the ejection port (see fig. 3-1).

Figure 3-1. Position of Hand.

- Grasp the charging handle with the index and middle fingers of the right hand.
- Pull the charging handle slightly to the rear and visually and physically inspect the chamber (see fig. 3-2). *Right-handed Marines*, insert one finger of the left hand into the ejection port and feel whether a round is present. *Left-handed Marines*, insert the thumb of the right hand into the ejection port and feel whether a round is present.

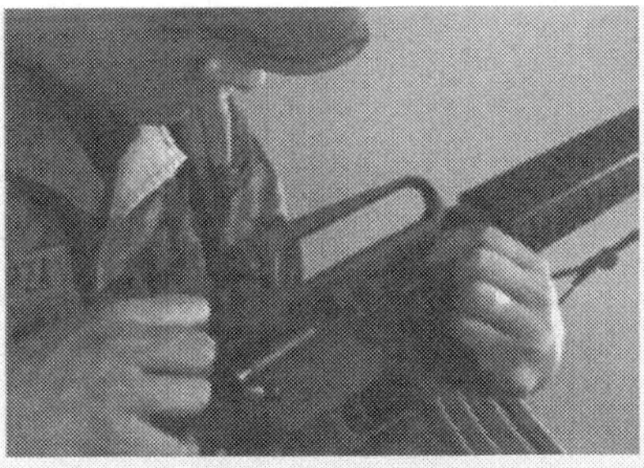

Figure 3-2. Chamber Check.

Note

The same procedure is used in daylight as during low visibility. A chamber check may be conducted at any time.

CAUTION

Pulling the charging handle too far to the rear while inspecting the chamber may cause double feed or ejection of one round of ammunition.

- Release the charging handle and observe the bolt going forward.
- Tap the forward assist.
- Close the ejection port cover (if time and the situation permit).
- Remove the magazine (if present) and observe if ammunition is present. If time permits, count the rounds. Reinsert the magazine into magazine well.

3004. Weapons Commands

Weapons commands dictate the specific steps required to load and unload the rifle. Six commands are used in weapons handling:

Load. This command is used to take the weapon from Condition 4 to Condition 3.

Make Ready. This command is used to take the weapon from Condition 3 to Condition 1.

Fire. This command is used to specify when a Marine may engage targets.

Cease-Fire. This command is used to specify when a Marine must stop target engagement.

Unload. This command is used to take the weapon from any condition to Condition 4.

Unload and Show Clear. This command is used when an observer must check the weapon to verify that no ammunition is present before the rifle is placed in Condition 4.

Loading the Rifle

Perform the following steps to load the rifle (take the rifle to Condition 3):

- Ensure the rifle is on safe.
- Withdraw the magazine from the magazine pouch.
- Observe the magazine to ensure it is filled.
- Fully insert the magazine in the magazine well. Without releasing the magazine, tug downward on the magazine to ensure it is seated.
- Close the ejection port cover.
- Fasten the magazine pouch.

Making the Rifle Ready

Perform the following steps to make the rifle ready for firing (take the rifle to Condition 1):

- Pull the charging handle to the rear and release. There are two methods of doing this:
- Grip the pistol grip firmly with the right hand and pull the charging handle with the left hand to its rearmost position and release (see fig. 3-3).

  Or grip the handguards firmly with the left hand and pull the charging handle with the right hand to its rearmost position and release (see fig. 3-4).

Figure 3-3. Pulling the Charging Handle with the Left Hand.

Figure 3-4. Pulling the Charging Handle with the Right Hand.

- To ensure ammunition has been chambered, conduct a chamber check (see para. 3003) to ensure a round has been chambered.
- Check the sights (to ensure proper battlesight zero [BZO] setting, correct rear sight aperture, etc.).
- Close ejection port cover.

Fire

On the command "Fire," aim the rifle, take the rifle off safe, and pull the trigger.

Cease-Fire

On the command "Cease Fire," perform the following steps:

- Place your trigger finger straight along the receiver.
- Place the weapon on safe.

Unloading the Rifle

Perform the following steps to unload the rifle (take the rifle to Condition 4):

- Ensure the weapon is on safe.
- Remove the magazine from the rifle and retain it on your person.

- Cup the left hand under the ejection port, rotate the weapon until the ejection port faces down.
- Pull the charging handle to the rear and catch the round in the left hand (see fig. 3-5).

Figure 3-5. Catching the Round.

- Lock the bolt to the rear.
- Put the weapon on safe if the selector lever would not move to safe earlier.
- Ensure the chamber is empty and that no ammunition is present.
- Depress the bolt catch and observe the bolt moving forward on an empty chamber (see fig. 3-6).

Figure 3-6. Observing the Chamber.

- Close the ejection port cover.
- Check the sights (for proper BZO setting, correct rear sight aperture, etc.).
- Place any ejected round into the magazine and return the magazine to the magazine pouch and close the magazine pouch.

Unloading and Showing the Rifle Clear

Perform the following steps to unload the rifle and show it clear to an observer (take the rifle to Condition 4). The Marine—

- Ensures the weapon is on safe.
- Removes the magazine from the rifle and retains it.
- Cups the left hand under the ejection port, rotates the weapon until the ejection port faces down.
- Pulls the charging handle to the rear and catches the round in the left hand.
- Locks the bolt to the rear and ensures the chamber is empty and that no ammunition is present.
- Has another Marine inspect the weapon to ensure no ammunition is present (see fig. 3-7).

Figure 3-7. Observer Inspection.

The observer—

- Visually inspects the chamber to ensure it is empty, no ammunition is present, and the magazine is removed.
- Ensures the weapon is on safe.
- Acknowledges the rifle is clear.

The Marine, after receiving acknowledgment that the rifle is clear—

- Depresses the bolt catch and observes the bolt moving forward on an empty chamber.
- Closes the ejection port cover.
- Checks the sights (for proper BZO setting, correct rear sight aperture, etc.).
- Places any ejected round into the magazine and returns the magazine to the magazine pouch and closes the magazine pouch.

3005. Filling, Stowing, and Withdrawing Magazines

Filling the Magazine with Loose Rounds

Perform the following steps to fill the magazine:

- Remove a magazine from the magazine pouch.
- Place a round on top of the follower.
- Press down until the round is held between the follower and magazine feed lips (see fig. 3-8).

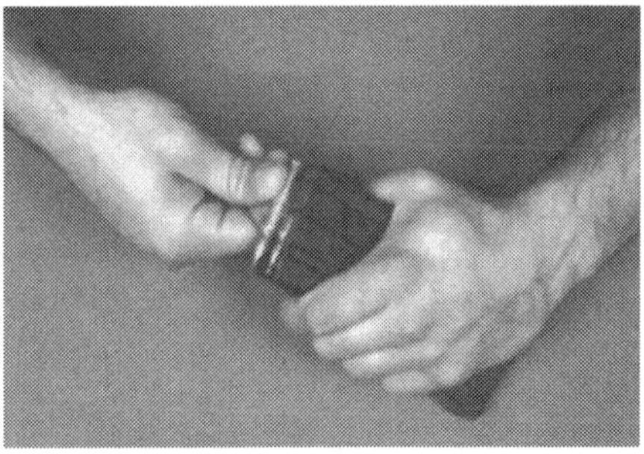

Figure 3-8. Filling the Magazine.

- Repeat until the desired number of rounds is inserted. The recommended number of rounds per magazine is 28 or 29. Thirty rounds in the magazine may prohibit the magazine from seating properly on a closed bolt.
- Tap the back of the magazine to ensure the rounds are seated against the back of the magazine.

Filling the Magazine Using a 10-round Stripper Clip and Magazine Filler

The magazine can also be filled quickly using a 10-round stripper clip and the magazine filler (see fig. 3-9).

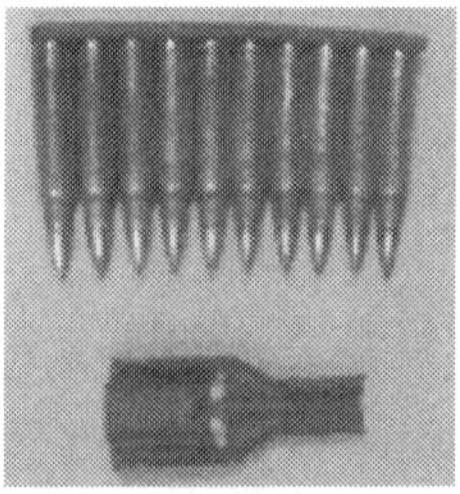

Figure 3-9. Magazine Filler and 10-round

Perform the following steps to fill the magazine with the 10-round stripper clip (see fig. 3-10):

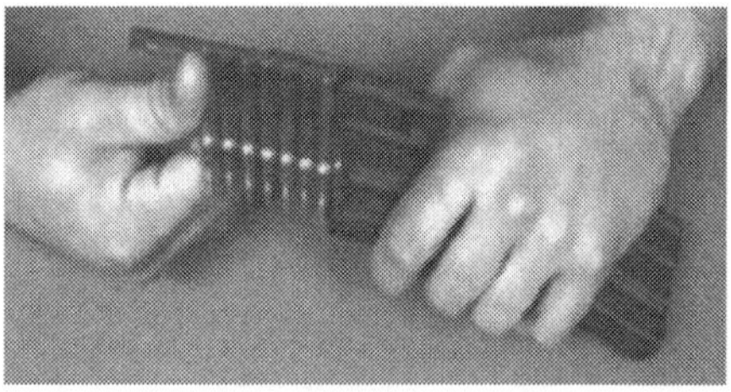

Figure 3-10. Filling the Magazine with a Stripper Clip and Magazine Filler.

- Remove a magazine from the magazine pouch.
- Slide the magazine filler into place.
- Place a 10-round stripper clip into the narrow portion of the magazine filler.
- Using thumb pressure on the rear of the top cartridge, press down firmly until all ten rounds are below the feed lips of the magazine.
- Remove the empty stripper clip while holding the magazine filler in place.
- Repeat until the desired number of rounds is inserted. The recommended number of rounds per magazine is 28 or 29. Thirty rounds in the magazine may prohibit the magazine from seating properly on a closed bolt.
- Remove magazine filler and retain it for future use.
- Tap the back of the magazine to ensure the rounds are seated against the back of the magazine.

Stowing Magazines

Magazine Pouch

In a magazine pouch, filled magazines are stored with rounds down and projectiles pointing away from the body.

Load-bearing Vest

In a load-bearing vest, filled magazines are stored with rounds down and projectiles pointing outboard.

Empty or Partially Filled Magazines

Empty or partially filled magazines are stored with the follower up to allow the selection of filled magazines by touch (i.e., at night).

Withdrawing Magazines

Magazine Pouch

With the right hand, withdraw magazines from the magazine pouch on the right side of the body. With the left hand, withdraw magazines from the magazine pouch on the left side of the body. To withdraw magazines from a magazine pouch—

- Use the thumb and index finger, pinch the magazine pouch release to open the magazine pouch.
- Slide the thumb over the magazines, feeling for a base plate indicating a filled magazine. Continue sliding the thumb until it rests on the back of the magazine.
- Grasp the magazine with the thumb, index finger, and middle finger and lift the magazine directly out of the pouch.
- Once the magazine is clear of the pouch, curl the ring finger and little finger underneath the magazine and rotate it up to observe rounds in the magazine.

Load-bearing Vest

With the right hand, withdraw magazines from the left side of the vest (see fig. 3-11). With the left hand, withdraw magazines from the right side of the vest (see fig. 3-12).

To withdraw magazines from a load-bearing vest:

- With the thumb and index finger, unfasten the snap on the vest pouch.
- Slide the thumb over the magazine, feeling for a base plate indicating a filled magazine.
- Rotate the hand over the magazine while sliding the thumb to the back of the magazine.
- Grasp the magazine with the thumb, index finger, and middle finger, while curling the ring finger and little finger on top of the base plate.
- Lift the magazine directly out of the pouch and rotate it up to observe the rounds in the magazine.

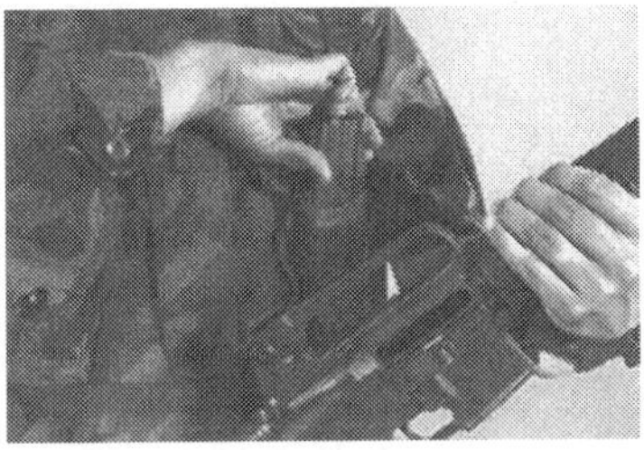

Figure 3-11. Withdrawing a Magazine with the Right Hand.

3006. Reloading the Rifle

Principles of Reloading

The first priority when performing a reload is to get the rifle reloaded and back into action. The second priority when performing a reload is to retain the magazine so when you move, the magazine moves with you. When time permits, retain magazines securely on your person (e.g., in magazine pouch, flak jacket, and cargo pocket). The combat situation may dictate dropping the magazine to the deck when performing a reload. This is acceptable as long as it is picked up before moving to another location.

Take cover before reloading. Always reload before leaving cover to take advantage of the protection provided by cover.

When moving, your focus should be on moving, therefore every effort should be made to not reload while on the move.

When reloading, your focus is on the magazine change.

When reloading, draw the weapon in close to your body so you can see what you are doing and retain positive control of the magazine.

When the new magazine is inserted, tug on it to ensure it is seated. Do not slam the magazine into the weapon hard enough to cause a round to partially pop out of the magazine. This action will cause a double feed and require remedial action.

Figure 3-12. Withdrawing a Magazine

Retain your empty magazines. When there is a lull in the action, refill those magazines so they will be available for future use.

During a lull in the action, replace your magazine when you know you are low on ammunition. This ensures a full magazine of ammunition in the rifle should action resume. Do not wait until the magazine is completely empty to replace it.

Condition 1 Reload

A condition 1 reload is performed when the weapon is in condition 1 by replacing the magazine before it runs out of ammunition. To perform a condition 1 reload, perform the following steps:

- Withdraw a filled magazine from the magazine pouch. With the same hand, press the magazine button and remove the partially filled magazine so it can be retained in the remaining fingers.
- Fully insert the filled magazine into the magazine well and tug downward on the magazine to ensure it is properly seated.
- Store the partially filled magazine in the magazine pouch with rounds up and projectiles pointing away from the body.
- Fasten the magazine pouch.

Dry Reload

A dry reload is required when the magazine in the weapon has been emptied and the bolt has locked to the rear. To perform a dry reload—

- Press the magazine release button.
- Remove the empty magazine and retain it on your person when time permits.
- Fully insert a filled magazine into the magazine well and tug downward on the magazine to ensure it is properly seated.
- Depress the bolt catch to allow the bolt carrier to move forward and observe the round being chambered. This places the rifle in Condition 1.

3007. Remedial Action

If the rifle fails to fire, a Marine performs remedial action. Remedial action is the process of investigating the cause of the stoppage, clearing the stoppage, and returning the weapon to operation.

Observe for Indicators

Once the rifle ceases firing, the Marine must visually or physically observe the ejection port to identify the problem before he can clear it. The steps taken to clear the weapon are based on observation of one of the following three indicators:

Indicator: The bolt is forward or the ejection port cover is closed. See figures 3-13 and 3-14.

Figure 3-13. Bolt Forward.

Figure 3-14. Ejection Port Cover Closed.

To return the weapon to operation—

- Seek cover if the tactical situation permits.
- Tap—Tap the bottom of the magazine.
- Rack—Pull the charging handle to the rear and release it.
- Bang—Sight in and attempt to fire.

Indicator: Brass is obstructing chamber area (usually indicating a double feed or failure to eject). See figure 3-15.

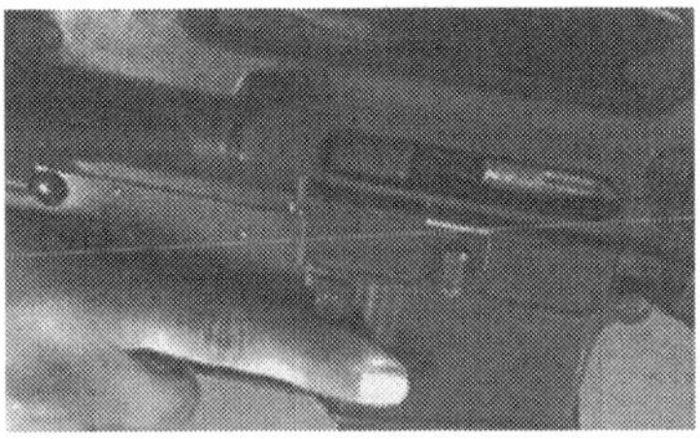

Figure 3-15. Brass Obstructing the Chamber.

To return the weapon to operation—

- Seek cover if the tactical situation permits.
- Attempt to remove the magazine.
- Attempt to lock the bolt to the rear

If the bolt will not lock to the rear, rotate the rifle so the ejection port is facing down; hold the charging handle to the rear as far as it will go and shake the rifle to free the round(s). If the rounds do not shake free, hold the charging handle to the rear and strike the butt of the rifle on the ground or manually clear the round. Conduct a reload. Sight in and attempt to fire.

Indicator: The bolt is locked to the rear. See figure 3-16. To clear return the weapon to operation—

Figure 3-16. Bolt Locked to the Rear.

Note

Although a dry weapon is not considered a true stoppage or mechanical failure, the Marine must take action to return the weapon to operation.

- Seek cover if the tactical situation permits.
- Conduct a dry reload.
- Sight in and attempt to fire.

Audible Pop or Reduced Recoil

An audible pop occurs when only a portion of the propellant is ignited. It is normally identifiable by reduced recoil and is sometimes accompanied by excessive smoke escaping from the chamber area. To clear the rifle in a combat environment:

- Place the rifle in Condition 4.
- Move take down pin from left to right as far as it will go to allow the lower receiver to pivot.
- Remove the bolt carrier group.
- Inspect the bore for an obstruction from the chamber end.
- Insert a cleaning rod into the bore from the muzzle end and clear the obstruction.
- Reassemble the rifle.
- Conduct a reload.
- Sight in and attempt to fire.

3008. Weapons Carries

Weapons carries provide an effective way to handle the rifle while remaining alert to enemy engagement. Weapons carries are tied to threat conditions and are assumed in response to a specific threat situation. The weapons carry assumed prepares the Marine, both mentally and physically, for target engagement. The sling provides additional support for the weapon when firing; therefore, the hasty sling should be used in conjunction with the carries.

Tactical Carry

A Marine carries the rifle at the tactical carry if no immediate threat is present. The tactical carry permits control of the rifle while a Marine is moving, yet it still allows quick engagement of the enemy. A Marine performs the following steps to assume the tactical carry:

- Place left hand on the handguards, right hand around the pistol grip, trigger finger straight along the receiver (see fig. 3-17), and right thumb on top of the selector lever (see fig. 3-18).

Figure 3-17. Straight Trigger Finger.

Figure 3-18. Thumb on Selector Lever.

- Place the buttstock along the side of the body at approximately hip level.
- Angle the muzzle upward about 45 degrees in a safe direction.
- Position the muzzle in front of the eyes, slightly below eye level (see fig. 3-19).
- Move the head and the eyes with the muzzle as it moves.

Figure 3-19. Tactical Carry.

Alert Carry

A Marine carries the rifle at the alert if enemy contact is likely. The alert is also used for moving in close terrain (e.g., urban, jungle). A Marine can engage the enemy faster from the alert than from the tactical carry. However, the alert is more tiring than the tactical carry and its use can be physically demanding. A Marine performs the following steps to assume the alert:

- Place the left hand on the handguards, the right hand around the pistol grip, the trigger finger straight along the receiver (see fig. 3-17), and the right thumb on top of the selector lever (see fig. 3-18).
- Place the buttstock in the shoulder.
- Angle the muzzle downward about 45 degrees and point it in a safe direction or the general direction of likely enemy contact (see fig. 3-20).

Figure 3-20. Alert Carry.

Ready Carry

A Marine carries the rifle at the ready if contact with the enemy is imminent. The ready allows immediate target engagement, but it is very tiring to maintain over a long period of time. A Marine performs the following steps to assume the ready:

- Place left hand on handguards, right hand around the pistol grip, the trigger finger straight along the receiver (see fig. 3-17), and the right thumb on top of the selector lever (see fig. 3-18).
- Place the buttstock in the shoulder.
- Point the muzzle in the direction of the enemy.

- Lower the sights to just below eye level so that a clear field of view is maintained so that a target may be identified (see fig. 3-21).

Figure 3-21. Ready Carry.

3009. Weapons Transports

Weapons transports are used to carry the rifle over the back or shoulders when moving for long periods; they provide a more relaxed position for walking. Weapons transports are used if no immediate threat is present. They are also used whenever one or both hands are needed for other work.

Strong Side Sling Arms Transport (Muzzle Up)

To assume the strong side sling arms (muzzle up) transport from the tactical carry, a Marine performs the following steps (see fig. 3-22):

- Release the hold on the pistol grip.
- Lower the buttstock and bring the rifle to a vertical position.
- With the right hand, grasp the sling above the left forearm.

Figure 3-22. Strong Side Sling Arms Transport (Muzzle Up).

- With the left hand, guide the rifle around the right shoulder.
- With the right hand, apply downward pressure on the sling. This stabilizes the rifle on the shoulder.

Weak Side Sling Arms Transport (Muzzle Down)

The weak side sling arms (muzzle down) transport can be used in inclement weather to keep moisture out of the rifle's bore. To assume this transport from the tactical carry, a Marine performs the following steps (see fig. 3-23):

- Release the hold on the pistol grip.
- With the left hand, rotate muzzle down and bring the rifle to a vertical position on the left side of the body. The pistol grip is pointed outboard.
- With right hand, place sling on left shoulder.
- Grasp sling above the waist with the left hand.
- With the left hand, apply downward pressure on the sling. This stabilizes the rifle on the shoulder.

Figure 3-23. Weak Side Sling Arms Transport (Muzzle Down.)

Cross Body Sling Arms Transport

A Marine uses the cross body sling arms transport if he requires both hands for work. The rifle is slung across the back with the muzzle up or down. Normally, the rifle is carried with the muzzle down to prevent pointing the muzzle in an unsafe direction.

To assume a cross body sling transport, a Marine performs the following steps from weak side sling arms (muzzle down) (see fig. 3-24):

Figure 3-24. Cross Body Sling Arms Transport (Muzzle Down).

- With the right hand, grasp the sling.
- With the left hand, grasp the handguards.
- Pull up on the rifle with both hands.
- Slide the sling over the head.
- Position the rifle so that it rests comfortably across the back.

To assume a cross body sling transport, a Marine performs the following steps from strong side sling arms (muzzle up) (see fig. 3-25):

- With the left hand, grasp the sling.
- With the right hand, grasp the pistol grip.
- Pull up on the rifle with both hands.
- Slide sling over head.
- Position the rifle so that it rests comfortably across the back.

Figure 3-25. Cross Body Sling Arms. Transport (Muzzle Up).

Note

Ensure the muzzle of the rifle is maintained in a safe direction when assuming this transport.

3010. Transferring the Rifle

Proper weapons handling is required every time a Marine picks up a weapon, passes a weapon to another Marine, or receives a weapon from another Marine. It is the responsibility of the Marine receiving or taking charge of a weapon to determine its condition. Depending on the situation, there are two procedures that can be used to transfer a rifle from one Marine to another Marine: show clear transfer and condition unknown transfer.

Show Clear Transfer

When time and the tactical situation permit, the Marine should transfer the rifle using the show clear transfer. To properly pass a rifle between Marines, the Marine handing off the rifle must perform the following procedures:

- Ensure the rifle is on safe.
- Remove the magazine if it is present.
- Lock the bolt to the rear.
- Visually inspect the chamber to ensure there is no ammunition present.
- Leave the bolt locked to the rear and hand the weapon to the other Marine.

The Marine receiving the weapon must:

- Ensure the rifle is on safe.
- Visually inspect the chamber to ensure there is no ammunition present.
- Release the bolt catch and observe the bolt going forward on an empty chamber.
- Close the ejection port cover.

Condition Unknown Transfer

There are times when time or the tactical situation does not permit a show clear transfer of the rifle. The procedures for the condition unknown transfer are conducted by a Marine when he takes charge of a rifle in any situation when the condition of the rifle is unknown (e.g., an unmanned rifle from a casualty, a rifle stored in a rifle rack). To properly take charge of a rifle when its condition is unknown, the Marine must perform the following procedures:

- Ensure the rifle is on safe.
- Conduct a chamber check to determine the condition of the weapon (see para. 3003).
- Remove the magazine and observe if ammunition is present. If time permits, count the rounds.
- Insert the magazine into the magazine well.
- Close the ejection port cover.

3011. Clearing Barrel Procedures

Purpose of a Clearing Barrel

The sole purpose of a clearing barrel is to provide a safe direction in which to point a weapon when loading; unloading; and unloading and showing clear. See figure 3-26 on page 3-12. Clearing barrel procedures are identical to the weapons handling procedures for the rifle for the loading; unloading; and unloading and showing clear.

Figure 3-26. Clearing Barrel.

Procedures for "Load"

On the command "Load," the Marine will perform the following steps to take the rifle from Condition 4 to Condition 3:

- With a straight trigger finger, point the rifle in the clearing barrel.
- Ensure the rifle is on safe.
- Withdraw a magazine from the magazine pouch.
- Observe the magazine to ensure it is filled.
- Fully insert the magazine into the magazine well.
- Without releasing the magazine, tug downward on the magazine to ensure it is seated.
- Fasten the magazine pouch.
- Close the ejection port cover.

Procedures for "Make Ready"

If standard operating procedures (SOP) or rules of engagement (ROE) require the rifle to be carried in Condition 1, the Marine will "Make Ready" at the clearing barrel. On the command "Make Ready," the Marine will perform the following steps to take the rifle from Condition 3 to Condition 1:

- Pull the charging handle to the rear and release. There are two methods of doing this:
  - Grip the pistol grip firmly with the right hand and pull the charging handle with the left hand to its rearmost position and release.
  - Or grip the handguards firmly with the left hand and pull the charging handle with the right hand to its rearmost position and release.
- To ensure ammunition has been chambered, conduct a chamber check.
- Check the sights (to ensure proper BZO setting, correct rear sight aperture, etc.).
- Close the ejection port cover.

Procedures for "Unload"

On the command "Unload," the Marine will perform the following steps to take the rifle from any condition to Condition 4:

- With a straight trigger finger, point the rifle in the clearing barrel.
- Ensure the weapon is on safe.
- Remove the magazine from the rifle and retain it on your person.
- While cupping the left hand under the ejection port, rotate weapon until the ejection port is facing down.
- Pull the charging handle to the rear and catch the round in the left hand (see fig. 3-17).
- Lock the bolt to the rear.
- Put weapon on safe if not already on safe.
- Ensure that the chamber is empty and that no ammunition is present.
- Release the charging handle and observe the bolt moving forward on an empty chamber.
- Close the ejection port cover.
- Check the sights (for proper BZO setting, correct rear sight aperture, etc.).
- Place any ejected round into the magazine and return the magazine to the magazine pouch and close the magazine pouch.

Procedures for "Unload and Show Clear"

On the command "Unload and Show Clear," the Marine will perform the following steps to take the rifle from any condition to Condition 4:

- With a straight trigger finger, point the rifle in the clearing barrel.
- Ensure the weapon is on safe.
- Remove the magazine from the rifle and retain it on your person.
- While cupping the left hand under the ejection port, rotate the weapon until the ejection port is facing down.
- Pull the charging handle to the rear and catch the round in the left hand.
- Lock the bolt to the rear.

- Ensure the chamber is empty and no ammunition is present.
- Have an observer inspect the weapon to ensure no ammunition is present. The observer:
  - Visually inspects chamber to ensure it is empty, no ammunition is present, and the magazine is removed.
  - Ensures the weapon is on safe.
  - Acknowledges the rifle is clear.
- After receiving acknowledgment from the observer that the rifle is clear, the Marine releases the bolt catch and observes the bolt moving forward on an empty chamber.
- Close the ejection port cover.
- Check the sights (for proper BZO setting, correct rear sight aperture, etc.).
- Place any ejected round into the magazine and return the magazine to the magazine pouch and close the magazine pouch.

CHAPTER 4. FUNDAMENTALS OF MARKSMANSHIP

The fundamentals of marksmanship are aiming, breathing, and trigger control. These techniques provide the foundation for all marksmanship principles and skills. For rifle fire to be effective, it must be accurate. A rifleman who merely sprays shots in the vicinity of the enemy produces little effect. The fundamentals of marksmanship, when applied correctly, form the basis for delivering accurate fire on enemy targets. These skills must be developed so that they are applied instinctively. During combat, the fundamentals of marksmanship must be applied in a time frame consistent with the size and the distance of the target. At longer ranges, the target appears to be smaller and a more precise shot is required to accurately engage the target. The fundamentals are more critical to accurate engagement as the range to the target increases. To be accurate at longer ranges, the Marine must take the time to slow down and accurately apply the fundamentals. At shorter ranges, the enemy must be engaged quickly before he engages the Marine. As the size of the target increases, and the distance to the target decreases, the fundamentals, while still necessary, become less critical to accuracy.

Note

☞ The procedures in this manual are written for right-handed Marines; left-handed Marines should reverse instructions as necessary.

4001. Aiming

Sight Alignment

Sight alignment is the relationship between the front sight post and rear sight aperture and the aiming eye. This relationship is the most critical to aiming and must remain consistent from shot to shot. To achieve correct sight alignment (see fig. 4-1):

- Center the tip of the front sight post vertically and horizontally in the rear sight aperture.
- Imagine a horizontal line drawn through the center of the rear sight aperture. The top of the front sight post will appear to touch this line.
- Imagine a vertical line drawn through the center of the rear sight aperture. The line will appear to bisect the front sight post.

Sight Picture

Sight picture is the placement of the tip of the front sight post in relation to the target while maintaining sight alignment. Correct sight alignment but improper sight placement on the target will cause the bullet to impact the target incorrectly on the spot where the sights were aimed when the bullet exited the muzzle.

To achieve correct sight picture, place the tip of the front sight post at the center of the target while maintaining sight alignment (see fig. 4-2). Center mass is

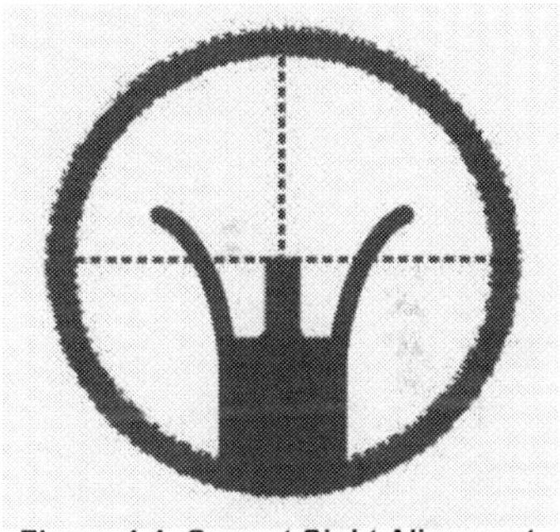

Figure 4-1. Correct Sight Alignment.

Figure 4-2. Correct Sight Picture.

Figure 4-3. Examples of Correct Sight Picture.

the correct aiming point so that point of aim/point of impact is achieved.

The sighting system for the M16A2 rifle is designed to work using a center mass sight picture.

In combat, targets are often indistinct and oddly shaped. The center mass hold provides a consistent aiming point (see fig. 4-3).

Importance of Correct Sight Alignment

A sight alignment error results in a misplaced shot. The error grows proportionately greater as the distance to the target increases. An error in sight picture, however, remains constant regardless of the distance to the target. See figure 4-4.

Factors Affecting Sight Alignment and Sight Picture

Stock Weld

Stock weld is the point of firm contact between the cheek and the stock of the rifle. See figure 4-5. The head should be as erect as possible to enable the aiming eye to look straight through the rear sight aperture. If the position of the Marine's head causes him to look across the bridge of his nose or out from under his eyebrow, the eye will be strained. The eye functions best in its natural forward position. Changing the placement of the cheek up or down on the stock from shot to shot may affect the zero on the rifle due to the perception of the rear sight aperture. A consistent and proper stock weld is critical to the aiming process because it provides consistency in eye relief, which affects the ability to align the sights.

Eye Relief

Eye relief is the distance between the rear sight aperture and the aiming eye. See figure 4-6. Normal eye relief is two to six inches from the rear sight aperture. The distance between the aiming eye and the rear sight aperture depends on the size of the Marine and the firing position. While eye relief varies slightly from one position to another, it is important to have the same eye relief for all shots fired from a particular position.

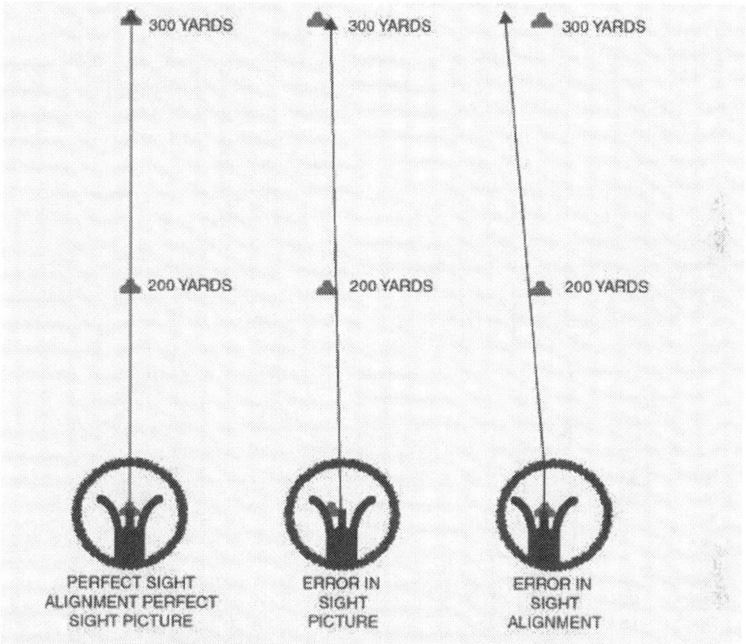

Figure 4-4. Importance of Correct Sight Alignment.

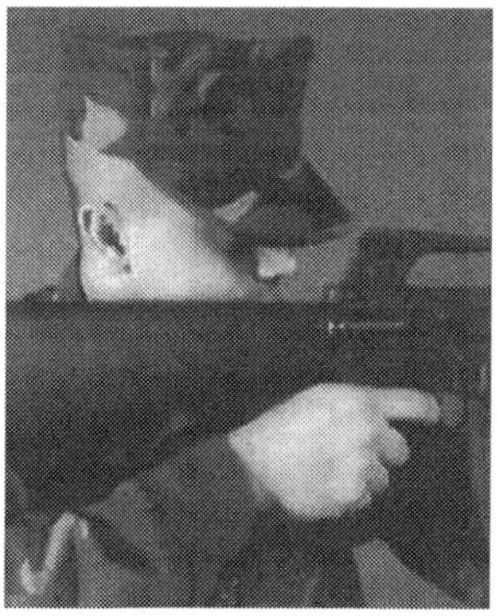

Figure 4-5. Stock Weld.

Figure 4-6. Proper Eye Relief.

If the eye is too close to the rear sight aperture, it will be difficult to line up the front sight post in the rear sight aperture (see fig. 4-7). Moving the eye back from the rear sight aperture will make the aperture appear smaller and allow the tip of the front sight post to be easily lined up inside the rear sight aperture.

Figure 4-7. Shortened Eye Relief.

If the eye is too far from the rear sight aperture, it will be difficult to acquire the target and to maintain a precise aiming point (see fig. 4-8).

Wearing of Glasses

Wearing glasses can alter the perception of sight alignment and sight picture. If wearing glasses, it is critical to look through the optic center of the lens.

Acquiring and Maintaining Sight Alignment and Sight Picture

The human eye can focus clearly on only one object at a time. For accurate shooting, it is important to focus on the tip of the front sight post. When the shot is fired, focus must be on the tip of the front sight post; peripheral vision will include the rear sight and the target. The rear sight and the target will appear blurry. Staring or fixing the vision on the front sight post for longer than a few seconds can distort the image, making it difficult to detect minute errors in sight alignment.

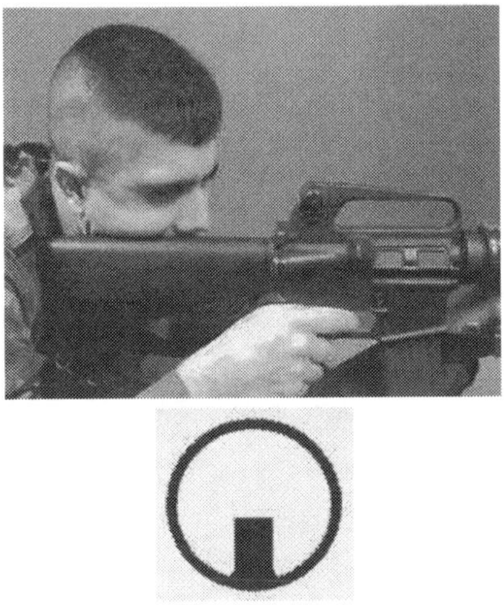

Figure 4-8. Extended Eye Relief.

Proper stock weld and placement of the rifle butt in the shoulder aid in establishing sight alignment quickly. The rifle butt's placement in the shoulder serves as the pivot point for presenting the rifle up to a fixed point on the cheek (stock weld). During combat, a Marine will look at the target as the rifle is presented. As rifle sights become level with the aiming eye, a Marine visually locates the target through the rear sight aperture. As the rifle settles, a Marine's focus shifts back to the front sight post to place the tip of the post on the target and obtain sight alignment and sight picture. To maintain sight alignment and sight picture, the Marine's focus should shift repeatedly from the front sight post to the target until correct sight alignment and sight picture are obtained. This enables the detection of minute errors in sight alignment and sight picture.

Size and Distance to the Target

During combat, the fundamentals of marksmanship must be applied in a time frame consistent with the size and the distance to the target.

Long-range Engagements

At longer ranges, the target appears smaller and a more precise shot is required to accurately engage the target. Sight alignment and sight picture are more critical to accurate engagement as the range to the target increases. To be accurate at longer ranges, the Marine must take the time to slow down and accurately apply the fundamentals.

As the distance to the target increases, the front sight post covers more of the target, making it difficult to establish a center of mass hold (see fig. 4-9). Since the Marine must see the target to engage it, there is a tendency to look at the target by lowering the tip of the front sight post. This causes shots to impact low or miss the target completely. A Marine must consciously aim at the center of mass and attempt to maintain a center mass sight picture.

Short-range Engagements

At shorter ranges, the enemy must be engaged quickly before he engages the Marine. As distance to the target decreases, the size of the target appears to increase, and sight alignment becomes less critical to accuracy. At very short ranges, a deviation in sight alignment can still produce accurate results as long as the tip of the front sight post is in the rear sight aperture and on the target (see fig. 4-10). The time required to engage a target is unique to each individual. Although a Marine must engage the target rapidly, some semblance of sight alignment is still required to be accurate.

4002. Breath Control

Proper breath control is critical to the aiming process. Breathing causes the body to move. This movement transfers to the rifle making it impossible to maintain proper sight picture. Breath control allows the Marine to fire the rifle at the moment of least movement.

Breath Control During Long-range or Precision Fire (Slow Fire)

It is critical that Marines interrupt their breathing at a point of natural respiratory pause before firing a long-range shot or a precision shot from any distance. A respiratory cycle lasts 4 to 5 seconds. Inhaling and exhaling each require about 2 seconds. A natural pause of 2 to 3 seconds occurs between each respiratory cycle. The pause can be extended up to 10 seconds. During the pause, breathing muscles are relaxed and the sights settle at their natural point of aim. To minimize movement, Marines must fire the shot during the natural respiratory pause. The basic technique is as follows:

- Breathe naturally until the sight picture begins to settle.
- Take a slightly deeper breath.
- Exhale and stop at the natural respiratory pause.
- Fire the shot during the natural respiratory pause.

Note

If the sight picture does not sufficiently settle to allow the shot to be fired, resume normal breathing and repeat the process.

Breath Control During All Other Combat Situations

A Marine in a combat environment may not have the time to fire a shot during the natural respiratory pause.

Figure 4-9. Sight Picture at Long-range Engagements.

Figure 4-10. Sight Picture at Short-range Engagements.

It may be necessary to take several deep breaths before holding the breath. A Marine should not make an exaggerated effort to perform breath control. A natural respiratory pause will help stabilize the shooter's sight picture. The basic technique is as follows:

- Take a deep breath filling the lungs with oxygen.

Note

It may be necessary to take several deep breaths quickly before holding the breath.

- Hold the breath and apply pressure to the trigger.
- Fire the shots.

4003. Trigger Control

Trigger control is the skillful manipulation of the trigger that causes the rifle to fire without disturbing sight alignment or sight picture. Controlling the trigger is a mental process, while pulling the trigger is a physical process.

Grip

A firm grip is essential for effective trigger control. The grip is established before starting the application of trigger control and it is maintained through the duration of the shot. To establish a firm grip on the rifle, position the "V" formed between the thumb and index finger on the pistol grip behind the trigger. The fingers and the thumb are placed around the pistol grip in a location that allows the trigger finger to be placed naturally on the trigger and the thumb in a position to operate the safety. Once established, the grip should be firm enough to allow manipulation of the trigger straight to the rear without disturbing the sights. See figure 4-11.

Trigger Finger Placement

Correct trigger finger placement allows the trigger to be pulled straight to the rear without disturbing sight alignment. The trigger finger should contact the trigger naturally. The trigger finger should not contact the rifle receiver or trigger guard.

Figure 4-11. Grip.

Types of Trigger Control

There are two techniques of trigger control: uninterrupted and interrupted

Uninterrupted Trigger Control

The preferred method of trigger control in a combat environment is uninterrupted trigger control. After obtaining sight picture, the Marine applies smooth, continuous pressure rearward on the trigger until the shot is fired.

Interrupted Trigger Control

Interrupted trigger control is used at any time the sight alignment is interrupted or the target is temporarily obscured. An example of this is extremely windy conditions when the weapon will not settle, forcing the Marine to pause until the sights return to his aiming point. To perform interrupted trigger control:

- Move the trigger to the rear until an error is detected in the aiming process.
- When this occurs, stop the rearward motion on the trigger, but maintain the pressure on the trigger, until sight picture is achieved.
- When the sight picture settles, continue the rearward motion on the trigger until the shot is fired.

Resetting the Trigger

During recovery, release the pressure on the trigger slightly to reset the trigger after the first shot is delivered (indicated by an audible click). Do not remove the finger from the trigger. This places the trigger in position to fire the next shot without having to reestablish trigger finger placement.

4004. Follow-Through/Recovery

Follow-Through

Follow-through is the continued application of the fundamentals until the round has exited the barrel. In combat, follow-through is important to avoid altering the impact of the round by keeping the rifle as still as possible until the round exits the barrel.

Recovery

It is important to get the rifle sights back on the target for another shot. This is known as recovery. Shot recovery starts immediately after the round leaves the barrel. To recover quickly, a Marine must physically bring the sights back on target as quickly as possible.

CHAPTER 5. RIFLE FIRING POSITIONS

In a combat environment, a Marine must be prepared to engage the enemy under any circumstance. There are four basic firing positions: prone, sitting, kneeling, and standing. These positions provide a stable foundation for effective shooting. Any firing position must provide stability, mobility, and observation of the enemy. During training, a Marine learns positions in a step-by-step process, guided by a series of precise movements until the Marine assumes a correct position. The purpose of this is to ensure that the Marine correctly applies all of the factors that assist him in holding the rifle steady. The Marine will gradually become accustomed to the feel of the positions through practice and eventually will be able to know instinctively whether his position is correct. In combat, it may not be possible to assume a textbook firing position due to terrain, available cover, engagement time, dispersion of targets, and other limiting factors. Modifications to the basic positions may have to be made to adjust to the combat environment. The Marine must strive to assume a position that offers stability for firing, maximum cover and concealment from the enemy, and maximum observation of the target.

Note

+The procedures in this manual are written for right-handed Marines; left-handed Marines should reverse instructions as necessary.

5001. Selecting a Firing Position

The selection of a firing position (prone, sitting, kneeling, standing) is based on terrain, available cover, dispersion of targets, and other limiting factors. A Marine must select a position that offers stability, mobility, and observation.

Stability

A firing position must provide a stable platform for accurate and consistent shooting. If the position is solid, the front sight can be held steady and the rifle sights should recover after recoil to the same position on the target. This allows for rapid reengagement of the enemy. The prone position provides the most stability for firing, while the standing position provides the least stability.

Mobility

A firing position must provide a Marine with the mobility required to move to new cover or to another area. The standing position permits maximum mobility, and it also allows the most lateral movement for engagement of widely dispersed targets. The prone position allows the least mobility and allows limited lateral movement.

Observation of the Enemy

A firing position must limit a Marine's exposure to the enemy, yet allow observation of the enemy. Manmade structures and terrain features (e.g., vegetation, earth contours) often dictate the shooting position. The standing position normally provides the best field of view, but it usually allows the most exposure to the enemy. The prone position normally allows the least exposure, but it usually provides a limited field of view.

5002. Types and Uses of the Rifle Web Sling

The rifle sling, when adjusted properly, provides maximum stability for the weapon, and helps hold the front sight still and reduce the effects of the rifle's recoil. Once a sling adjustment is found that provides maximum control of the weapon, the same sling adjustment should be maintained. Varying the sling tension extensively will affect the strike of the bullet, which will make maintaining a BZO difficult. Using the same sling adjustment will ensure the accuracy of rounds on target. There are two basic types of rifle sling adjustments: the hasty sling and loop sling.

Note

In training situations, the parade sling may be used. See paragraph 5008.

Hasty Sling

Application

The hasty sling is used in all firing positions. The hasty sling is advantageous in combat because it can be acquired quickly and it provides added stability to the rifle. The same sling setting can be used for all firing positions. If properly adjusted, the hasty sling supports the weight of the weapon, provides stability for the rifle, and reduces the effects of the rifle's recoil. When using the hasty sling, controlled muscle tension is applied to offer resistance against the sling, enabling the rifle sights to be held steady.

Donning the Hasty Sling

A Marine performs the following steps to form a hasty sling:

- Hold the rifle vertical with the barrel pointing upward.
- Unhook the J-hook from the lower sling swivel.
- Loosen the sling keeper.
- Adjust the sling until the J-hook hangs below the rifle butt. (The distance will vary based on the individual Marine, but the J-hook will usually hang approximately 3 to 10 inches below the rifle butt.) Secure the sling keeper. See figure 5-1.

Figure 5-1. J-hook Location.

- Turn the sling a half turn outboard to allow the sling to lay flat against the arm.
- Attach the J-hook to the lower sling swivel so the open end of the J-hook faces outboard, away from the rifle. See figure 5-2.
- While holding the rifle with the right hand, place the left arm through the sling near the lower sling swivel. Slide the arm up through the sling below the half twist. The sling makes contact low on the arm just below the triceps, above the elbow. The sling lies flat on the back of the arm.

Figure 5-2. J-hook Turned Outboard.

- With the left hand, grasp the handguard by pinching it in the "V" formed by the thumb and forefinger. The sling lies flat against the back or side of the wrist or on the arm near the wrist. See figure 5-3.

Figure 5-3. Sling Against the Wrist.

- Move the left hand as required to level the rifle with the line of sight. Placement of the forward hand controls the tension on the sling between the back of the wrist or arm and the upper sling swivel (see fig. 5-4). This hand placement, with a straight locked wrist, will cause the sling to pull straight under the handguards and serves to stabilize the front sight of the rifle.

- Move the feed end of the sling in or out of the sling keeper to adjust the hasty sling. Sling tension is in-

Figure 5-4. Position of Forward Hand.

creased by pushing the elbow outboard. See figure 5-5. (This enables one sling setting to fit all positions.) It is important for the hasty sling to be adjusted so it supports the rifle. The sling setting must allow the left elbow to push outboard against the sling so the elbow is not inverted under the rifle.

Figure 5-5. Position of Left Elbow.

- Locate the sling keeper near the feed end of the sling and secure so the backside or flat end of the sling keeper is against the arm.

Loop Sling

Application

The loop sling provides the greatest amount of stability during firing. This stability allows the Marine to perfect marksmanship fundamentals. A loop sling takes longer to don or remove than a hasty sling. Therefore, it has limited combat application and is best used where stability of hold is needed for a precision or long-range shot. The loop sling is used in the prone, sitting, and kneeling positions.

Donning the Loop Sling

- Place the rifle butt on the right hip and cradle the rifle in the right arm.
- Disconnect the J-hook from the lower sling swivel.
- With the M-buckle near the hook, feed the sling through the top of the M-buckle to form a loop large enough to slip over the arm. See figure 5-6.
- Give the loop a half turn outboard and insert the left arm through the loop, positioning the loop above the biceps. The loop is high on the left arm above the biceps muscle in such a position that it does not transmit pulse beat to the rifle.
- Position the M-buckle on the outside of the left arm. See figure 5-7.
- Tighten the loop on the left arm, ensuring the M-buckle moves toward the center of the arm as the loop tightens. The sling must pull from the center of the arm to be properly positioned. In this way, as tension is applied to the sling in the firing position, the loop will tighten.

To adjust the sling for the proper length, loosen the sling keeper and pull up or down (toward or away) from the loop. This adjustment varies with every individual and every firing position. The loop should not be tightened excessively on the arm. If blood flow is restricted, excessive pulse beat is transmitted through the rifle sling to the rifle and causes a noticeable, rhythmic movement of the rifle sights. When this occurs, a stable hold at the desired aiming point is impossible to achieve.

Figure 5-6. Forming a Loop.

Figure 5-7. Position of M-buckle.

Tension on the rifle sling is correct when it causes the rifle butt to be forced rearward into the pocket of the shoulder. This serves to keep the butt plate in the shoulder pocket during recoil. To increase tension on the rifle sling, the sling must be shortened. To lessen the tension, the rifle sling must be lengthened.

Move the sling keeper toward the left arm and secure it. The sling keeper should be positioned near the feed end of the sling.

Place the left hand over the sling from the left side and under the rifle. The rifle handguard should rest in the "V" formed between the thumb and forefinger and across the palm of the hand.

Move the left hand as required to achieve desired sight picture. Adjust the length of the sling for proper sling tension and support. See figure 5-8.

Figure 5-8. Loop Sling Donned.

5003. Factors Common to All Shooting Positions

There are seven factors that are common to all shooting positions that affect the ability to hold the rifle steady, maintain sight alignment, and control the trigger. The way these factors are applied differs slightly for each position, but the principles of each factor remain the same.

Left Hand

Placement of the left hand affects placement of left elbow, eye relief, stock weld, and sling tension.

Hasty Sling

In a hasty sling configuration, the sling is attached to the upper and lower sling swivels of the rifle. When the left arm is placed in the hasty sling, tension created by the sling travels from side to side. The tension created by the sling affects how the position is established. There are fundamental differences between the application of the seven factors when using the hasty sling. The most obvious of these is placement of the left hand and the left elbow.

To maximize the support provided by the hasty sling, the left elbow should not be inverted and under the rifle. Instead, the elbow should be pushed outboard against the sling. To achieve this, the position of the shooter's body must almost face the target as opposed to being perpendicular to the target. In addition, the hasty sling must be loosened to allow the elbow to push out against the sling far enough so that the elbow is not under the rifle.

The tension on the sling created by the hasty sling causes the center of balance to change on the rifle. When the elbow is under the rifle with the hasty sling donned, the sling pulls down on the sling swivel disrupting the center of balance and causing the muzzle to drop. Therefore, the elbow must be pushed outboard.

Outboard tension on the sling by the elbow drives the buttstock into the pocket of the shoulder. To enable this, the sling must make contact on the arm just below the triceps, above the elbow. See figure 5-9.

Figure 5-9. Position of Left Elbow with Hasty Sling.

To stabilize the front sight of the rifle, the forward hand, wrist, and forearm should be straight with the wrist locked in place; the hand should be rotated up so the rifle rests in the "V" formed by the thumb and index finger; the fingers will not curl around the handguards. Instead, they will pinch the handguard slightly to keep the hand from slipping on the handguard during recoil. See figure 5-10.

Figure 5-10. Position of Left Hand with Hasty Sling.

When the wrist of the left hand is straight and locked, it creates resistance on the sling close to the muzzle. The sling is in contact with the back or side of the wrist or on the arm near the wrist. This resistance allows the front sight to be stabilized.

In contrast, when the rifle rests across the palm of the hand, the only resistance created is where the sling meets the triceps. Since the resistance is further from the muzzle of the rifle, it makes stabilizing the front sight more difficult.

Loop Sling

With the loop sling donned, the handguard of the rifle rests in the "V" formed by the thumb and index finger of the left hand. The rifle rests across the heel of the hand. The left elbow should be positioned under the weapon to create bone support and a consistent resistance to recoil. The fingers can curl around the handguard, but should apply only the minimum amount of pressure to prevent the hand from slipping on the handguard. See figure 5-11.

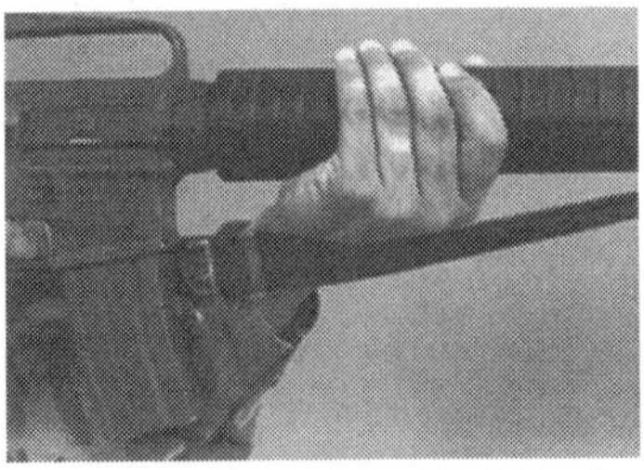

Figure 5-11. Position of Left Hand with Loop Sling.

Rifle Butt in the Pocket of the Shoulder

The rifle butt placed firmly in the pocket formed in the right shoulder provides resistance to recoil, helps steady the rifle, and prevents the rifle butt from slipping in the shoulder during firing. Consistent placement of the rifle butt in the shoulder pocket is essential to maintaining a BZO and firing tight shot groups.

Hasty Sling

With the hasty sling donned, the butt of the rifle is placed in the pocket of the shoulder. The body is squared to the target to provide a pocket for the butt of the weapon.

Outboard tension on the sling by the left elbow drives the buttstock into the pocket of the shoulder.

The rifle butt should be placed high in the shoulder to achieve a proper stock weld, allowing the Marine to bring the stock up to his head, rather than lower his head to the stock, which can degrade acquisition of sight alignment and sight picture. See figure 5-12.

Figure 5-12. Buttstock in the Shoulder with Hasty Sling.

Loop Sling

With the loop sling donned, the toe of the rifle butt is placed in the pocket of the shoulder. See figure 5-13.

Figure 5-13. Buttstock in the Shoulder with Loop Sling.

Grip of the Right Hand

Grasp the pistol grip with the right hand and place the forefinger on the trigger, with the thumb and remaining fingers wrapped around the pistol grip. See figure 5-14.

Figure 5-14. Grip of the Right Hand.

Firm rearward pressure should be exerted to help keep the rifle butt firmly in the shoulder, reducing the effects of recoil (this type of pressure is particularly true with the loop sling).

The trigger finger should be placed naturally on the trigger and care should be taken to ensure that the trigger finger can move independently without dragging on the side of the receiver. Proper placement of the right hand high on the pistol grip allows the trigger to be moved straight to the rear without disturbing sight alignment.

Right Elbow

The right elbow should be positioned naturally to provide balance to the position and create a pocket in the shoulder for the rifle butt. If the elbow is correctly positioned, it helps to form a pocket in the right shoulder where the rifle butt rests. The exact placement of the elbow varies with each shooting position but should remain consistent from shot to shot, ensuring the resistance to recoil remains constant. See figure 5-15.

Figure 5-15. Right Elbow.

Stock Weld

The placement of the shooter's cheek against the stock should remain firm and consistent from shot to shot. Consistency of stock weld is achieved through proper placement of the rifle butt in the pocket of the shoulder. A firm contact between the cheek and the stock enables consistent eye relief and enables the head and rifle to recoil as a single unit. Stock weld provides quick recovery between rapid fire shots, keeps the aiming eye centered in the rear sight aperture, and prevents the head from bouncing off the stock during recoil. The head should remain erect to allow the aiming eye to look straight through the rear sight aperture. See figure 5-16.

Figure 5-16. Stock Weld.

Breathing

Breathing causes chest movement and a corresponding movement in the rifle and its sights. Applying breath control will minimize this movement and the effect it has on aiming.

Muscular Tension/Relaxation

Muscular Tension–Hasty Sling

With the hasty sling donned, the shooter must apply an amount of controlled muscular tension in the left arm to keep the sling taut and stabilize the weapon sights. Resistance against the hasty sling controls the point at which the rifle sights will settle. The muscular tension is applied outward against the sling rather than in an effort to hold the rifle up. However, muscular tension should not be excessive to cause the shooter to shake, tremble, or experience fatigue.

Muscular Relaxation–Loop Sling

When using the loop sling, the muscles should be relaxed. Relaxation prevents undue muscle strain and reduces excessive movement. If proper relaxation is achieved, natural point of aim and sight alignment are more easily maintained.

5004. Elements of a Good Shooting Position

There are three elements of a good shooting position that apply when using a loop sling: bone support, muscular relaxation, and natural point of aim. The three elements of a shooting position applied with the loop sling do not apply in the same way as when firing with a hasty sling. While some degree of bone support is still achieved with the hasty sling, muscular **tension** is applied rather than muscular **relaxation**. Natural point of aim, however, applies to both the loop sling and the hasty sling.

Bone Support

The body's skeletal structure provides a stable foundation to support the rifle's weight. A weak shooting position will not withstand a rifle's repeated recoil when firing at the sustained rate or buffeting from wind. To attain a correct shooting position, the body's bones must support as much of the rifle's weight as possible. Proper use of the sling provides additional support.

The weight of the weapon should be supported by bone rather than muscle because muscles fatigue whereas bones do not.

By establishing a strong foundation for the rifle utilizing bone support, the Marine can relax as much as possible while minimizing weapon movement due to muscle tension.

Muscular Relaxation

Once bone support is achieved, muscles are relaxed. Muscular relaxation helps to hold the rifle steady and increase the accuracy of the aim. Muscular relaxation also permits the use of maximum bone support to create a minimum arc of movement and consistency in resistance to recoil. Muscular relaxation cannot be achieved without bone support. During the shooting process, the muscles of the body must be relaxed as much as possible. Muscles that are tense will cause excessive movement of the rifle, disturbing the aim. When proper bone support and muscular relaxation are achieved, the rifle will settle onto the aiming point, making it possible to apply trigger control and deliver a well-aimed shot.

Natural Point of Aim

The point at which the rifle sights settle when in a firing position is called the natural point of aim.

Since the rifle becomes an extension of the body, it may be necessary to adjust the position of the body until the rifle sights settle naturally on the desired aiming point on the target.

When in a shooting position with proper sight alignment, the position of the tip of the front sight post will indicate the natural point of aim. When completely relaxed, the tip of the front sight post should rest on the desired aiming point.

One method of checking for natural point of aim is to aim in on the target, close the eyes, take a couple of breaths, and relax as much as possible. When the eyes are opened, the tip of the front sight post should be positioned on the desired aiming point while maintaining sight alignment.

For each shooting position, specific adjustments will cause the rifle sights to settle center mass, achieving a natural point of aim.

In all positions, the natural point of aim can be adjusted by—

- Varying the placement of the left hand in relation to the handguards.
- Moving the left hand forward on the handguards to lower the muzzle of the weapon, causing the sights to settle lower on the target.
- Moving the left hand back on the handguards to raise the muzzle of the weapon, causing the sights to settle higher on the target.
- Varying the placement of the stock in the shoulder.
- Moving the stock higher in the shoulder to lower the muzzle of the weapon, causing the sights to settle lower on the target.
- Moving the stock lower in the shoulder to raise the muzzle of the weapon, causing the sights to settle higher on the target.
- Natural point of aim can be adjusted right or left by adjusting body alignment in relation to the target.

In the prone position, if the natural point of aim is above or below the desired aiming point, move the body slightly forward or back using the left elbow as a pivot and by digging the toes in.

- Pushing the body forward causes the sights to settle lower on the target.
- Pulling the body backward causes the sights to settle higher on the target.

In the kneeling and sitting positions, natural point of aim can be adjusted by varying the placement of the left elbow on the knee.

- Moving the left elbow forward on the knee lowers the muzzle of the weapon, causing the sights to settle lower on the target.
- Moving the left elbow back on the knee raises the muzzle of the weapon, causing the sights to settle higher on the target.

5005. Prone Position

Application

The prone position provides a very steady foundation for shooting and presents a low profile for maximum concealment. However, the prone position is the least mobile of the shooting positions and may restrict a Marine's field of view for observation. In this position, the Marine's weight is evenly distributed on the elbows, providing maximum support and good stability for the rifle.

Assuming the Prone Position

The prone position can be assumed by either moving forward or dropping backward into position, depending on the combat situation.

Moving Forward into Position

To move forward into the prone position, the Marine performs the following steps:

- With the hasty sling donned, stand erect and face the target, keeping the feet a comfortable distance apart (approximately shoulder width).
- Place the left hand on the handguard, the right hand on the pistol grip.
- Lower yourself into position by dropping to both knees (see fig. 5-17).

Figure 5-17. Dropping to Both Knees.

- Then shift the weight forward to lower the upper body to the ground using the right hand to break the forward motion. See figure 5-18.

Figure 5-18. Moving Forward Into Position.

Dropping Back into Position

It may be necessary to drop backward into position to avoid crowding cover, or to avoid covering uncleared terrain. To drop back into the prone position, the Marine performs the following steps:

- Face the target.
- Place the left hand on the handguard, the right hand on the pistol grip.
- Squat to the ground and break the fall with either hand (see fig. 5-19).
- Kick both legs straight to the rear (see fig. 5-20).
- If the fall was broken using the left hand, reestablish the hasty sling.

Figure 5-19. Breaking the Fall.

Figure 5-20. Kicking Back Into Position.

Straight Leg Prone Position with the Hasty Sling

Apply the seven factors to this position (para. 5003). To assume the straight leg prone position with the hasty sling, either move forward or drop back into position (see figs. 5-21, 5-22, and 5-23):

- Once on the ground, extend your left elbow in front of you. Stretch your legs out behind you. Spread the feet a comfortable distance apart with the toes pointing outboard and the inner portion of the feet in contact with the ground.
- As much of the body mass should be aligned directly behind the rifle as possible.
- If body alignment is correct, weapon recoil is absorbed by the whole body and not just the shoulder.
- Grasp the pistol grip with the right hand and place the rifle butt in the right shoulder pocket.
- Lower the head and place the cheek firmly against the stock to allow the aiming eye to look through the rear sight aperture.
- Rotate the left hand up, pinching the handguard between the thumb and forefinger.
- Slide both elbows outboard on the ground so there is outboard tension against the sling (moving the elbows out tightens the sling) and both shoulders are level. The elbows should provide a tripod of support with the body.
- Adjust the position of the left hand on the handguard to allow the sling to support the weapon and the front sight to be centered in the rear sight aperture. To adjust for elevation:
  - Move the left hand rearward or forward on the handguards (moving the hand rearward elevates the muzzle).
  - Open or close "V" of the left hand for small adjustments. (closing the "V" elevate the muzzle).
- To adjust for a minor cant in the rifle, rotate the handguard left or right in the "V" formed by the thumb and forefinger by rotating the pistol grip left or right.

Straight Leg Prone Position with the Hasty Sling

Figure 5-21. Front View.

Figure 5-22. Left View.

Figure 5-23. Right View.

Straight Leg Prone Position with the Loop Sling

Figure 5-24. Front View.

Figure 5-25. Left View.

Figure 5-26. Right View.

Straight Leg Prone Position with the Loop Sling

Apply the three elements and seven factors to this position (para. 5004). To assume straight leg prone position with the loop sling, either move forward or drop back into position (see figs. 5-24, 5-25, and 5-26):

- Once on the ground, roll the body to the left side as you extend and invert the left elbow on the ground. Stretch your legs out behind you. Spread the feet a comfortable distance apart with the toes pointing outboard and the inner portion of the feet in contact with the ground.
- As much of the body mass should be aligned directly behind the rifle as possible.
- If body alignment is correct, weapon recoil is absorbed by the whole body and not just the shoulder.
- Grasp the rifle butt with the right hand and place the rifle butt into the right shoulder pocket.
- Grasp the pistol grip with the right hand.
- Rotate the body to the right while the elbow is lowered to the ground so the shoulders are level. The right hand pulls and holds the rifle in the shoulder.
- Lower the head and place the cheek firmly against the stock to allow the aiming eye to look through the rear sight aperture.
- Move the left hand to a location under the handguard, which provides maximum bone support and stability for the weapon. This may require that you remove the rifle from the shoulder to reposition the left hand.

Cocked Leg Prone Position with the Hasty Sling

Figure 5-27. Front View.

Figure 5-28. Left View.

Figure 5-29. Right View.

Cocked Leg Prone Position with the Hasty Sling

Apply the seven factors to this position (para. 5003). To assume the cocked leg prone position with the hasty sling, either move forward or drop back into position (see figs. 5-27, 5-28, and 5-29):

- Once on the ground, roll the body to the left side. The left leg is stretched out behind you, almost in a straight line. This allows the mass of the body to be placed behind the rifle to aid in absorbing recoil.
- Turn the toe of the left foot inboard so the outside of the left leg and foot are in contact with the ground. Bend the right leg and draw it up toward the body to a comfortable position. Turn the right leg and foot outboard so the inside of the right boot is in contact with the ground. Cocking the leg will raise the diaphragm, making breathing easier.
- Grasp the pistol grip with the right hand and place the rifle butt in the right shoulder pocket.
- Lower the head and place the cheek firmly against the stock to allow the aiming eye to look through the rear sight aperture.
- Rotate the left hand up, pinching the handguard between the thumb and forefinger.
- Roll the body to the right while lowering the right elbow to the ground. Slide both elbows outboard on the ground so there is outboard tension against the sling (moving the elbows out tightens the sling). The right shoulder is higher than the left shoulder in the cocked leg position.
- Adjust the position of the left hand on the handguard to allow the sling to support the weapon and the front sight to be centered in the rear sight aperture. To adjust for elevation—
  - Move the left hand rearward or forward on the handguards (moving the hand rearward elevates the muzzle).
  - Open or close "V" of the left hand for small adjustments (closing the "V" elevates the muzzle).

To adjust for a minor cant in the rifle, rotate the handguard left or right in the "V" formed by the thumb and forefinger by rotating the pistol grip left or right.

Cocked Leg Prone Position with the Loop Sling

Apply the three elements and seven factors to this position (para. 5004). To assume the cocked leg prone position with the loop sling, either move forward or drop back into position (see figs 5-30, 5-31, and 5-32):

- Once on the ground, roll the body to the left side and extend and invert the left elbow on the ground. The left leg is stretched out behind you, almost in a straight line. This allows the mass of the body to be placed behind the rifle to aid in absorbing recoil.
- Turn the toe of the left foot inboard so the outside of the left leg and foot are in contact with the ground. Then bend the right leg and draw it up toward the body to a comfortable position. Turn the right leg and foot outboard so the inside of the right boot is in contact with the ground. Cocking the leg will raise the diaphragm, making breathing easier.
- Grasp the rifle butt with the right hand and place the rifle butt into the right shoulder pocket.

Cocked Leg Prone Position with the Loop Sling

Figure 5-30. Front View.

Figure 5-31. Left View.

Figure 5-32. Right View.

- Grasp the pistol grip with the right hand.
- Roll the body to the right while lowering the right elbow to the ground. The right shoulder is higher than the left shoulder in the cocked leg position.
- Lower the head and place the cheek firmly against the stock to allow the aiming eye to look through the rear sight aperture.
- Move the left hand to a location under the handguard which provides maximum bone support and stability for the weapon.

5006. Sitting Position

There are three variations of the sitting position: crossed ankle, crossed leg, and open leg. Experiment with all the variations and select the position that provides the most stability for firing.

Although the sitting position provides an extremely stable base, it limits lateral movement and maneuver ability. It has several variations that can be adapted to the individual Marine. The sitting position provides greater elevation than the prone position while still having a fairly low profile.

Crossed Ankle Sitting Position with the Hasty Sling

The crossed ankle sitting position is an extremely stable shooting position. This position places most of the body's weight behind the weapon and aids in quick shot recovery. Apply the seven factors to this position (para. 5003). To assume the crossed ankle sitting position with the hasty sling (see figs. 5-33, 5-34, 5-35)—

- Square the body to the target.
- Grasp the handguard with the left hand.
- Bend at knees and break the fall with the right hand.
- Push backward with the feet to extend the legs and place the buttocks on the ground.
- Cross the left ankle over the right ankle.
- Grasp the pistol grip with the right hand and place the rifle butt in the right shoulder pocket.
- Lower the head and place the cheek firmly against the stock to allow the aiming eye to look through the rear sight aperture.
- Rotate the left hand up, pinching the handguard between the thumb and forefinger.
- Bend forward at the waist and place the flat portion of the back of the left arm, just above the elbow, on the left leg just below the knee.
- Place the right elbow on the inside of the right knee.
- Adjust the position to adjust sling tension. To tighten the sling: square the body more to the target, or move the left elbow out, or draw the feet up slightly toward the body.
- Adjust the position of the left hand to allow the sling to support the weapon and the front sight to be centered in the rear sight aperture.

To adjust for elevation:

- Move left hand rearward or forward on handguards (moving the hand rearward elevates the muzzle).
- Open or close the "V" of the left hand for small adjustments (closing the "V" elevates the muzzle).

To adjust for a minor cant in the rifle, rotate the handguard left or right in the "V" formed by the thumb and forefinger by rotating the pistol grip left or right.

Crossed Ankle Sitting Position with the Hasty Sling

Figure 5-33. Left View.

Figure 5-34. Front View.

Figure 5-35. Right View.

Crossed Ankle Sitting Position with the Loop Sling

Figure 5-36. Right View.

Figure 5-37. Front View.

Crossed Ankle Sitting Position with the Loop Sling

Apply the three elements and seven factors to this position (para. 5004). To assume crossed ankle sitting position with the loop sling (see figs. 5-36 and 5-37)—

- Position the body at approximately a 30-degree angle to the target.
- Place the left hand under the handguard.
- Bend at knees and break the fall with the right hand.
- Push backward with the feet to extend the legs and place the buttocks on the ground.
- Cross the left ankle over the right ankle.
- Bend forward at the waist and place the left elbow on the left leg below the knee.
- Grasp the rifle butt with the right hand and place the rifle butt into the right shoulder pocket.
- Grasp the pistol grip with the right hand.
- Lower right elbow to the inside of the right knee.
- Lower the head and place the cheek firmly against the stock to allow the aiming eye to look through the rear sight aperture.
- Move the left hand to a location under the handguard, which provides maximum bone support and stability of the weapon.

Crossed Leg Sitting Position with the Hasty Sling

The crossed leg sitting position provides a medium base of support and places some of the body's weight behind the weapon for quick recovery after each shot. Apply the seven factors to this position (para. 5003). To assume the crossed leg sitting position with the hasty sling (see figures 5-38, 5-39, and 5-40)—

- Square the body to the target.
- Grasp the handguard with the left hand.
- Bend at the knees while breaking the fall with the right hand.
- Place the buttocks on the ground as close to the feet as you comfortably can.
- Cross the left leg over the right leg.
- Grasp the pistol grip with the right hand and place the rifle butt in the right shoulder pocket.

Crossed Leg Sitting Position with the Hasty Sling

Figure 5-38. Left View.

Figure 5-39. Front

Figure 5-40. Right View.

- Lower the head and place the cheek firmly against the stock to allow the aiming eye to look through the rear sight aperture.
- Rotate the left hand up, pinching the handguard between the thumb and forefinger.
- Bend forward at the waist while placing the left elbow into the bend of the knee or placing the flat portion of the back of the left arm, just above the elbow, in front of the knee.
- Place the right elbow on the inside of the right knee.
- Adjust the position to adjust sling tension. Squaring the body more to the target or drawing the feet closer together tightens the sling by forcing the left elbow outboard.
- Adjust the position of the left hand to allow the sling to support the weapon and the front sight to be centered in the rear sight aperture.

To adjust for elevation:

- Move left hand rearward or forward on handguards (moving the hand rearward elevates the muzzle).
- Open or close the "V" of the left hand for small adjustments (closing the "V" elevates the muzzle).

To adjust for a minor cant in the rifle, rotate the handguard left or right in the "V" formed by the thumb and forefinger by rotating the pistol grip left or right.

Crossed Leg Sitting Position with the Loop Sling

Apply the three elements and seven factors to this position (para. 5004). To assume crossed leg sitting position with loop sling (see figs. 5-41, 5-42, and 5-43)—

- Position body at a 45- to 60-degree angle to target.
- Place the left hand under the handguard.
- Cross the left leg over the right leg.
- Bend at the knees while breaking the fall with the right hand.
- Place the buttocks on the ground as close to the crossed legs as you comfortably can.
- Bend forward at the waist while placing the left elbow on the left leg into the bend of the knee.
- Grasp the rifle butt with the right hand and place the rifle butt into the right shoulder pocket.
- Grasp the pistol grip with the right hand.
- Lower right elbow to the inside of the right knee.
- Lower the head and place the cheek firmly against the stock to allow the aiming eye to look through the rear sight aperture.
- Move the left hand to a location under the handguard which provides maximum bone support and stability of the weapon.

Open Leg Sitting Position with the Hasty Sling

The open leg sitting position provides a medium base of support and is most commonly used when firing from a forward slope. Apply the seven factors to this position (para. 5003). To assume the open leg sitting position with the hasty sling (see figs. 5-44, 5-45, and 5-46)—

- Square the body to the target.
- Place the feet approximately shoulder width apart.
- Grasp the handguard with the left hand.
- Bend at the knees while breaking the fall with the right hand.
- Push backward with the feet to extend the legs and place the buttocks on the ground.
- Grasp the pistol grip with the right hand and place the rifle butt in the right shoulder pocket.
- Lower the head and place the cheek firmly against the stock to allow the aiming eye to look through the rear sight aperture.

Crossed Leg Sitting Position with the Loop Sling

Figure 5-41. Left View.

Figure 5-42. Front View.

Figure 5-43. Right View.

Open Leg Sitting Position with the Hasty Sling

Figure 5-44. Left View.

Figure 5-45. Front View.

Figure 5-46. Right View.

- Rotate the left hand up, pinching the handguard between the thumb and forefinger.
- Place the flat portion of the back of the left arm, just above the elbow, in front of the left knee.
- Place the right elbow on the inside of the right knee or place the flat portion of the back of the right arm, just above the elbow, in front of the knee.
- Adjust the position to adjust sling tension. Widening the stance tightens the sling by forcing the left elbow outboard. There must be some controlled muscular tension in the legs to hold them up and offer resistance to recoil.
- The position of the left hand to allow the sling to support the weapon and the front sight to be centered in the rear sight aperture.

To adjust for elevation:

- Move left hand rearward or forward on handguards (moving the hand rearward elevates the muzzle).
- Open or close the "V" of the left hand for small adjustments (closing the "V" elevates the muzzle).

To adjust for a minor cant in the rifle, rotate the handguard left or right in the "V" formed by the thumb and forefinger by rotating the pistol grip left or right.

Open Leg Sitting Position with the Loop Sling

Apply the three elements and seven factors to this position (para. 5004). To assume the open leg sitting position with the loop sling (see figs. 5-47, 5-48, and 5-49 on page 5-16)—

- Position the body at approximately a 30-degree angle to the target.
- Place the feet approximately shoulder width apart.
- Place the left hand under the handguard.
- Bend at the knees while breaking the fall with the right hand.
- Push backward with the feet to extend the legs and place the buttocks on the ground.
- Place the left elbow on the inside of the left knee.
- Grasp the rifle butt with the right hand and place the rifle butt into the right shoulder pocket.
- Lower right elbow to the inside of the right knee.
- Lower the head and place the cheek firmly against the stock to allow the aiming eye to look through the rear sight aperture.
- Move the left hand to a location under the handguard which provides maximum bone support and stability of the weapon.

5007. Kneeling Position

Description

The kneeling position is quick to assume and easy to maneuver from. It is usually assumed after initial engagement has been made from a standing position. It can easily be adapted to available cover. A tripod is formed by the left foot, right foot, and right knee when the Marine assumes the position, providing a stable foundation for shooting. The kneeling position also presents a higher profile to facilitate a better field of view as compared to the prone and sitting positions.

Assuming the Kneeling Position

The kneeling position can be assumed by either moving forward or dropping back into position, depending on the combat situation. For example, it may be necessary to drop back into position to avoid crowding cover, or to avoid covering uncleared terrain.

Open Leg Sitting Position with the Loop Sling

Figure 5-47. Left View.

Figure 5-48. Front View.

Figure 5-49. Right View.

Moving Forward into Position

To move forward into the kneeling position, the Marine steps forward toward the target with his left foot and kneels down on his right knee.

Dropping Back into Position

To drop back into the kneeling position, the Marine leaves his left foot in place and steps backward with his right foot and kneels down on his right knee.

High Kneeling Position with the Hasty Sling

Apply the seven factors to this position (para. 5003). To assume the high kneeling position with the hasty sling, either move forward or drop back into position (see figs. 5-50, 5-51, and 5-52):

- Square the body to the target.
- Keep the right ankle straight, with the toe of the boot in contact with the ground and curled under by the weight of the body.
- Place the right portion of the buttocks on or over the right heel. Contact with the heel provides more stability to the position, however, it is not mandatory that the buttocks make contact.
- Place the left foot forward to a point that allows the shin to be vertically straight. For the shin to be vertical, the heel should be directly under the knee. The left foot must be flat on the ground. To provide a wider base of support, slide the right knee and left foot outboard to form a tripod with the right foot.
- Grasp the pistol grip with the right hand and place the rifle butt in the right shoulder pocket.
- Lower the head and place the cheek firmly against the stock to allow the aiming eye to look through the rear sight aperture.
- Rotate the left hand up, pinching the handguard between the thumb and forefinger. The left hand will not grasp the slip ring or the magazine. The magazine must be on the inside of the left arm.
- Place the flat portion of the back of the upper left arm, just above the elbow, on the left knee or against the inside of the left knee so it is in firm

High Kneeling Position with the Hasty Sling

Figure 5-50. Left View.

Figure 5-51. Front View.

Figure 5-52. Right View.

contact. The upper portion of the triceps or the armpit will not rest on the knee.

- Bend the right elbow to provide the least muscular tension possible, lowering it to a natural position.
- Adjust the position to adjust sling tension. Widening the stance by moving the left foot and knee outboard will allow the sling to be tightened.
- Adjust the position of the left hand to allow the sling to support the weapon and the front sight to be centered in the rear sight aperture.

To adjust for elevation:

- Move left hand rearward or forward on handguards (moving the hand rearward elevates the muzzle).
- Open or close the "V" of the left hand for small adjustments (closing the "V" elevates the muzzle).

To adjust for a minor cant in the rifle, rotate the handguard left or right in the "V" formed by the thumb and forefinger by rotating the pistol grip left or right.

High Kneeling Position with the Loop Sling

Apply the three elements and seven factors to this position (para. 5004). To assume the high kneeling position with the loop sling, either move forward or drop back into position (see figs. 5-53, 5-54, and 5-55):

- Position the body at a 45-degree angle to the target.
- Place the left hand under the handguard.
- Kneel down on right knee so right lower leg is approximately parallel to the target (45 to 90 degrees).
- Keep the right ankle straight, with the toe of the boot in contact with the ground and curled under by the weight of the body.
- Place the right portion of the buttocks on or over the right heel; contact with the heel provides more stability to the position.
- Place the left foot forward to a point that allows the shin to be vertically straight. For the shin to be vertical, the heel should be directly under the knee. The left foot must be flat on the ground since it will be supporting the majority of the weight.
- Place the flat portion of the back of upper left arm, just above the elbow, on the left knee so it is in firm contact with the flat surface formed on top of the bent knee. The point of the left elbow will extend just slightly past the left knee. The upper portion of the triceps or the armpit will not rest on the knee.
- Lean slightly forward into the sling for support.
- Grasp the rifle butt with the right hand and place the rifle butt into the right shoulder pocket.
- Grasp the pistol grip with the right hand.
- Bend the right elbow to provide the least muscular tension possible and lower it to a natural position.
- Lower the head and place the cheek firmly against the stock to allow the aiming eye to look through the rear sight aperture.
- Move the left hand to a location under the handguard, which provides maximum bone support and stability for the weapon.

Medium Kneeling Position

This is also referred to as the bootless kneeling position. Assume the medium kneeling position in the same way as the high kneeling position with the ex-

High Kneeling Position with the Loop Sling

Figure 5-53. Left View.

Figure 5-54. Front View.

Figure 5-55. Right View.

ception of the right foot. The right ankle is straight and the foot is stretched out with the bootlaces in contact with the ground. The buttocks are in contact with the heel of the right foot. See figure 5-56.

Figure 5-56. Medium Kneeling Position.

Low Kneeling Position

The low kneeling position is most commonly used when firing from a forward slope. Assume the low kneeling position in the same way as the high kneeling position with the exception of the placement of the right foot. Turn the right ankle so the outside of the foot is in contact with the ground and the buttocks are in contact with the inside of the foot. See figure 5-57.

Figure 5-57. Low Kneeling Position.

5008. Standing Position

Description

The standing position is the quickest position to assume and the easiest to maneuver from. It allows greater mobility than other positions. The standing position is often used for immediate combat engagement. The standing position is supported by the shooter's legs and feet and provides a small area of contact with the ground. In addition, the body's center of gravity is high above the ground. Therefore, maintaining balance is critical in this position.

Standing Position with the Hasty Sling

Apply the seven factors to this position (para. 5003). To assume the standing position with the hasty sling (see figs. 5-58, 5-59, and 5-60)—

- Square the body to the target.
- Spread the feet apart to a comfortable distance with the left foot slightly in front of the right foot. This distance may be wider than shoulder width.
- Distribute the weight evenly over both feet and hips. Balance will shift forward slightly to reduce recovery time and increase the stability of the hold. The legs should be slightly bent for balance.
- Grasp the pistol grip with the right hand and place the rifle butt in the right shoulder pocket.
- Bring the rifle sights up to eye level instead of lowering the head to the sights and place the cheek firmly against the stock. Ensure the head is erect so the aiming eye can look through rear sight aperture.
- Rotate the left hand up, pinching the handguard between the thumb and forefinger.
- The left hand will be under the handguard with the thumb on the outboard side of the handguard.The left hand will not grasp the slip ring or magazine. The magazine must be on the inside of the left arm.
- Hold the right elbow in a natural position.
- Adjust the position of the left hand on the handguard to allow the sling to support the weapon and front sight to be centered in the rear sight aperture.

To adjust for elevation:

- Move left hand rearward or forward on handguards (moving the hand rearward elevates the muzzle).
- Open or close the "V" of the left hand for small adjustments (closing the "V" elevates the muzzle).
- Adjust the position to adjust sling tension. Moving the left elbow out or squaring the body more to the target tightens the sling.

To adjust for a minor cant in the rifle, rotate the handguard left or right in the "V" formed by the thumb and forefinger by rotating the pistol grip left or right.

Standing Position with the Hasty Sling

Figure 5-58. Left View.

Figure 5-59. Front View.

Figure 5-60. Right View.

Standing Position with the Parade Sling

The parade sling is used to emphasize marksmanship fundamentals while firing from the standing position on KD courses during entry-level training. To achieve proper positioning of the parade sling, perform the following steps:

- Attach sling to rifle by placing feed end of sling down through the upper sling swivel.
- Place feed end of sling through sling keeper and lock into place.
- Attach J-hook to lower sling swivel.
- Pull feed end of sling through sling keeper until sling is taut.
- Move sling keeper down near feed end of sling.
- Lock sling keeper into place.

Perform the following steps to assume the standing position with a parade sling (figure 5-61):

- Stand erect.
- Face approximately 90 degrees to the right of the line of fire.
- Place feet approximately shoulder width apart.
- Place left hand under handguard.
- Grasp pistol grip with right hand.
- Place rifle butt into right shoulder pocket.
- Invert left elbow across rib cage.
- Rest left arm naturally against rib cage.
- Lower right elbow to a natural position.
- Place cheek firmly against stock to obtain a firm stock weld.

Figure 5-61. Standing Position with the Parade Sling.

CHAPTER 6. USE OF COVER AND CONCEALMENT

Note

+The procedures in this manual are written for right-handed Marines; left-handed Marines should reverse instructions as necessary.

6001. Cover and Concealment

In a combat environment, a Marine must be prepared to fire from any type of cover or concealment. Cover is anything that protects a Marine from enemy fire. Cover may be an existing hole, a hastily dug shelter, or a well-prepared fighting position with overhead protection. Concealment is anything that hides a Marine from enemy view, but it may not afford protection. Concealment can be obtained from buildings, trees, crops, and skillful use of ground contours. A Marine can use any object or terrain feature that protects him from enemy fire, hides him from enemy view, allows him to observe the enemy, and provides support for a firing position.

Types of Cover

Frontal Cover

A firing position should have frontal cover that provides protection from small arms fire and indirect fire fragments. Ideally, frontal cover should be thick enough to stop small arms fire, high enough to provide protection from enemy fire, and wide enough to provide cover when firing to the left or right edge of a sector of fire.

Ideal Cover

The ideal cover provides:

- Overhead, flank, and rear protection from direct and indirect fire.
- Free use of personal weapons.
- Concealment from enemy observation.
- A concealed route in and out.
- Unobstructed view of the sector of fire.

Common Cover Materials

Any material that protects a Marine from small arms fire can be used for cover. Some common materials include sandbags, trees, logs, and building debris. Table 6-1 presents some common materials and their minimum thickness required for protection from small arms fire.

Table 6-1. Minimum Thickness for Protection Against Small Arms.

Material	Minimum Thickness (in inches)
Concrete	7
Broken stone (rubble)	20
Dry sand	24
Wet sand	35
Logs (oak)	40
Earth (packed)	48

Sandbags

Cover can be improved and positions can be fortified by filling sandbags with dirt/sand and placing them around the position. Sandbags should be tightly packed because bullets can easily penetrate moist or loosely packed sandbags. Overlapping sandbags increase protection and decrease the bullet's ability to penetrate the sandbag. A minimum thickness of three sandbags is required to stop small arms fire.

Trees/Logs

Wood is a relatively dense material and offers good cover and protection. Bullets have a tendency to fragment when they penetrate wood. Live trees have a greater resistance to bullet penetration than dead trees. Wood that has been treated with creosote, such as telephone poles and railroad ties, offers better protection from projectiles than untreated wood, but it still does not ensure protection from small arms fire.

Cinder Bocks

Cinderblocks are not impenetrable cover. Although they are made of a dense material, the composition of a cinderblock is so brittle that a bullet can shatter the block upon impact. This can cause injury to a Marine by secondary fragmentation.

Firing From Specific Types of Cover

Effective cover allows a Marine to engage enemy targets while protecting himself from enemy fire. Several types of cover provide support, protection, and concealment and do not interfere with target engagement. A Marine must adapt firing positions to the type of cover available.

Fighting Hole

A Marine should use fighting holes if available. See figure 6-1. After a Marine enters the fighting hole, he adds or removes dirt, sandbags or other supports to fit his height. To assume a firing position, a Marine performs the following steps:

- Place the right foot to the rear as a brace.
- Lean forward until the chest is against the forward wall of the fighting hole.
- Extend the left arm and elbow over the forward side of the fighting hole so the left forearm rests against the back of the parapet.
- Place the rifle butt into the pocket of the right shoulder and grasp the pistol grip with the right hand.
- Place the right elbow on solid support using the elbow rest of the fighting hole or sandbags placed around the fighting hole.

Figure 6-1. Fighting Hole Position.

Rooftop

If possible, a Marine's entire body should be positioned behind the apex of the rooftop, using the apex to support the rifle. See figure 6-2. If the body cannot be positioned behind the apex, place the left arm over the apex of the roof to hold the weight of the body. Expose as little of the head and shoulders as possible. See figure 6-3.

Figure 6-2. Rooftop Position Behind the Apex.

Figure 6-3. Rooftop Position Supported by the Apex.

Window

The Marine can establish a supported or unsupported position from a window.

Unsupported. A Marine can establish an unsupported position back from the opening of the window so that the muzzle does not protrude and interior shadows provide concealment so as not to provide a silhouette to the enemy. See figure 6-4.

Figure 6-4. Unsupported Window Position.

Supported. When additional stability is needed, a Marine can establish a supported position by placing the rifle handguards or his forearm in the "V" formed by the side and bottom of the windowsill. A drawback to this technique is that the muzzle of the weapon and the Marine may be exposed to view. See figures 6-5 and 6-6.

Vehicle

In many combat situations, particularly in urban environments, a vehicle may be the best form of cover. When using a vehicle for cover, the engine block provides the most protection from small arms fire. The Marine should establish a position behind the front wheel so the engine block is between him and the target (see figs. 6-7 and 6-8). From this position, the Marine may fire over, under or around the vehicle. This is a particularly effective position for larger vehicles that are high off the ground.

The Marine can establish additional support for the rifle by positioning himself behind the doorjamb (frame of door) and placing the rifle against the "V" formed by the open door and doorframe (see figure 6-9). From this position, the Marine may fire over the hood of the vehicle while using the engine block for protection.

Figure 6-5. Supported Window Position (Handguards on Support).

Figure 6-6. Supported Window Position (Forearm on Support).

Figure 6-7. Firing Around Front of Vehicle.

Figure 6-8. Firing Over Front of Vehicle.

Figure 6-9. Establishing a Supported Position in a Vehicle.

However, this position limits lateral mobility and it is more difficult to maneuver from.

At the back of the vehicle, only the axle and the wheel provide cover. If the Marine must shoot from the back of the vehicle, he must position himself directly behind the wheel as much as possible (see fig. 6-10).

Figure 6-10. Using the Back of a Vehicle for Cover.

6002. Supported Firing Positions

Supports are foundations for positions; positions are foundations for the rifle. To maximize the support the position provides, the firing position should be adjusted to fit or conform to the shape of the cover. Elements of a sound firing position, such as balance and stability, must be incorporated and adjusted to fit the situation and type of cover. A supported firing position should minimize exposure to the enemy, maximize the stability of the rifle, and provide protection from enemy observation and fires. A Marine can use any available support (e.g., logs, rocks, sandbags or walls) to stabilize his firing position. The surrounding combat environment dictates the type of support and position used.

Considerations Using Cover and Concealment

Cover and concealment considerations are similar, regardless of the combat environment (e.g., urban, desert, jungle).

Adjusting the Shooting Position

The type of cover can dictate which shooting position (e.g., standing, kneeling, sitting, and prone) will be the most effective. For example, a Marine's height in relation to the height of the cover aids in the selection of a firing position.

The firing position selected should be adjusted to fit the type of cover to:

- Provide stability. The position should be adjusted to stabilize the rifle sights and allow the management of recoil to recover on target.
- Permit mobility. The position should be adjusted to permit lateral engagement of dispersed targets and movement to other cover.
- Allow observation of the area/enemy while minimizing exposure to the enemy.

The firing position is adjusted to fit the type of cover by adjusting the seven factors (i.e., left hand, pocket of shoulder, right elbow, stock weld, grip of right hand) to support the rifle or the position.

Keeping the Entire Body Behind Cover

A Marine should minimize exposure of any part of his body to fire. Be especially aware of the head, right elbow, knees or any other body part that may extend beyond the cover.

Firing From the Right or Left Side of Cover

To minimize exposure and maximize the cover's protection, a right-handed Marine should fire from the right side of cover and a left-handed Marine should fire from the left side, if possible (see fig. 6-11).

Figure 6-11. Firing from the Right Side of Cover.

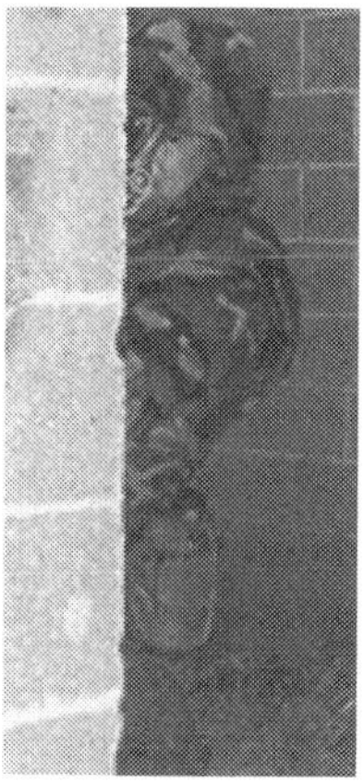

Figure 6-12. Firing from the Left Side of Cover.

If, however, a right-handed Marine must fire from the left side of cover, he fires right-handed but adjusts his position behind cover and uses the rollout technique (see para. 6003) to engage the target. See figure 6-12.

Firing Over the Top of Cover

Firing over the top of cover provides a wider field of view and lateral movement. When firing over the top of cover, the position may be supported and stabilized by resting the handguard or the left forearm on the cover (see fig. 6-13). The Marine should keep as low a profile as possible; the rifle should be as close to the top of cover as possible.

Figure 6-13. Firing Over the Top of Cover.

Maintaining Muzzle Awareness

When firing over the top of cover, a Marine must remember that the sights are higher than the barrel and remain aware of the location of his muzzle. Therefore, a Marine must maintain a position that ensures the muzzle is high enough to clear the cover (e.g., window sill, top of wall) as he obtains sight alignment/sight picture on the target (see figs. 6-14a and 6-14b).

Clearing the Ejection Port

Ensure the cover does not obstruct the ejection port. If the ejection port is blocked, the obstruction can interfere with the ejection of the spent cartridge case and cause a stoppage.

Figures 6-14a and 6-14b. Clearing Cover with the Muzzle.

Resting the Magazine

The bottom, front or side of the rifle magazine can rest on or against support to provide additional stability (see figs. 6-15, 6-16, and 6-17).

CAUTION

The back of the magazine should not be pulled back against support because it can cause a stoppage by not allowing a round to feed from the magazine.

Seven Factors

The seven factors (left hand, rifle in shoulder pocket, stock weld, right elbow, grip of the right hand, breathing, muscular tension) are applied when firing from cover, however, some may have to be modified slightly to accommodate the artificial support provided the rifle and position.

Left Hand

The support should be used to help stabilize both the firing position and the rifle to enable the Marine to maintain sight alignment and sight picture.

The forearm or left hand can contact the support to stabilize the weapon. Rest the forearm or the meaty portion of the bottom of the left hand on the support and rest the rifle in the "V" formed by the thumb and forefinger of the left hand (see fig. 6-18).

The rifle's handguards may rest on the support, but the barrel may not (see fig. 6-19). Placement of left hand on the handguard may have to be adjusted forward or back to accommodate the cover and the additional support provided by the rifle resting on the cover. If the handguards are resting on the cover, the left hand

Figure 6-15. Bottom of Magazine on Support.

Figure 6-16. Front of Magazine on Support.

Figure 6-17. Side of Magazine on Support.

can pull down on the handguards to further stabilize the weapon.

Rifle in Shoulder Pocket

Regardless of the combat situation or the height of the cover, the rifle butt must remain in the shoulder pocket to manage the effects of recoil, stabilize the rifle, and maintain the rifle's battlesight zero. The firing position must be adjusted behind cover to enable the rifle to be placed in the shoulder.

Stock Weld

Regardless of the cover, the firing position must be adjusted to allow stock weld to be achieved. Proper stock weld provides quick recovery between shots and keeps the aiming eye centered in the rear sight aperture.

Right Elbow

The right elbow can be placed on or against the support to stabilize the weapon and the position. Ensure the elbow is not extended beyond cover revealing the position to the enemy.

Grip of the Right Hand

If rifle handguards, forward hand, or forearm rest on cover for support, the grip of the right hand should pull back and down on the pistol grip to further stabilize the weapon in the shoulder and on the support.

Breathing

Breathing does not change when firing from a supported position.

Muscular Tension

To create balance and support for the position, the Marine may shift his body weight into or against the support (see fig. 6-20). This enables the Marine to use cover to support his body weight, reducing the need for muscular tension.

Figure 6-18. Forearm Resting on Cover.

Figure 6-19. Handguards Resting on Cover.

Figure 6-20. Shifting Body Weight into Cover.

Types of Supported Positions

Supported Prone

If possible, a Marine should use the supported prone position when firing from behind cover. This position is the steadiest, provides the lowest silhouette, and provides maximum protection from enemy fire.

- Support the position by placing the handguards, the forearm, or the magazine on or against support (see fig. 6-21).

Figure 6-21. Supported Prone.

- The prone position can be assumed behind a tree, a wall, a log, or almost any type of cover. It is flexible and allows shooting from all sides of cover and from cover of various sizes.
- The body must be adjusted to conform to the cover. For example, if the cover is narrow, keep the legs together. The body should be in line with the rifle and directly behind the rifle (see figs. 6-22 and 6-23). This presents a smaller target to the enemy and provides more body mass to absorb recoil.

Figure 6-22. Supported Prone Behind Narrow Cover—Rear View.

Supported Kneeling

When the prone position cannot be used because of the height of the support, the supported kneeling position may be appropriate. The kneeling position provides additional mobility over the prone position.

- The kneeling position allows shooting from all sides and from cover of varying sizes. This position may be altered to maximize the use of cover or support by assuming a variation of the kneeling position (high, medium or low).
- In the kneeling position, the Marine must not telegraph his position behind the cover with his knee. When shooting around the sides of cover, the Marine should strive to keep his right knee in line with his left foot so as not to reveal the position to the enemy. See figure 6-24.
- Support the position by placing the handguards, the forearm, or the magazine on or against support. In addition, the position (e.g., a knee, the side of the body) may rest against support (see fig. 6-25).
- If the rifle is resting on support, the Marine may not need to stabilize the weapon by placing his left elbow on his knee (see fig. 6-26).

Supported Sitting

A supported sitting position may be used to fire over the top of cover when mobility is not as critical. A sitting position can be comfortably assumed for a longer period of time than a kneeling position and it can conform to higher cover when a prone position cannot be used.

Figure 6-23. Supported Prone Behind Narrow Cover—Front View.

Figure 6-24. Supported Kneeling.

Figure 6-25. Supported Kneeling with Body Against Support.

- Support the position by placing the handguards, the forearm, or the magazine on or against support (see fig. 6-27).

- If the rifle is resting on support, the Marine may not need to stabilize the weapon by placing his left or right elbows on his legs (see fig. 6-28).

Supported Standing

The supported standing position provides greater mobility than the other positions and usually provides greater observation of the enemy. The supported standing position is effectively used behind high cover (e.g., window, over a wall) or narrow cover (e.g., tree, telephone pole).

- To assume the supported standing position, the Marine leans his body forward or against support to stabilize the weapon and the position.

Figure 6-26. Supported Kneeling with Rifle Resting on Support.

Figure 6-27. Supported Sitting.

Figure 6-28. Supported Sitting with Rifle on Support.

- Support the position by placing the handguards, the forearm, or the magazine on or against support. In addition, the position (e.g., the side of the body) may rest against support (see fig. 6-29).

Figure 6-29. Supported Standing.

6003. Searching for and Engaging Targets From Behind Cover

To locate targets when behind cover or to ensure the area is clear before moving, the Marine must expose as little of himself as possible to the enemy. Additionally, the Marine must be ready to fire if a target is located. Two techniques that can be used to locate and engage targets from behind cover are: the pie and rollout. These techniques minimize the Marine's exposure to enemy fire while placing the Marine in a position to engage targets or to move to another location if necessary. These techniques are also used to enter a building or structure. Both techniques are used in the kneeling and standing positions. To be accurate in engaging targets using either technique, the seven factors must be applied. See page 6-6 for a discussion of seven factors.

Pie Technique

To perform the pie technique:

- Staying behind cover, move back and away from the leading edge of the cover. The surroundings and situation will dictate the distance you should move back and away from the cover. Generally, the further back the Marine is from cover, the greater his area of observation; staying too close to cover decreases the area of observation. However, if the Marine is too far back from the leading edge of cover, he may unknowingly expose himself to the enemy.
- Assume a firing position and lower the rifle sights enough to have a clear field of view, orienting the muzzle on the leading edge of the cover. (In a building, the baseboards serve as a reference point for the muzzle of the rifle when searching for targets.)
- Taking small side steps, slowly move out from behind the cover, covering the field of view with the aiming eye and muzzle of the weapon. Wherever the eyes move, the muzzle should move (eyes, muzzle, target). The muzzle should remain on the leading edge of cover, serving as a pivot point when moving out.
- Continue taking small side steps and moving out from cover until a target is identified or the area is found to be clear.
- When a target is identified, sweep the safety, place the finger on the trigger, and engage the target.

Rollout Technique

To perform the rollout technique:

- Staying behind cover, move back and position the body so it is in line with the leading edge of the cover, ensuring that no part of the body extends beyond the cover.
- Assume a firing position and come to the Ready, ensuring the muzzle is just behind the cover.
- Canting the head and weapon slightly, roll the upper body out to the side just enough to have a clear field of view and allow the muzzle to clear the cover. Keeping the feet in place, push up on the ball of one foot to facilitate rolling out.
- Continue rolling out from cover until a target is identified or the area is found to be clear.
- When a target is identified, sweep the safety, place the finger on the trigger, and engage the target.
- If a target is identified before moving out from cover, the rifle should be taken off safe before moving out using the rollout technique.

Combining Techniques

In some situations, it may be necessary to combine the pie and rollout techniques to search an entire area for targets (e.g., corner of a building, a doorway). Changing from one technique to another may permit the Marine to minimize his exposure to enemy observation and fires.

6004. Moving Out From Behind Cover

A Marine must be constantly aware of his surroundings and available cover, should a threat appear. He should avoid obvious danger areas and move quickly through danger areas that cannot be avoided.

When moving from cover to cover, the Marine should select the next covered location and plan his route before moving from his present position. This is done by quickly looking from behind cover to ensure the area is clear, ensuring the head and eyes are exposed for as short a time as possible.

If necessary, the Marine should conduct a Condition 1 reload before moving from cover.

Once the Marine is committed to moving, all focus should be on moving until cover is reassumed.

CHAPTER 7. RIFLE PRESENTATION

Note

☞The procedures in this manual are written for right-handed Marines; left-handed Marines should reverse instructions as necessary.

7001. Presentation of the Rifle

In a combat environment, targets may present themselves with little or no warning. To maintain an advantage, the Marine carries his weapon in a position appropriate to the threat level that permits the rifle to be both easily carried and presented as quickly as possible. A carry is also established based on the situation such as moving in a close quarter environment, moving over or under objects, etc.

Presenting the Rifle From the Tactical Carry

The Marine uses the Tactical Carry when no immediate threat is present. This carry permits the rifle to be easily carried for long periods of time, but it does not permit the quickest presentation to a target. If the situation changes and a target presents itself, a Marine performs the following steps to present the rifle from the Tactical Carry once a target appears:

- Extend the rifle toward the target keeping the muzzle slightly up so the buttstock clears all personal equipment. Continue to look at the target.
- At the same time, place the rifle in Condition 1. Two methods can be used to place rifle in Condition 1 if it is in Condition 3:
  - Grip the pistol grip firmly with the right hand. Pull the charging handle with the left hand to its rearmost position and release (see fig. 7-1).

Figure 7-1. Pulling Charging Handle with Left Hand.

  - Grip the handguards firmly with the left hand. Pull the charging handle with the right hand to its rearmost position and release (see fig. 7-2).

Figure 7-2. Pulling Charging Handle with Right Hand.

- As the rifle is being presented, take the rifle off safe and place the trigger finger on the trigger (see figs. 7-3a, 7-3b, and 7-3c on page 7-2).
- Level the rifle while pulling it firmly into the pocket of the shoulder to obtain proper stock weld. Do not move the head down to meet the stock of the rifle.

Note

If the rifle is in the shoulder properly, the aiming eye will be able to look through the rear sight as soon as the stock makes contact with the cheek.

- As the sights become level with the aiming eye, visually locate the target through the rear sight aperture. As the rifle sights settle, shift the focus back to the front sight post to obtain sight alignment, and place the tip of the post center mass on the target to obtain sight picture.

Presenting the Rifle From the Alert Carry and From the Ready Carry

The Marine uses the Alert Carry when enemy contact is likely. The Alert is also used for moving in close terrain (e.g., urban, jungle).

The Marine uses the Ready Carry when enemy contact is imminent.

Figures 7-3a, 7-3b, and 7-3c. Sweeping the Safety.

To present the rifle from the Alert and from the Ready, a Marine performs the following steps once a target appears:

- While looking at the target, bring the muzzle up by raising the left hand, allowing the rifle butt to pivot in the shoulder. At the same time, pull the rifle firmly into the pocket of the shoulder.
- As the rifle is being presented, take the rifle off safe and place the trigger finger on the trigger (see figs. 7-3a, 7-3b, and 7-3c).
- As the stock makes contact with the cheek, level the rifle to obtain proper stock weld.
- Do not move the head down to meet rifle stock.

Note

If the rifle is in the shoulder properly, the aiming eye will be able to look through the rear sight as soon as the stock makes contact with the cheek.

- As the sights become level with the aiming eye, visually locate the target through the rear sight aperture. As the rifle sights settle, shift the focus back to the front sight post to obtain sight alignment, and place the tip of the post center mass on the target to obtain sight picture.

Presenting the Rifle From the Strong Side Sling Arms Transport

Once a target appears, a Marine performs the following to present the rifle from Strong Side Sling Arms:

- While looking at the target, lean forward slightly to facilitate removal of the rifle from the shoulder.
- Reach under the right arm with the left hand between the sling and the body and grasp the handguards. See figure 7-4). At the same time, pull down on the sling and raise right elbow out and parallel to the deck.

Figure 7-4. Grasping the Handguards.

- Roll the right shoulder forward and release the sling from the right hand once the handguards have cleared the elbow. At the same time, pull rifle forward off the shoulder with left hand.
- Continue pulling the rifle forward with the left hand while rotating the rifle parallel to the deck; when the right arm is free of the sling and the rifle clears all personal gear, grasp the charging handle with the right hand (see fig. 7-5) and pull it to its rearmost position and release.
- Establish a firing grip with right hand while keeping the trigger finger straight along the receiver.
- Take rifle off safe and place the trigger finger on the trigger (see figs. 7-3a, 7-3b, and 7-3c).
- Level the rifle while pulling it firmly into the pocket of the shoulder to obtain proper stock weld. Do not move the head down to meet the stock of the rifle.

Figure 7-5. Clearing Gear and Grasping the Charging Handle.

Note

If the rifle is in the shoulder properly, the aiming eye will be able to look through the rear sight as soon as the stock makes contact with the cheek.

- As the sights become level with the aiming eye, visually locate the target through the rear sight aperture. As the rifle sights settle, shift the focus back to the front sight post to obtain sight alignment, and place the tip of the post center mass on the target to obtain sight picture.

Presenting the Rifle From the Weak Side Sling Arms Transport

The hasty sling should be maintained while presenting the rifle from this transport. To present the rifle from Weak Side Sling Arms, a Marine performs the following steps once a target appears:

- While looking at the target, lean forward slightly to facilitate removal of rifle from the shoulder.
- Grasp the sling with the right hand to prevent the rifle from falling off the shoulder.
- Grasp handguards with left hand (the index finger points toward the muzzle). See figure 7-6.
- Rotate the rifle counterclockwise while extending the muzzle toward the target.
- Continue extending the rifle toward the target to ensure the rifle clears all personal gear.
- Grasp the charging handle with the right hand and pull it to its rearmost position and release (see fig. 7-7).
- Establish a firing grip with right hand while keeping the trigger finger straight along the receiver.
- Take the rifle off safe and place the trigger finger on the trigger (see figures 7-3a, 7-3b, and 7-3c).
- Level the rifle while pulling it firmly into the pocket of the shoulder to obtain proper stock weld. Do not move the head down to meet the stock of the rifle.

Figure 7-6. Grasping the Handguards.

Note

If the rifle is in the shoulder properly, the aiming eye will be able to look through the rear sight as soon as the stock makes contact with the cheek.

Figure 7-7. Clearing Gear and Grasping the Charging Handle.

- As the sights become level with the aiming eye, visually locate the target through the rear sight aperture. As the rifle sights settle, shift the focus back to the front sight post to obtain sight alignment, and place the tip of the post center mass on the target to obtain sight picture.

7002. Search and Assess

After a Marine engages a target, he must immediately search the area and assess the results of his engagement. Searching and assessing enables the Marine to avoid tunnel vision that can restrict the focus so that an indication of other targets is overlooked.

Purpose

The Marine searches the area for additional targets or for cover. The Marine assesses the situation to determine if he needs to re-engage a target, engage a new target, take cover, assume a more stable position, cease engagement, etc.

Technique

To search and assess, a Marine performs the following steps:

- Keeping the buttstock in the shoulder, lower the muzzle of the rifle slightly to look over the sights.
- Place the trigger finger straight along the receiver (see fig. 7-8).
- Search the area and assess the situation/threat by moving the head, eyes, and rifle left and right (approximately 45 degrees from center) to cover the immediate area. The muzzle moves with the head and eyes in one fluid motion while searching. Keep both eyes open to increase the field of view.

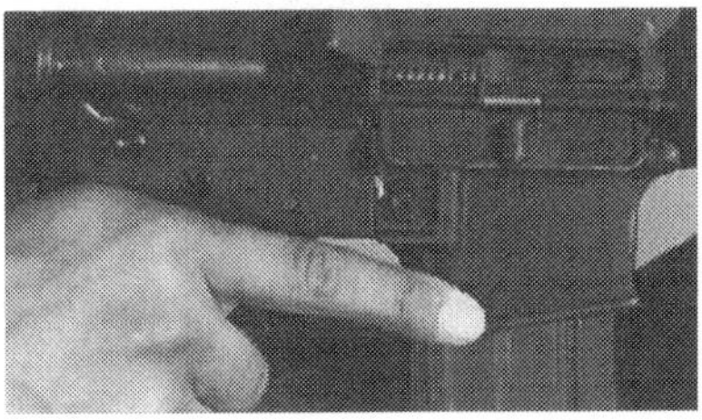

Figure 7-8. Straight Trigger Finger.

- Once a Marine determines the area is clear of enemy threat, he places the rifle on safe.

Searching and Assessing to a Higher Profile

Depending on the tactical situation, the Marine may choose to increase his area of observation by searching and assessing to a higher profile position.

Prone to Kneeling

After searching and assessing at the prone position, move to a kneeling position by performing the following steps:

- While maintaining control of the pistol grip, lower the rifle butt out of the shoulder.
- Drop the left hand to the deck and, bringing it back, push up off the deck to both knees (see figs. 7-9a and 7-9b).
- Grasp the handguard with the left hand and place the rifle butt in the pocket of the shoulder.
- Assume a kneeling position and search and assess (see fig. 7-10).

Sitting to Kneeling

After searching and assessing at the sitting position, move to a kneeling position by performing the following steps:

- Maintain control of the rifle with the rifle butt in the pocket of the shoulder.

Figures 7-9a and 7-9b. Pushing Up Off the Deck to Both Knees.

- Uncross the legs to an open leg position.
- Tuck the right foot underneath the left thigh, as close to the buttocks as possible (see fig. 7-11).
- Lean forward and to the right and roll on to the right knee to a kneeling position and search and assess (see fig. 7-12).

Note

It may be necessary to release the rifle with the right hand and push off the deck with the right hand to assist in rolling up to a kneeling position.

Kneeling to Standing

After searching and assessing at the kneeling position, maintain control of the rifle with the rifle butt in the pocket of the shoulder, and stand while continuing to search and assess.

Figure 7-10. Kneeling Search and Assess.

Figure 7-11. Tucking the Right Foot.

Figure 7-12. Rolling Up to a Kneeling Position.

Chapter 8. Effects of Weather

Wind, temperature, and precipitation can affect the trajectory of the bullet. In addition, all weather conditions have a physical and psychological effect on Marines. Marines must use techniques to offset the effects of wind, light, temperature, and precipitation (snow, sleet, and rain). Through proper training, Marines can develop the confidence required to reduce the physical and psychological effects of weather during combat situations.

8001. Physical Effects of Wind on the Bullet

Physical Effects

The weather condition that presents the greatest problem to shooting is the wind. Wind affects a bullet's trajectory. The effect of wind on the bullet as it travels down range is referred to as deflection. The wind deflects the bullet laterally in its flight to the target (see fig. 8-1).

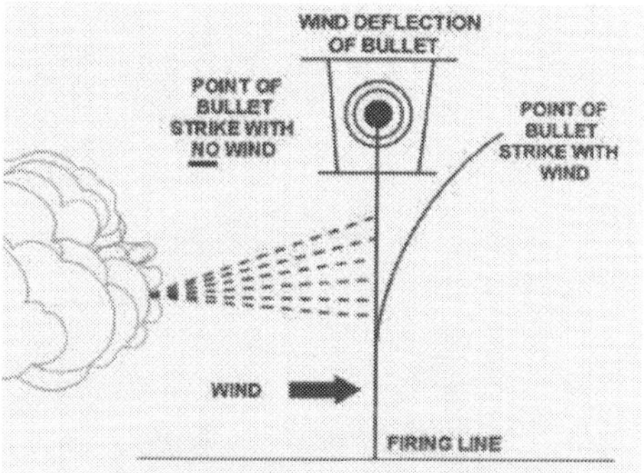

Figure 8-1. Deflection of a Bullet.

The bullet's exposure time to the wind determines the amount the bullet is deflected from its original trajectory. Deflection increases as the distance to the target increases. There are three factors that affect the amount of deflection of the bullet:

- Velocity of the wind—The greater the velocity of the wind, the more the bullet will be deflected.
- Range to the target—As the distance to the target increases, the speed of the bullet slows allowing the wind to have a greater effect on shot placement.
- Velocity of the bullet—A bullet with a high muzzle velocity will not be affected by the wind as much as a bullet with a low muzzle velocity.

Determining Windage Adjustments to Offset Wind Effects

The velocity and direction of the wind in relationship to the bullet must be determined to offset the wind's effects. If Marines can classify wind values and determine velocity within 5 mph, they can effectively engage targets in windy conditions.

Wind Direction

Determine wind direction by observing direction vegetation is moving, by feeling the wind blow against the body, or by observing direction of a flag (in training).

Wind Value Classifications

Winds are classified according to the direction from which they are blowing in relation to the direction of fire. The clock system indicates wind direction and value (see fig. 8-2). Winds can be classified as half

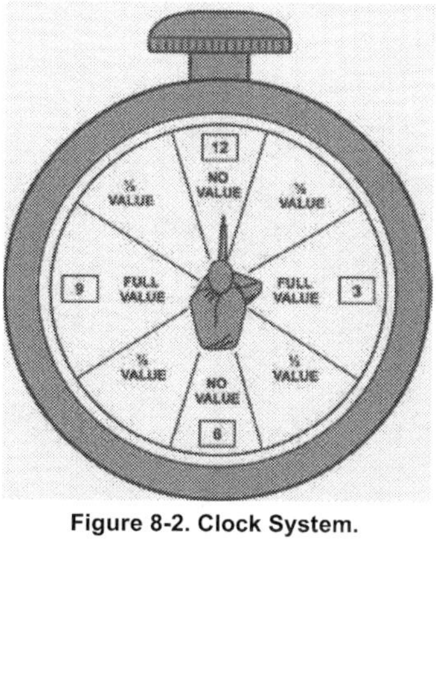

Figure 8-2. Clock System.

value, full value, or no value. The target is always located at 12 o'clock.

Wind Velocity

There are two methods used to determine wind velocity: observation and flag. The flag method is used as a training tool on the known distance (KD) range to learn the observation method. This method teaches Marines to relate the effect a given wind condition has on the natural surroundings in order to develop the base of knowledge used during the observation method. The observation method is the primary method used to estimate wind velocity and direction in a tactical situation. The following are guidelines used during the observation method:

- Under 3 miles per hour (mph) the wind can hardly be felt on the face. The presence of a slight wind can be detected by drifting smoke.
- 3 to 5 mph winds can be felt lightly on the face.
- 5 to 8 mph winds keep tree leaves in a constant motion.
- 8 to 12 mph winds raise dust and loose paper.
- 12 to 15 mph winds cause small trees to sway.
- 16 to 25 mph winds cause large trees to sway.

Flag Method

The flag method is primary method used on the KD range. To estimate wind velocity in miles per hour:

Estimate the angle created between the flagpole and the flag in degrees.

- Divide the angle by four to estimate wind velocity in miles per hour. See figure 8-3.

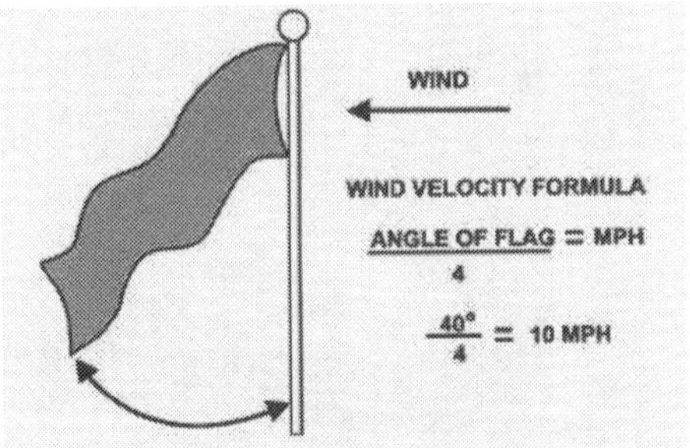

Figure 8-3. Flag Method.

Note

Information given is based on a dry flag. A wet flag is heavy and gives a false reading.

Windage Adjustments

After identifying wind direction, wind classification, and wind velocity, windage adjustments needed to enable the bullet to strike the target are estimated in the following ways:

Observation Method. Using the windage chart provided in figure 8-4, match the wind velocity, wind direction, and range to the target to the information in the chart to estimate the correct number of clicks to apply to the windage knob.

Flag Method. Using the windage chart provided in figure 8-5, match the wind velocity, wind direction, and range to the target to the information in the chart to determine the correct number of clicks to apply to the windage knob.

Once the number of windage clicks is determined, turn the windage knob causing the rear sight aperture to move into the direction of the wind. (See chapter 9.)

8002. Physical Effects of Temperature and Precipitation on the Bullet and the Rifle

Temperature

Extreme changes in temperature cause fluctuation in the rifle's chamber pressure. This fluctuation is caused by changes in the propellant's temperature. In cold weather, as rifle chamber pressure decreases, the bullet exits the muzzle at a lower velocity, and the bullet impacts the target below the point of aim. In extreme heat, the rifle's chamber pressure increases causing the bullet to exit the muzzle at a higher velocity and impact the target above the point of aim. Hot air is less dense than cool air and provides less resistance to the bullet; this allows the bullet to travel faster and experience less deflection from the wind. Cold air is dense and provides the bullet with more resistance; this causes the bullet to travel slower and experience greater deflection from the wind.

	WINDS CAN BE FELT LIGHTLY ON FACE AND LEAVES ARE IN CONSTANT MOTION		WINDS RAISE DUST AND LOOSE PAPER 8-12 MPH		WINDS CAUSE SMALL TREE TO SWAY 12-15 MPH		WINDS CAUSE LARGE TREES TO SWAY 16-20 MPH		21-25 MPH	
RANGE YARDS	WIND VALUE		WIND VALUE		WIND VALUE		WIND VALUE		WIND VALUE	
	FULL	HALF	FULL	HALF	FULL	HALF	FULL	HALF	FULL	HALF
200	2	1	3	1	5	2	6	3	8	4
300	3	1	6	3	10	5	13	6	16	8
400	4	2	9	4	13	6	18	9	22	11
500	6	3	12	6	18	9	24	12	30	15

Figure 8-4. Windage Click Chart for the Observation Method.

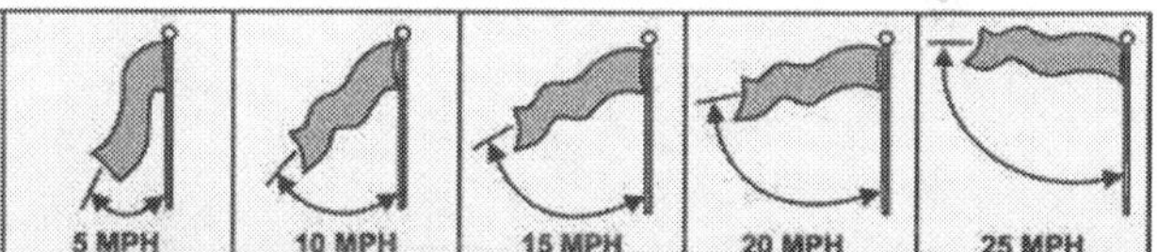

RANGE FLAG ANGLES	5 MPH		10 MPH		15 MPH		20 MPH		25 MPH	
RANGE YARDS	WIND VALUE		WIND VALUE		WIND VALUE		WIND VALUE		WIND VALUE	
	FULL	HALF	FULL	HALF	FULL	HALF	FULL	HALF	FULL	HALF
200	2	1	3	1	5	2	6	3	8	4
300	3	1	6	3	10	5	13	6	16	8
400	4	2	9	4	13	6	18	9	22	11
500	6	3	12	6	18	9	24	12	30	15

Figure 8-5. Windage Click Chart for the Flag Method.

Once the rifle is zeroed, a change in temperature of 20 degrees or more can cause the bullet to strike above or below the point of aim. Therefore, if the temperature changes 20 degrees or more, a Marine should re-zero the rifle.

If the rifle is exposed to below freezing temperatures, it should not be brought immediately into a warm location. Condensation may form on and in the rifle, and it may freeze if re-exposed to the cold. Ice that forms inside the rifle may cause it to malfunction. Ice can form on the rear sight aperture due to condensation, making it impossible to acquire sight alignment.

Precipitation

Freezing rain and other types of precipitation may make the rifle difficult to handle, foul the rifle and cause stoppages, or build up in the barrel or compensator and cause erratic shots. Care should be taken to keep the barrel and muzzle free of water. If the rifle has been submerged, ensure the bore is drained before firing. To drain the bore, pull the charging handle slightly to the rear and hold for a few seconds while the muzzle points down. Once the barrel has been drained, turn the rifle muzzle up to allow the water to drain out of the stock.

8003. Physical and Psychological Effects of Weather on Marines

Wind

Marines can shoot effectively in windy conditions if they apply a few basic techniques and develop the proper mental attitude. The Marine can combat the wind in a number of ways:

- Make subtle changes to the basic shooting positions, such as increasing muscular tension, to reduce movement of the rifle sights.
- Select a more stable firing position.
- Seek support to stabilize the rifle.
- Hold the shot and apply the fundamentals during a lull in the wind.

Temperature

Extreme Heat

In extreme heat, a Marine may experience rapid fatigue. Heat can cause muscle cramps, heat exhaustion, heat stroke, blurred vision, and reduced concentration levels that result in inaccurate shooting. Increased fluid intake and good physical condition can offset the effects of extreme heat. Sweat running into the eyes can cause irritation and make it difficult to see the sights. Extreme heat also can create ground mirages that cause a target to appear indistinct and to drift from side to side. Heat waves or mirages may also distort the target shape or the appearance of the front sight post. A mirage created by the heat of the barrel reduces a Marine's ability to see the sight clearly. To overcome the effects of heat and accurately engage a target, a Marine should maintain a center of mass hold.

Extreme Cold

Extreme cold may affect a Marine's ability to concentrate. If a Marine's hands are numb, he will have difficulty holding a frigid rifle and executing effective trigger control. To protect the hands in a cold environment, a Marine should wear arctic mittens or gloves. To operate the rifle while wearing arctic mittens or gloves, a Marine depresses the trigger guard plunger to open the trigger guard. This allows easier access to the trigger. See figure 8-6. The hasty sling can assist in holding the hand in place on the hand guards so the hand does not slip while wearing mittens.

Precipitation

Precipitation (rain, snow, hail, sleet) can affect target engagement, a Marine's comfort level, and a Marine's ability to concentrate. The amount and type of precipitation may obscure or completely hide the target and it may reduce a Marine's ability to establish an accurate sight picture. Precipitation collecting on rear sight aperture can make it difficult to establish sight alignment and sight picture. Protect sights as much as possible

Figure 8-6. Open Trigger Guard.

during periods of precipitation. It is easy to lose concentration when wet and uncomfortable. Proper dress reduces the effects of precipitation on the Marine.

Light

Light conditions can change the appearance of a target. Light affects each Marine differently. Light can affect range estimation, visual acuity, or the placement of the tip of the front sight on the target. By maintaining a center of mass hold, the effects of light can be reduced.

Bright Light

Bright light conditions exist under a clear blue sky with no fog or haze present to filter the sunlight. Bright light can make a target appear smaller and farther away. As a result, it is easy to overestimate range. Maintaining a center of mass hold, regardless of how indistinct the target appears, ensures the best chances for an effective shot.

Overcast

An overcast condition exists when a solid layer of clouds blocks the sun. The amount of available light changes as the overcast thickens. Overcast conditions can make a target appear larger and closer. As a result, it is easy to underestimate range. During a light overcast, the target appears very distinct and the rifle sights appear very distinct, making it easy to establish sight alignment. As the overcast thickens, it becomes difficult to identify the target from its surroundings.

Haze

Hazy conditions exist when fog, dust, humidity or smoke is present. Hazy conditions can make a target appear indistinct making it difficult to establish sight picture.

CHAPTER 9. ZEROING

To be combat effective, it is essential for the Marine to know how to zero his rifle. Zeroing is adjusting the sights on the weapon to cause the shots to impact where the Marine aims. This must be done while compensating for the effects of weather and the range to the target. It is critical that Marines can zero their rifles and make the sight adjustments required to engage targets accurately.

See the appropriate technical manual for procedures on boresighting and zeroing with supplemental aiming devices (e.g., laser, night vision devices).

Note

☞The procedures in this manual are written for right-handed Marines; left-handed Marines should reverse instructions as necessary.

9001. Elements of Zeroing

There are five basic elements involved in zeroing a rifle: line of sight, aiming point, centerline of the bore, trajectory, and range. See figure 9-1.

Line of Sight

The line of sight is a straight line, which begins with the shooter's eye, proceeds through the center of the rear sight aperture, and passes across the tip of the front sight post to a point of aim on a target.

Aiming Point

The aiming point is the precise point where the tip of the front sight post is placed in relationship to target.

Centerline of the Bore

Centerline of the bore is an imaginary straight-line beginning at the chamber end of the barrel, proceeding out of the muzzle, and continuing indefinitely.

Trajectory

In flight, a bullet does not follow a straight line but travels in a curve or arc, called trajectory. Trajectory is the path a bullet travels to the target. As the bullet exits the muzzle, it travels on an upward path, intersecting the line of sight (because the sights are above the muzzle). As the bullet travels farther, it begins to drop and intersects the line of sight again.

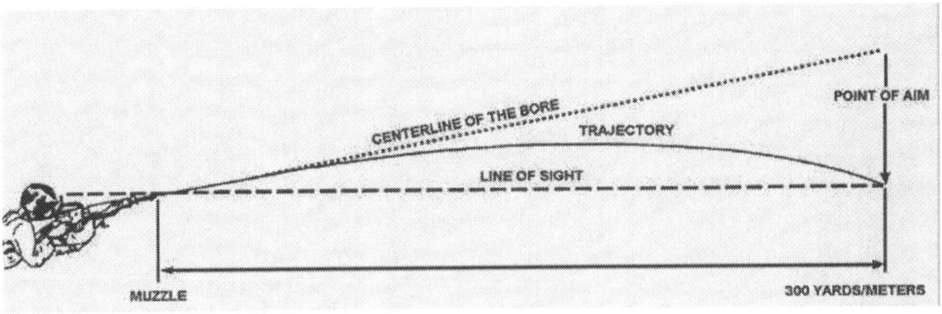

Figure 9-1. Elements of Zeroing.

Range

Range is the KD from the rifle muzzle to the target.

9002. Types of Zeros

Battlesight Zero (BZO)

A BZO is the elevation and windage settings required to place a single shot, or the center of a shot group, in a predesignated location on a target at 300 yards/meters, under ideal weather conditions (i.e., no wind). A BZO is the sight settings placed on your rifle for combat. In combat, your rifle's BZO setting will enable engagement of point targets from 0–300 yards/meters in a no-wind condition.

Zero

A zero is the elevation and windage settings required to place a single shot, or the center of a shot group, in a predesignated location on a target at a specific range, from a specific firing position, under specific weather conditions.

True Zero

A true zero is the elevation and windage settings required to place a single shot, or the center of a shot group, in a predesignated location on a target at a specific range other than 300 yards/meters, from a specific firing position, under ideal weather conditions (i.e., no wind).

9003. M16A2 Sighting System

The sighting system of the M16A2 service rifle consists of a front sight post and two rear sight apertures windage and elevation knob. Scales of the sighting system may be applied accurately to both yard and meter measurements. For example, a rear sight elevation setting of 8/3 may be used for 300 yards/meters.

Front Sight

The front sight post is used to adjust for elevation. The front sight consists of a square, rotating sight post with a four-position, spring-loaded detent (see fig. 9-2). To adjust for elevation, use a pointed instrument (or the tip of a cartridge) to depress the detent and rotate the front sight post (see fig. 9-3). To raise the strike of the bullet, rotate the post clockwise (in the direction of the arrow marked UP) or to the right. To lower the strike of the bullet, rotate the post counter-clockwise (in the opposite direction of the arrow) or to the left.

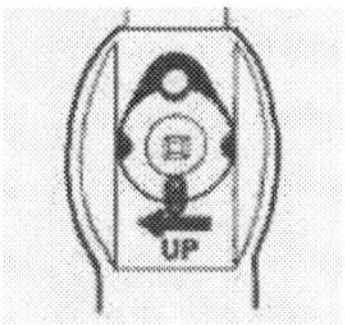

Figure 9-2. Front Sight.

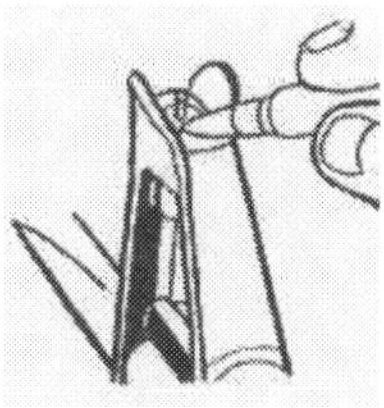

Figure 9-3. Front Sight Adjustment.

Rear Sight

The rear sight consists of two sight apertures, a windage knob, and an elevation knob. See figure 9-4. The large aperture marked 0-2 is used for target engagement during limited visibility, when a greater field of view is desired, or for engagements of targets closer than 200 yards/meters. The unmarked aperture (small aperture) is used for zeroing and normal firing situations.

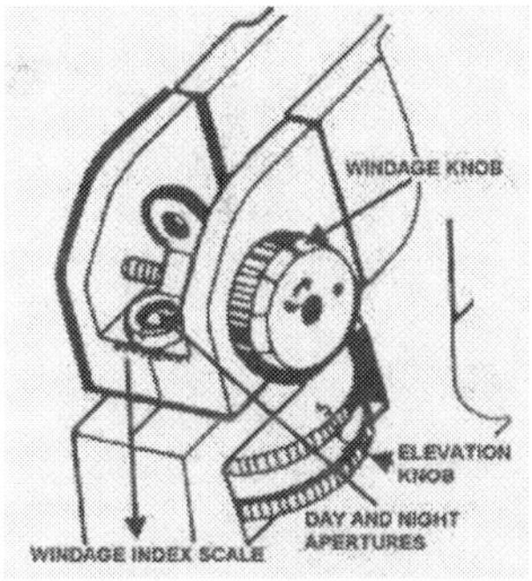

Figure 9-4. Rear Sight.

Elevation Knob

The rear sight elevation knob is used to adjust the sight for a specific range to the target. The elevation knob is indexed as shown in figure 9-5. Each number on the knob represents a distance from the target in 100-yard/meter increments. To adjust for range to the target, rotate the elevation knob so the desired setting is aligned with the index on the left side of the receiver.

Figure 9-5. Elevation Knob.

If the elevation knob is turned so the number 8/3 aligns with the elevation index line, the 3 indicates 300 yards/meters (see fig. 9-6).

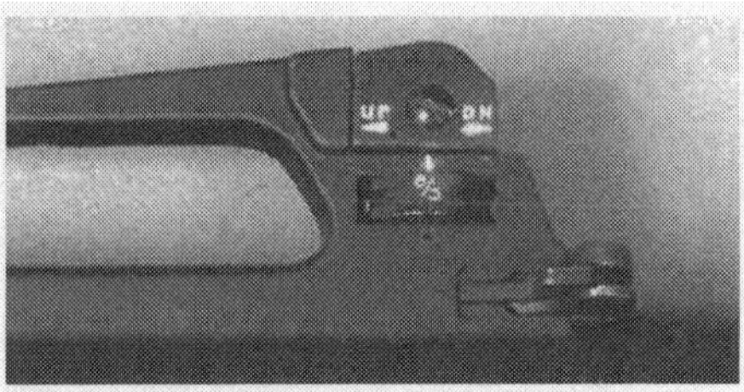

Figure 9-6. Rear Sight Elevation Knob Set for 300 Yards/Meters.

When the rear sight elevation knob is set on 8/3 for 800 yards/meters, there will be a considerable gap (about a quarter inch) between the rear sight housing and the upper receiver (see fig. 9-7).

Figure 9-7. Rear Sight Elevation Knob Set for 800 Yards/Meters.

A hasty sight setting is the setting placed on the rear sight elevation knob to engage targets beyond 300 yards/meters. Hasty sight settings for ranges of 400 to 800 yards/meters are applied by rotating the rear sight elevation knob to the number that corresponds with the engagement distance of the enemy. Aligning the number 4, 5, 6 or 7 with the elevation index line places the elevation at 400, 500, 600 or 700 yards/meters, respectively. If a clockwise rotation is continued, the number 8/3 appears for the second time on the elevation index line and indicates an 800 yard/meter elevation.

Windage Knob

The windage knob is used to adjust the strike of the round right or left. The windage knob is marked with an arrow and the letter R that shows the direction the strike of the round is being moved. See figure 9-8.

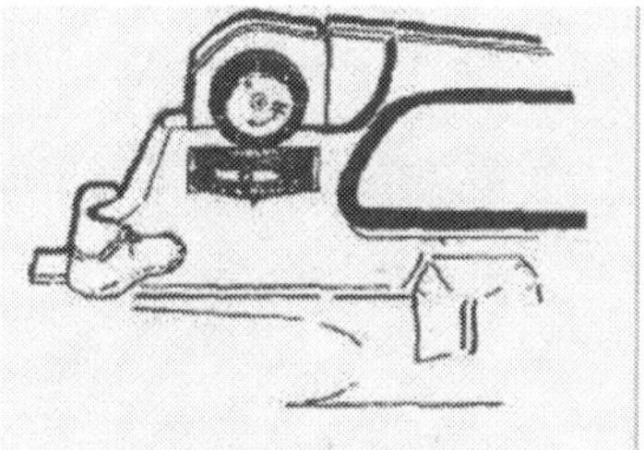

Figure 9-8. Windage Knob.

To move the strike of the round to the right, rotate the windage knob clockwise (in the direction of the arrow marked R).

To move the strike of the round to the left, rotate the windage knob counterclockwise.

9004. Windage and Elevation Rules

Moving the front sight post, elevation knob or windage knob one graduation or notch is referred to as moving one "click" on the sight. The windage and elevation rules define how far the strike of the round will move on the target for each click of front and rear sight elevation or rear sight windage for each 100 yards/meters of range to the target.

Front Sight Elevation Rule

One click of front sight elevation adjustment will move the strike of the round on target approximately 1.25 inches for every 100 yards of range to the target or 3.5 centimeters for every 100 meters of range to the target.

Rear Sight Elevation Rule

One click of rear sight elevation adjustment will move the strike of the round on the target approximately 1 inch for every 100 yards of range to the target or 2.5 centimeters for every 100 meters of range to the target.

Windage Rule

One click of windage adjustment will move the strike of the round on the target approximately 0.5 inch for every 100 yards of range to the target or 1.25 centimeters for every 100 meters of range to the target.

9005. Initial Sight Settings

Initial sight settings are those settings that serve as the starting point for initial zeroing from which all sight adjustments are made. (If the Marine already has a BZO established on his rifle, he may begin the zeroing process by using the previously established BZO sight settings.) To set the sights to initial sight settings:

Front Sight Post

To set the front sight post to initial sight setting, depress the front sight detent and rotate the front sight post until the base of the front sight post is flush with the front sight housing.

Rear Sight Elevation Knob

To set the elevation knob at the initial sight setting, perform the following:

- Rotate the rear sight elevation knob counterclockwise until the moveable rear sight housing is bottomed out on the upper receiver. See figure 9-9.

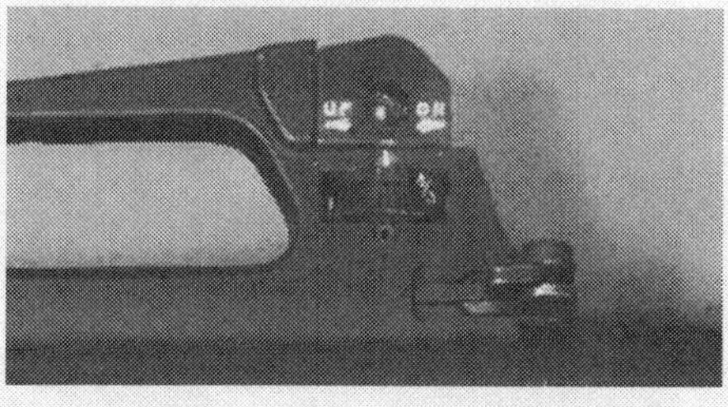

Figure 9-9. Bottoming Out Elevation Knob.

Note

Once bottomed out, the rear sight elevation knob should be three clicks counterclockwise from 8/3. If the sight fails to move three clicks counterclockwise from 8/3, it must be adjusted by a qualified armorer.

- Rotate rear sight elevation knob clockwise until the number 8/3 aligns with the index mark located on the left side of the upper receiver. See figure 9-10.

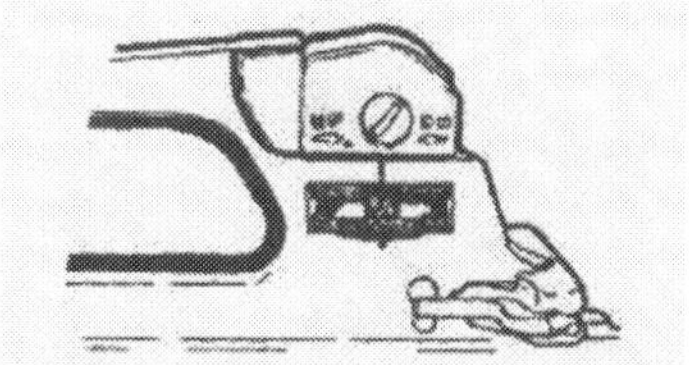

Figure 9-10. Elevation Knob Set at 8/3.

Windage Knob

To set the windage knob to initial sight setting, rotate windage knob until the index line located on the top of the large rear sight aperture aligns with the centering on the windage index scale located on the moveable base of the rear sight assembly. See figure 9-11.

Figure 9-11. Aligning Index Line.

9006. Zeroing Process

During the zeroing process, all elevation adjustments are made on the front sight post. Once a BZO is established, the front sight post should never be moved, except when rezeroing the rifle. (The rear sight elevation knob is used for dialing in the range to the target.) Zeroing is conducted at a range of 300 yards/meters. To prepare a rifle for zeroing, the rifle sights must be adjusted to the initial sight settings as outlined in paragraph 9005. Perform the following steps to zero the rifle:

- Fire a 3-shot group.
- Triangulate the shot group to find the center. See figure 9-12.
- Determine the vertical distance in inches from the center of the shot group to the center of the target.
- Make elevation adjustments on the front sight post to move the center of the shot group to the center of the target.
- Determine the horizontal distance from the center of the shot group to the center of the target.
- Make lateral adjustments on the windage knob to move the center of the shot group to the center of the target.
- Repeat preceding steps until shot group is centered.
- Fire a 4-round shot group to confirm sight setting.

Once the sight setting is confirmed, determine the value and direction of the wind and remove the number of clicks added to the windage knob (if necessary) to compensate for current wind conditions. This becomes the BZO setting for the rifle.

9007. Battlesight Zero

Zeroing is conducted at a range of 300 yards/meters. If a 300-yard/-meter range is not available, a field expedient BZO can be established at a reduced range of 36 yards/30 meters. When a rifle is zeroed for 300 yards/meters, the bullet crosses the line of sight twice. It first

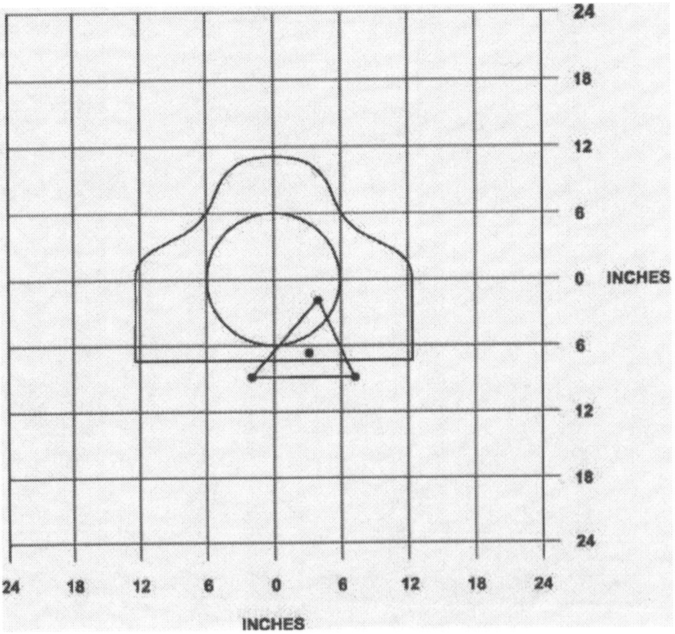

Figure 9-12. Triangulating Shot Group.

crosses the line of sight on its upward path of trajectory at 36 yards/30 meters, and again farther down range at 300 yards/meters (see fig. 9-13). Therefore a rifle's BZO may be established at a distance of 36 yards/30 meters and the same BZO will be effective at 300 yards/meters.

To establish a field expedient BZO at 36 yards or 30 meters when a 300-yard/-meter range is not available, a Marine performs the same steps as the zeroing process outlined in paragraph 9006. However, since wind does not affect the round at 36 yards/30 meters, windage is not added nor is it removed from the windage knob after confirming the BZO. To be accurate, the target must be placed exactly 36 yards (or 30 meters) from the muzzle of the rifle.

9008. Factors Causing a BZO to be Reconfirmed

Marines are responsible for maintaining a BZO on their rifles at all times. Many factors influence the BZO of a rifle. Atmospheric conditions, humidity, and temperature can cause BZOs to change on a daily basis. If operating in a combat environment, Marines should confirm their BZO as often as possible. To confirm a BZO, a Marine may begin the zeroing process by using the previously established BZO sight settings rather than placing the sights at initial sight setting. The following factors cause a BZO to be reconfirmed.

Maintenance

It is possible for the BZO to change if ordnance personnel perform maintenance on a rifle. If maintenance was performed, it is critical that the rifle be rezeroed as soon as possible.

Temperature

An extreme change in temperature (i.e., 20 degrees or more) will cause the elevation BZO to change. Changes in temperature cause chamber pressure to increase when hot and decrease when cold. This causes shots to impact the target high in hot temperatures and low in cold temperatures.

Climate

Changing climates (i.e., moving from a dry climate to a tropical climate) can mean changes in air density, moisture content, temperature or barometric pressure. Any of these elements can affect the rifle's BZO.

Ammunition

Inconsistencies in the production of ammunition lots can change a rifle's BZO.

Ground Elevation

Drastic changes in ground elevation can create changes in air density, moisture content, temperature or barometric pressure. Any of these elements can affect the rifle's BZO.

Uniform

If Marines zero their rifles in utility uniform and fire in full battle gear, their BZOs will change. The wearing of full battle gear changes eye relief, placement of the rifle in the shoulder pocket, and the way the rifle is supported on the handguard. Marines must establish their BZOs while wearing the uniform and equipment they will be wearing while engaging targets.

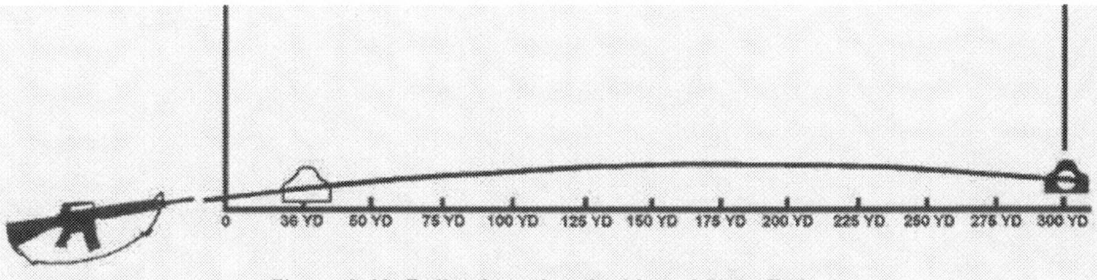

Figure 9-13. Bullet Crossing the Line of Sight Twice.

9009. Factors Affecting the Accuracy of a BZO

Anything the Marine changes from shot to shot affects the accuracy of his BZO. It further affects the accuracy of shot placement. The following factors, when applied inconsistently, diminish the accuracy of a BZO:

- Any of the seven factors (left hand, rifle butt in the pocket of the shoulder, grip of the right hand, right elbow, stock weld, breathing, and muscular relaxation/tension).
- Stability of hold.
- Sling tension.
- Trigger control.
- Sight picture.

CHAPTER 10. ENGAGEMENT TECHNIQUES

A Marine must maintain the ability to react instinctively in a combat environment—day or night. He must possess a combat mindset that eliminates hesitation, fear, or uncertainty of action. A Marine's combat mindset allows him to engage the enemy rapidly and focus on the actions required to fire well-aimed shots. He must remember that speed alone does not equate to effective target engagement. He should fire only as fast as he can fire accurately. He should never exceed his physical capability to engage a target effectively. To be effective in combat, a Marine must train to perfect the physical skills of target engagement (such as presenting the weapon and assuming a shooting position) until they become instinctive. A Marine must also employ effective engagement techniques that enable acquisition and engagement of a variety of targets in diverse combat conditions.

10001. Target Detection

To be proficient, a Marine rifleman must be able to detect targets, determine the range to targets, and accurately engage the targets. There are many variables affecting a Marine's ability to detect and determine the range to combat targets. Enemy targets on the battlefield may be single or multiple, stationary or moving, or completely hidden from view. Success in locating an enemy target will depend upon the observer's position, his skill in searching an area, and his ability to recognize target indicators.

Target Indicators

Most combat targets are detected at close range by smoke, flash, dust, noise or movement, and are usually seen only momentarily. Target indicators are anything that reveals an individual's position to the enemy. These indicators are grouped into three general areas: movement, sound, and improper camouflage.

Movement

The human eye is attracted to movement, especially sudden movement. The Marine need not be looking directly at an object to notice movement. The degree of difficulty in locating moving targets depends primarily on the speed of movement. A slowly moving target will be harder to detect than one with quick, jerky movements.

Sound

Sound can also be used to detect an enemy position. Sound may be made by movement, rattling equipment, or talking. Sound provides only a general location of the enemy, making it difficult to pinpoint a target by sound alone. However, sound can alert the Marine to the presence of a target and increase his probability of locating it through other indicators.

Improper Camouflage

There are three indicators caused by improper camouflage: shine, outline, and contrast with the background. Most targets on the battlefield are detected due to improper camouflage. However, many times an observation post or enemy firing position will blend almost perfectly with the natural background. Only through extremely careful, detailed searching will these positions be revealed.

Shine. Shine is created from reflective objects such as metal or glass. It may also come from pools of water and even the natural oils from the skin. Shine acts as a beacon to the target's position.

Outline. Most enemy soldiers will camouflage themselves, their equipment, and their positions. The outline of objects such as the body, head and shoulders, weapons, and web gear are recognizable even from a distance. The human eye will often pick up a recognizable shape and concentrate on it even if the object cannot be identified immediately. The reliability of this indicator depends upon visibility and the experience of the observer.

Contrast With the Background. Indicators in this category include objects that stand out against (contrast with) a background because of differences in color, surface, and shape. For instance, a target wearing a dark uniform would be clearly visible in an area of snow or sand. Geometric shapes, such as helmets or rifle barrels, can be easy to detect in a wooded area. Fresh soil around a fighting hole contrasts with the otherwise unbroken ground surface. While observing an area, take note of anything that looks out of place or unusual and study it in more detail. This will greatly increase your chances of spotting a hidden enemy.

Identifying Target Location

Observation Position

A good position is one that offers maximum visibility of the area while affording cover and concealment. The optimal observation position should allow the Marine to scan all the areas of observation and offer enough concealment to prevent his position from being detected.

The Marine should avoid positions that are obvious or stand out, such as a lone tree in a field or a pile of rocks on a hill. These positions may be ideal points for easy observation, but they will also make it easier for the enemy to locate the Marine.

Methods for Searching an Area

In searching an area, the Marine will be looking for target indicators. There are two techniques for searching an area: the hasty search and the detailed search.

Hasty Search. When a Marine moves into a new area, he must quickly check for enemy activity that may pose an immediate danger. This search is known as the hasty search and should take about 30 seconds, depending on the terrain.

Quickly glance at various points throughout the area rather than sweeping the eyes across the terrain in one continuous movement. The Marine should search the area nearest him first since it poses the greatest potential for danger.

This method of search is effective because it takes advantage of peripheral vision. Peripheral vision enables the detection of any movement in a wide area around the object being observed. For this technique to be effective, the eyes must be focused briefly on specific points (i.e., areas that may provide cover or concealment for the enemy).

Detailed Search. A detailed search is a systematic examination of a specific target indicator or of the entire observation area. A detailed search should be conducted immediately on target indicators located during the hasty search. The detailed search should be made from top to bottom or side to side, observing the entire object in exact detail. If multiple indicators were observed during the hasty search, the detailed search should begin with the indicator that appears to pose the greatest threat.

After a thorough search of target indicators, or if no indicators were located during the hasty search, a detailed search should be made of the entire observation area. The 50-meter overlapping strip method is normally used.

Normally, the area nearest the observer offers the greatest potential danger and should be searched first. Begin the search at one flank, systematically searching the terrain at the front in 180-degree arcs, searching everything in exacting detail, 50 meters in depth. See figure 10-1.

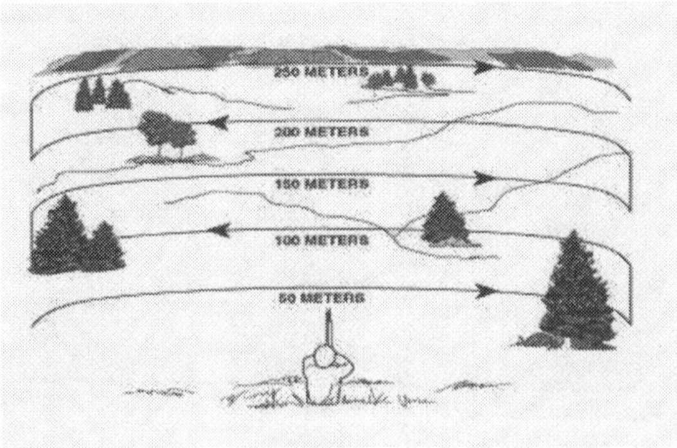

Figure 10-1. Searching the Terrain in Overlapping Strips.

After reaching the opposite flank, systematically cover the area between 40 and 90 meters from your position. The second search of the terrain includes about 10 meters of the area examined during the first search. This technique ensures complete coverage of the area.

Continue the overlapping strip search method for as far as you can see.

Maintaining Observation

Method

The combat situation will dictate the method of maintaining observation of an area. Generally, the observation method will include a combination of hasty and detailed searches.

Sequence of Observation

Observation is often conducted by a two-man team. One team member should constantly observe the entire area using the hasty search technique and the other team member should conduct a detailed overlapping strip search. If you are observing as an individual, devise a plan to ensure that the area of observation is completely covered. When entering a new area, immediately conduct a hasty search. Since a hasty search may fail to detect some indicators, periodically conduct a detailed search of the area. A detailed search should also be conducted any time your attention has been diverted from the search area.

Remembering Target Location

Most targets are seen only briefly and most areas contain multiple targets. Once you have located a target indicator, you will need to remember its location to engage it successfully. To help remember the location of a target, select a known feature and use it as a reference point to determine the distance and general direction to the target.

10002. Range Estimation

To engage targets at unknown distances, a Marine must determine the distance from his location to a known point. This is known as range estimation. The ability to determine range is a skill that must be developed if a Marine is to successfully engage targets at unknown distances. Precise range estimation enhances accuracy, enhances the chance of survival, and determines if a target can be effectively engaged using the rifle's existing BZO or if a new sight setting or point of aim is required.

Range Estimation Methods

Unit of Measure Method

To use this method, a Marine visualizes a distance of 100 meters on the ground, and then estimates how many of these units will fit between him and the target. This determines the total distance to the target. See figure 10-2.

Figure 10-2. Unit of Measure Method.

The greatest limitation of this method is that its accuracy is related to the amount of visible terrain. For example, if a target appears at a range of 500 meters or more and only a portion of the ground between a Marine and the target can be seen, it becomes difficult to use the unit of measure method to estimate range accurately. A Marine must practice this method frequently to be proficient. Whenever possible, a Marine should select an object, estimate the range, and then verify the actual range by either pacing or using another accurate measurement.

Rifle Front Sight Post Method

The area of the target covered by the rifle's front sight post can be used to estimate range to a target. A Marine notes the appearance of the front sight post on a known-distance target. A Marine then uses this as a guide to determine range over an unknown distance. Because the apparent size of the target changes as the distance to the target changes, the amount of the target covered by the front sight post varies based on the range. In addition, a Marine's eye relief and perception of the front sight post affect the amount of the target that is visible. To use this method, a Marine must apply the following guidelines:

- The front sight post covers the width of a man's chest or body at approximately 300 meters.
- If the target is less than the width of the front sight post, the target is in excess of 300 meters and the rifle's BZO cannot be used effectively.
- If the target is wider than the front sight post, the target is less than 300 meters and can be engaged point of aim/point of impact using the rifle's BZO.

See figure 10-3.

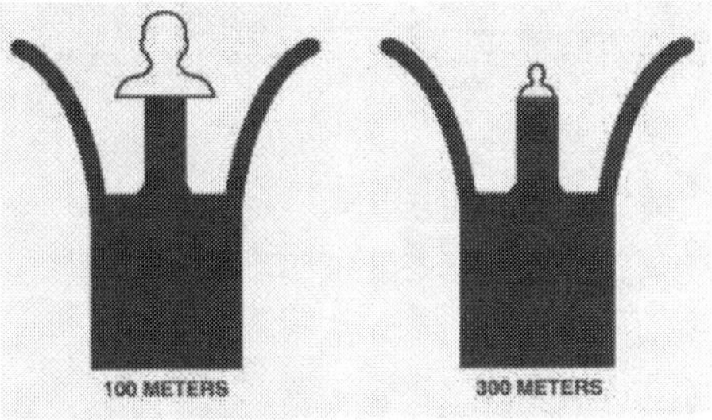

Figure 10-3. Front Sight Post Method.

Visible Detail Method

The amount of detail seen at various ranges can provide a Marine with an estimate of the target's distance. To use this method, a Marine must be familiar with the size and various details of personnel and equipment at known distances. Visibility (such as weather, smoke or darkness) limits the effectiveness of this method. A Marine should observe a man while he is standing, kneeling, and in the prone position at known ranges of 100 to 500 meters. He should note the man's size, characteristics/size of his uniform and equipment, and any other pertinent details. The Marine then uses this as a guide to determine range over an unknown distance. A Marine also should study the appearance of other familiar objects such as rifles and vehicles. To use this method, a Marine applies the following general guidelines:

- At 100 meters, the target is clearly observed in detail and facial features are distinguished
- At 200 meters, the target is clearly observed. There is a loss of facial detail. The color of the skin and equipment are still identifiable.
- At 300 meters, the target has a clear body outline, face color usually remains accurate, but remaining details are blurred.
- At 400 meters, the body outline is clear, but remaining detail is blurred.
- At 500 meters, the body shape begins to taper at the ends, and the head becomes indistinct from the shoulders.
- At 600 meters, the body appears wedge-shaped and headless.

Bracketing Method

This method of range estimation estimates the shortest possible distance and the greatest possible distance to the target. For example, a Marine estimates that a target may be as close as 300 meters but it could be as far away as 500 meters. The estimated distances are averaged to determine the estimated range to the target. For example, the average of 300 meters and 500 meters is 400 meters.

Halving Method

To use the halving method, a Marine estimates the distance halfway between him and the target, then doubles that distance to get the total distance to the target. A Marine must take care when judging the distance to the halfway point; any error made in judging the halfway distance is doubled when estimating the total distance.

Combination Method

The methods previously discussed require optimal conditions with regard to the target, terrain, and visibility in order to obtain an accurate range estimation. A Marine should estimate the range using two methods and then compare the estimates, or two Marines can compare their estimates. The average of the two estimates should be close to the actual range to target.

Factors Affecting Range Estimation

The following specific factors will affect the accuracy of estimation. A Marine must be aware of these factors and attempt to compensate for their effects.

Nature of the Target

- An object with a regular outline such as a steel helmet, rifle, or vehicle on a clear day will appear to be closer than one with an irregular outline such as a camouflaged object.

- A target that contrasts with its background will appear to be closer than a target that blends in with its background.
- A partially exposed object will appear to be farther away than it is.
- A target will appear to be farther away if the target is smaller than the objects surrounding it.

Nature of the Terrain

- Terrain that slopes upward gives the illusion of shorter distance.
- Terrain that slopes downward gives the illusion of greater distance.
- Terrain with dead space makes the target appear to be closer.
- Smooth terrain such as sand, water, or snow gives the illusion of greater distance.

Limited Visibility

The position of the light source significantly affects a Marine's ability to estimate range. Other factors that affect range estimation include smoke, fog, rain, angled light, or anything that obscures the battlefield. Generally, when the sun is bright, a target appears further away. When the sky is overcast, a target generally appears closer. Target contrast is another factor to consider when estimating range as well as obstacles located between the shooter and the target. If there is contrast (e.g., color variation) between the target and the background, the target will appear closer. If there is little or no contrast between the target and the background, the target will appear farther away. If there is an object between the target and the shooter that will distract the shooter (e.g., a tree), the target will appear farther away.

10003. Offset Aiming

The conditions of rifle fire in combat may not permit mechanical adjustments of the sights. To engage a target during combat, a Marine may be required to aim his rifle at a point on the target other than center mass. This is known as offset aiming. Offset aiming is used to compensate for the distance and size of the target, wind, and speed and angle of a moving target. There are two primary techniques for offset aiming: point of aim technique and known strike of the round.

Point of Aim Technique

The point of aim technique is the shifting of the point of aim (sight picture) to a predetermined location on or off the target to compensate for a known condition (i.e., wind, distance, and movement). Each predetermined location is known as a point of aim. Figure 10-4 illustrates points of aim for elevation.

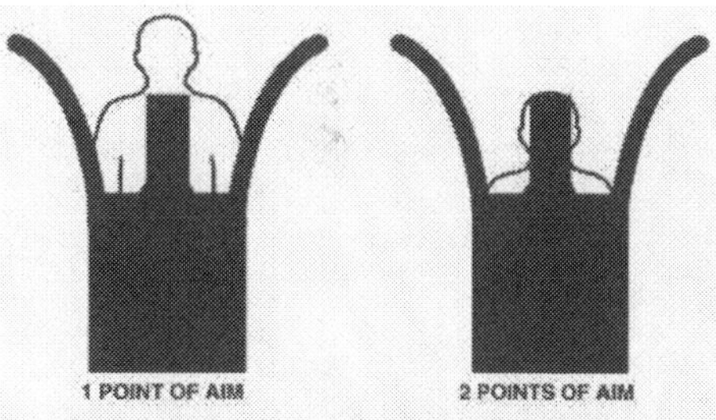

Figure 10-4. Points of Aim for Elevation.

Elevation

Predetermined points of aim sector the target horizontally. The tip of the front sight post held at shoulder level is considered one point of aim; the tip of the front sight post held at the top of the target's head is considered two points of aim. A Marine uses these points of aim to compensate for the elevation required to engage a target beyond the BZO capability of the rifle or to engage a small target (e.g., head shot) inside the BZO of the weapon.

Beyond the BZO. To use the point of aim technique to engage a target beyond the BZO of the rifle, a Marine must apply the following guidelines:

- When range to the target is estimated to be beyond 300 meters out to 400 meters, hold one point of aim.
- When the range to the target is estimated to be beyond 400 meters out to 500 meters, hold two points of aim.

Note

It is better to apply a hasty sight setting at ranges beyond the rifle's BZO. Points of aim are only guidelines at these distances because the front sight will mask the target when the front sight is held above center mass, making it difficult to acquire sight picture.

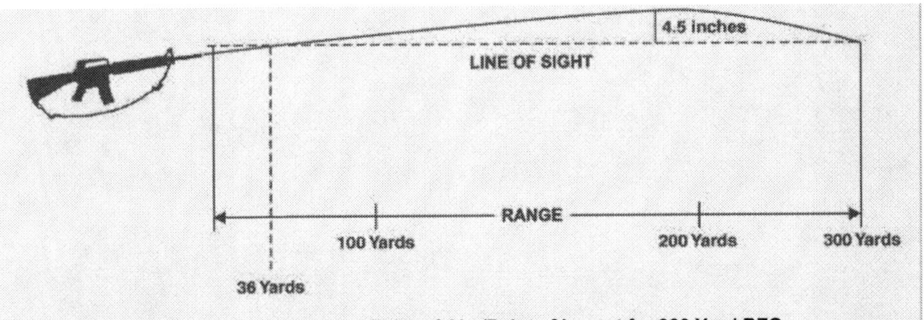

Figure 10-5. Trajectory and Point of Aim/Point of Impact for 300 Yard BZO.

Inside the BZO. If the rifle is properly zeroed for 300 yards/meters, the trajectory (path of the bullet) will rise approximately 4.5 inches (11 centimeters) above the line of sight at a distance of approximately 175 yards (160 meters). At other distances, the strike of the bullet will be less than 4.5 inches above the point of aim. Only at 36 yards/30 meters and 300 yards/meters does the point of impact coincide with the point of aim. If only a portion of the target is visible (e.g., the head of an enemy soldier), the trajectory of the bullet may have to be taken into consideration when firing at a distance less than 300 yards/meters. If a Marine does not consider trajectory, he may shoot over the top of the target if the target is small and at a distance less than 300 yards/meters. See figure 10-5.

Windage

Predetermined points of aim sector the target vertically (see fig. 10-6). The tip of the front sight post centered on the edge of the target into the wind is considered one point of aim; the trailing edge of the front sight post held on the edge of the target into the wind is considered two points of aim. The same units of measure are applied off the target for holds of additional points of aim. These points of aim are used to compensate for wind affecting the strike of the round and when there is no time to adjust the rifle's sights, or when a lead is required to engage a moving target (points of aim for moving targets are discussed in paragraph 10007). Table 10-1 provides points of aim for full value winds.

Known Strike of the Round

This offset aiming technique shifts the aiming point (sight picture) to compensate for rounds that strike off target center. The known strike of the round method is used if the strike of the round is known. To engage a target using this method, a Marine aims an equal distance from center mass opposite the known strike of the round. For example, if the round strikes high and left, a Marine aims an equal and opposite distance low and right.

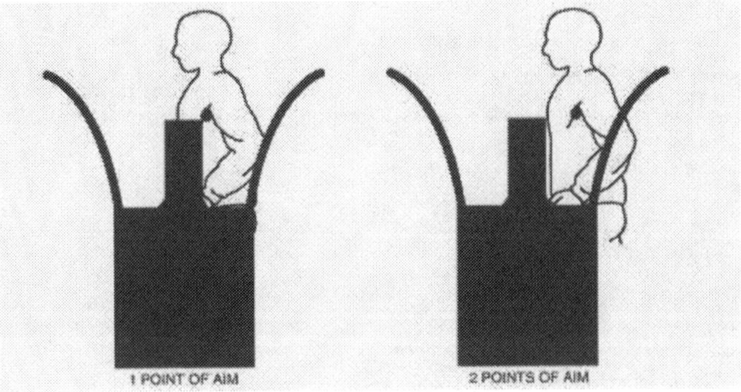

Figure 10-6. Points of Aim.

Table 10-1. Points of Aim for Full Value Winds.

Wind	Distance (yards/meters)	Points of Aim
Light	0-300	0
Medium	200-300	1
Strong	0-200	1
Strong	200-300	2

10004. Techniques of Fire

The size and distance to the target should dictate the technique of fire.

Two-Shot Technique

In combat, an effective technique for eliminating a threat is to rapidly fire more than one shot on the target. Two shots fired in rapid succession will increase the trauma (i.e., shock, blood loss) on the target, increasing the Marine's chances of quickly eliminating the threat. Firing two shots enables the Marine to break out of the tunnel vision often associated with firing in combat and then assess the situation to determine follow-on action. To execute the two-shot technique, the Marine acquires sight picture for each shot fired (each pull of the trigger).

Single Shot Technique

If the target is at a long range or if the target is small (i.e., partially exposed), it can best be engaged with a single, precision shot. A Marine's stability of hold and sight alignment are more critical to accurate engagement of long-range or small targets. To engage a target with the single shot technique, the Marine must slow down the application of the fundamentals and place one well-aimed shot on target.

Sustained Rate of Fire

An effective method for delivering suppressive fire is to fire at the sustained rate of 12 to 15 rounds per minute. Management of recoil is critical to bring the sights back on target after the shot is fired.

Three-Round Burst Technique

When set on burst, the design of the rifle permits three shots to be fired from a single trigger pull. The rounds fire as fast as the weapon will function and cause the muzzle to climb during recoil. The ability to manage recoil is extremely important when firing the rifle on burst. To achieve the desired effect (i.e., 3 rounds on target), the Marine must control the jump angle of the weapon to maintain the sights on target. At short ranges (i.e., 25 meters or less), firing on three-round burst can be an effective technique to place rounds on a man-sized target quickly to increase trauma on the target. To execute the three-round burst technique, the Marine places the selector lever on burst, aims center mass, and acquires sight picture once for the single trigger pull.

10005. Engaging Immediate Threat Targets

Immediate threat target engagement is characterized by short-range engagement (i.e., less than 50 meters) with little or no warning that requires an immediate response to engage an enemy. This type of engagement is likely in close terrain (e.g., urban, jungle). If this type of engagement is likely, the large rear sight aperture (0-2) could be placed up to provide a wider field of view and detection of targets. Marksmanship skills include quick presentation and compression of the fundamentals (i.e., quick acquisition of sight picture, uninterrupted trigger control). At close ranges, perfect sight alignment is not as critical to accuracy on target. However, the front sight post must be in the rear sight aperture; proper sight alignment is always the goal.

10006. Engaging Multiple Targets

When engaging multiple targets, a Marine must prioritize each target and carefully plan his shots to ensure successful target engagement. Mental preparedness and the ability to make split-second decisions are the key to successful engagement of multiple targets. The proper mindset allows a Marine to react instinctively and to control the pace of the battle rather than just reacting to the threat.

After the first target is engaged, a Marine must immediately engage the next target and continue to engage targets until they are eliminated. While engaging multiple targets, a Marine must be aware of his surroundings and not fixate on just one target. He must rapidly prioritize the targets, establish an engagement sequence, and engage the targets. A Marine also must maintain constant awareness and continuously search the terrain for additional targets.

Prioritizing Targets

The combat situation will usually dictate the order of multiple target engagement. Target priority is based on factors such as proximity, threat, and opportunity; no two situations will be the same. The principal method is to determine the level of threat for each target so all may be engaged in succession from the most threatening to the least threatening. The target that poses the greatest threat (e.g., closest, greatest firepower) should be engaged first. Prioritizing targets is an ongoing process. Changes in threat level, proximity, or the target itself may cause a Marine to revise his priorities. Therefore, a Marine must remain alert to changes in a target's threat level and proximity and other target opportunities as the battle progresses.

Technique of Engagement

To engage multiple targets, the Marine performs the following steps:

- The first target with two rounds.
- The recoil of the rifle can be used to direct the recovery of the weapon on to the next target. As the weapon is coming down in its recovery, the Marine physically brings the sights onto the desired target. Pressure is maintained on the trigger throughout recovery and trigger control is applied at a rate consistent with the Marine's ability to establish sight picture on the desired target.
- When possible, such as when all targets are of equal threat, the Marine should engage targets in a direction that maximizes his ability to support and control the weapon.
- The preceding steps are repeated until all targets have been engaged.

Firing Position

The selection and effective use of a firing position are critical to the successful engagement of multiple targets. A Marine should make a quick observation of the terrain and select a firing position that provides good cover and concealment, as well as the flexibility to engage multiple targets. If enemy targets are widely dispersed, the selected position must provide the Marine with flexibility of movement. The more restrictive the firing position, the longer it will take a Marine to eliminate multiple targets.

Prone

The prone position limits left and right lateral movement and is, therefore, not recommended for engaging short-range dispersed targets. Because the elbows are firmly placed on the ground in the prone position, upper body movement is restricted.

Sitting

Like the prone position, the sitting position allows limited lateral movement. This makes engagement of widely-dispersed multiple targets difficult. To ease engagement, the forward arm can be moved by pivoting on the elbow, but this movement disturbs the stability of the position.

Kneeling

The kneeling position provides a wider, lateral range of motion since only one elbow is used for support. A Marine moves from one target to another by rotating at the waist to move the forward arm in the direction of the target, either right or left.

Standing

The standing position allows maximum lateral movement. Multiple targets are engaged by rotating the upper body to a position where the sights can be aligned on the desired target. If severe or radical adjustments are required to engage widely dispersed targets, a Marine moves his feet to establish a new position rather than give up maximum stability of the rifle. This avoids poorly placed shots that can result from an unstable position.

10007. Engaging Moving Targets

In combat, it is unlikely that a target will remain stationary. The enemy will move quickly from cover to cover, exposing himself for the shortest possible time. Therefore, a Marine must quickly engage a moving target before it disappears.

Types of Moving Targets

There are two types of moving targets: steady moving targets and stop and go targets. Steady moving targets move in a consistent manner and remain in a Marine's field of vision. A walking or running man is an example of a steady moving target. A stop and go target appears and disappears during its movement. A stop and go target will present itself for only a short time before it reestablishes cover. An enemy moving from cover to cover is an example of a stop and go target. This target is most vulnerable to fire at the beginning and end of its movement to new cover because the target must gain momentum to exit its existing cover and then slow down to occupy a new position.

Leads

When a shot is fired at a moving target, the target continues to move during the time the bullet is in flight. Therefore, a Marine must aim in front of the target; otherwise, the shot will fall behind the target. This is called leading a target. Lead is the distance in advance of the target that the rifle sights are placed to accurately engage the target when it is moving.

Amount of Lead Required

Factors that affect the amount of lead are the target's range, speed, and angle of movement.

Range. Lead is determined by the rifle's distance to the target. When a shot is fired at a moving target, the target continues to move during the time the bullet is in flight. This time of flight could allow a target to move out of the bullet's path if the round was fired directly at the target. Time of flight increases as range to the target increases. Therefore, the lead must be increased as the distance to the target increases.

Speed. If a man is running, a greater lead is required because the man will move a greater distance than a walking man will while the bullet is in flight.

Angle of Movement. The angle of movement across the line of sight relative to the flight of the bullet determines the type (amount) of lead.

Three Types of Leads: Full, Half, No

- **Full Lead**. The target is moving straight across a Marine's line of sight with only one arm and half the body visible. This target requires a full lead because it will move the greatest distance across a Marine's line of sight during the flight of the bullet.
- **Half Lead**. The target is moving obliquely across a Marine's line of sight (at a 45-degree angle). One arm and over half the back or chest are visible. This target requires half of a full lead because it will move half as far as a target moving directly across a Marine's line of sight during the flight of the bullet.
- **No Lead**. The target is moving directly toward or away from a Marine and presents a full view of both arms and the entire back or chest. No lead is required. A Marine engages this target as if it were a stationary target because it is not moving across his line of sight.

Point of Aim Technique

Predetermined points of aim sector the target vertically. The tip of the front sight post centered on the leading edge of the target is considered one point of aim; the trailing edge of the front sight post held on the leading edge of the target is considered two points of aim. The same units of measure are applied off the target for holds of additional points of aim. These points of aim are used to compensate when a lead is required to engage a moving target. The following guidelines apply if a Marine uses the point of aim technique to establish a lead for a moving target at various ranges and speeds (see fig. 10-7, on page 10-10). These guidelines do not consider wind or other effects of weather. Body width in these examples is considered to be 12 inches (30 centimeters) (side view of the target).

For a slow walking target (approximately 2 to 2.5 mph/3.2 to 4 kph) moving directly across the line of sight (full lead)—

- At a range of 200 meters or less, no lead is required.
- At a range of 300 meters, hold one point of aim in the direction the target is moving.

For a fast walking target (approximately 4 mph/6.4 kph) moving directly across the line of sight (full lead)—

- At a range of 200 meters or less, hold one point of aim in the direction the target is moving.
- At a range of 300 meters, hold two points of aim in the direction the target is moving.

For a target running (approximately 6 mph/9.7 kph) directly across the line of sight (full lead)—

- At a range of 100 meters or less, hold one point of aim in the direction the target is moving.

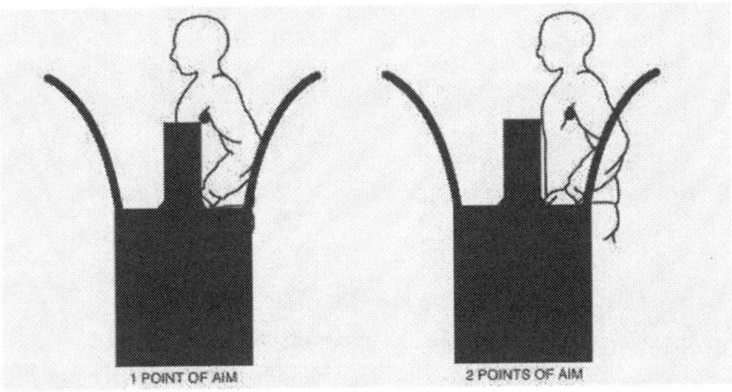

Figure 10-7. Points of Aim.

- At a range of 200 meters, hold two points of aim in the direction the target is moving.

For a target moving at a 45-degree angle (an oblique target) across the line of sight, the lead is one half of the lead that is required for a target moving directly across the line of sight.

Engagement Methods

Moving targets are the most difficult targets to engage. However, they can be engaged successfully by using the tracking or the ambush method.

The Tracking Method

The tracking method is used for a target that is moving at a steady pace over a well-determined route. If a Marine uses the tracking method, he tracks the target with the rifle's front sight post while maintaining sight alignment and a point of aim on or ahead of (leading) the target until the shot is fired. When establishing a lead on a moving target, rifle sights will not be centered on the target and instead will be held on a lead in front of the target. See figure 10-8. A Marine performs the following steps to execute the tracking methods:

- Present the rifle to the target.
- Swing rifle muzzle through the target (from the rear of the target to the front) to the desired lead (point of aim). The point of aim may be on the target or some point in front of the target depending upon the target's range, speed, and angle of movement.
- Track and maintain focus on the front sight post while acquiring the desired sight picture. It may be necessary to shift the focus between the front sight post and the target while acquiring sight picture, but the focus must be on the tip of the front sight post when the shot is fired.
- Engage the target once sight picture is acquired while maintaining the proper lead.
- Follow-through so the lead is maintained as the bullet exits the muzzle.
- Continue to track in case a second shot needs to be fired on the target.

The Ambush Method

The ambush method is used when it is difficult to track the target with the rifle, as in the prone, sitting, or any supported position. The lead required to effectively engaging the target determines the engagement point. With the sights settled, the target moves into the predetermined engagement point and creates the desired sight picture. See figure 10-9.

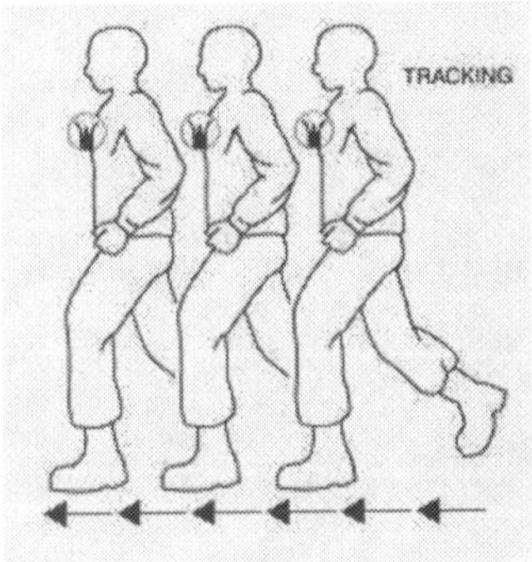

Figure 10-8. Tracking Method.

The trigger is pulled simultaneously with the establishment of sight picture. To execute the ambush method, a Marine performs the following steps:

- Select an aiming point ahead of the target.
- Obtain sight alignment on the aiming point.
- Hold sight alignment until the target moves into vision and the desired sight picture is established.
- Engage the target once sight picture is acquired.
- Follow-through so the rifle sights are not disturbed as the bullet exits the muzzle.

A variation of the ambush method can be used when engaging a stop and go target. A Marine should look for a pattern of exposure (e.g., every 15 seconds). Once a pattern is determined, a Marine establishes a lead by aiming at a point in front of the area in which the target is expected to appear, then he fires the shot at the moment the target appears.

Marksmanship Fundamentals

Engaging moving targets requires concentration and adherence to the fundamentals of marksmanship. The following modifications to the fundamentals of marksmanship are critical to engagement of moving targets.

Sight Picture

Typically, sight picture is the target's center of mass. If a Marine engages a moving target, he bases his sight picture on the target's range, speed, and angle of movement, i.e., sight alignment may be established on a point of aim in front of the target.

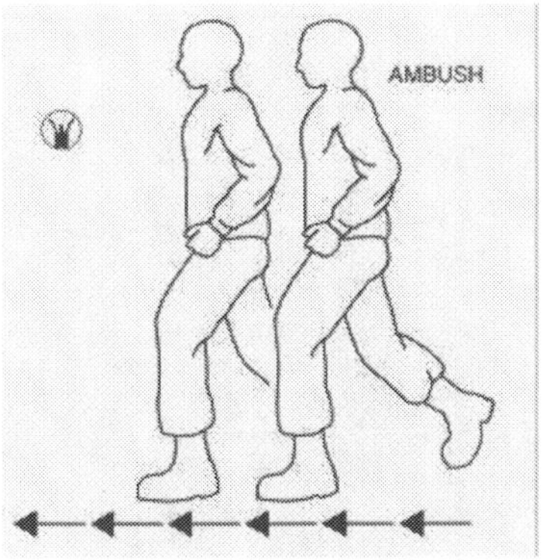

Figure 10-9. Ambush Method.

Trigger Control

As with any target engagement, trigger control is critical to the execution of shots. A Marine can apply pressure on the trigger prior to establishing sight picture, but there should be no rearward movement of the trigger until sight picture is established. Interrupted trigger control is not recommended because the lead will be lost or have to be adjusted to reassume proper sight picture. When using the tracking method, continue tracking as trigger control is applied to ensure the shot does not impact behind the moving target.

Follow-through

If a Marine uses the tracking method to engage moving targets, he continues to track the target during follow-through so the desired lead is maintained as the bullet exits the muzzle. Continuous tracking also enables a second shot to be fired on target if necessary.

Stable Position

To engage moving targets using the tracking method, the rifle must be moved smoothly and steadily as the target moves. A stable position steadies the rifle sights while tracking. Additional rearward pressure may be applied to the pistol grip to help steady the rifle during tracking and trigger control. Elbows may be moved from the support so the target can be tracked smoothly.

10008. Engaging Targets at Unknown Distances

Hasty Sight Setting

While a BZO is considered true for 300 meters, a Marine must be capable of engaging targets beyond this distance. The rifle's sighting system allows sight settings for distances out to 800 meters in 100-meter increments. If a Marine must establish a BZO for extended ranges, it is referred to a hasty sight setting. To achieve a hasty sight setting, a Marine dials the appropriate range numeral on the rear sight elevation knob that corresponds to the range to the target. For example, if the rear sight elevation knob is set at 8/3 and a target appears at 500 meters, rotate the knob to the 5 setting.

Note

Upon completion of firing with a hasty sight setting for extended ranges, return the rear sight to the BZO setting.

Point of Aim Technique

If the distance to the target is beyond the BZO capability of the rifle and time does not permit adjustment of the sights, a Marine can use offset aiming techniques to apply a point of aim for elevation to engage the target. See paragraph 10003.

10009. Engaging Targets During Low Light and Darkness

Combat targets are frequently engaged during periods of darkness or under low-light conditions. Although basic marksmanship fundamentals do not change, the principles of night vision must be applied and target detection is applied differently. During periods of darkness or low light, a Marine's vision is extremely limited. A Marine must apply the techniques of night observation in order to detect potential targets, and he must develop skills that allow him to engage targets under these conditions.

Night Vision

A Marine can improve his ability to see during periods of darkness or low light by obtaining and maintaining night vision. Since adapting to night vision is a slow and gradual process, steps should be taken to protect night vision once it is obtained.

Obtaining Night Vision

There are two methods used to obtain night vision. The first method is to remain in an area of darkness for about 30 minutes. This area can be indoors or outdoors. The major disadvantage of this approach is that an individual is not able to perform any tasks while acquiring night vision in total darkness. The second method is to remain in a darkened area under low intensity red light (similar to the light used in a photographer's darkroom) for about 20 minutes, followed by about 10 minutes in darkness without the red light. This method produces almost complete night vision adaptation while permitting the performance of some tasks during the adjustment period.

Maintaining Night Vision

Because the eyes take a long time to adjust to darkness, it is important to protect night vision once it is acquired. To maintain night vision—

- Avoid looking at any bright light. Bright light will eliminate night vision and require readaptation.
- Shield eyes from parachute flares, from spotlights or from headlights.
- When using a flashlight to read a map or any other written material:
  - Put one hand over the glass to limit the area illuminated and the intensity of the light. Keeping one eye shut will reduce the amount of night vision lost.
  - Cover the light with a red filter to help reduce the loss of night vision.
  - Minimize the time spent using a flashlight.

Factors Affecting Night Vision

Some physical factors may affect your night vision and reduce your ability to see as clearly as possible in low light or darkness. These factors include—

- Fatigue.
- Lack of oxygen.
- Long exposure to sunlight.
- Heavy smoking.
- Drugs.
- Headaches.
- Illness.
- Consumption of alcohol within the past 48 hours.
- Improper diet.

Searching Methods

Once night vision has been acquired, the Marine can locate targets. Some daylight observation techniques (e.g., searching for target indicators) also apply during periods of darkness or low light.

Off-center Vision

Off-center vision is the technique of keeping the attention focused on an object without looking directly at it (see fig. 10-10). To search for targets using off-center vision, never look directly at the object you are observing. You will see the object much better by using off-center vision. Look slightly to the left, right, above or below the object. Experiment and practice to find the best off-center angle for you. For most people, it is about 6 to 10 degrees away from the object, or about a fist's width at arm's length.

Note

Staring at a stationary object in the dark may make it appear to be moving. This occurs because the eye has nothing on which

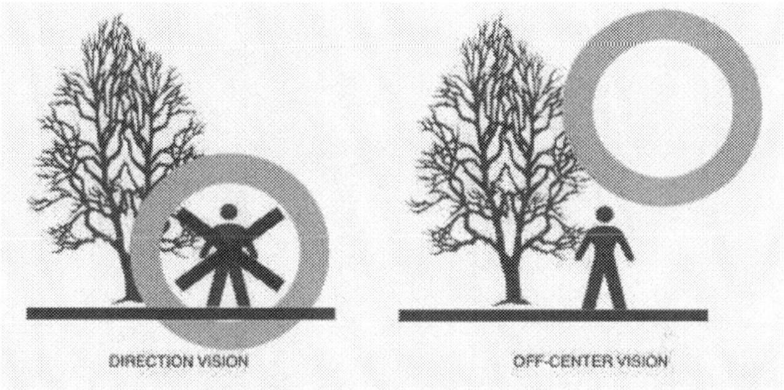

Figure 10-10. Off-center Vision.

to reference the exact position of the object. This illusion can be prevented by visually aligning the object against something else, such as a finger at arm's length.

Scanning/Figure Eight Scan

Scanning is the use of off-center vision to observe an area or object and involves moving the eyes in a series of separate movements across the objective area.

A common method of scanning is to move the eyes in a figure eight pattern (see fig. 10-11). A Marine moves the eyes in short, abrupt, irregular movements over and around the area. Once a target indicator is detected, focus is concentrated in that area, but not directly at it. Pause a few seconds at each point of observation

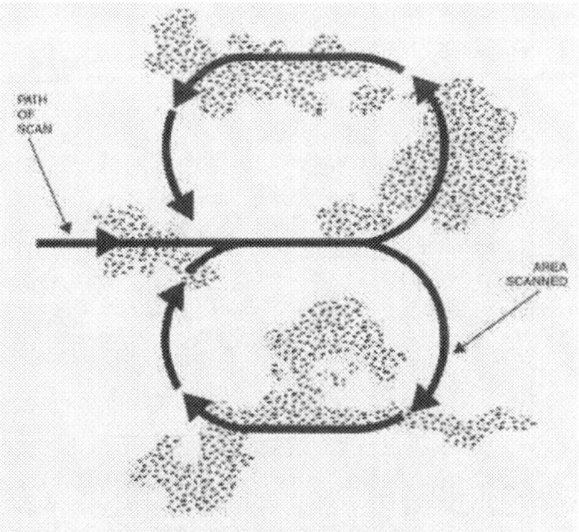

Figure 10-11. Figure Eight Scan.

since the eyes cannot focus on a still object while in motion. Rest the eyes frequently when scanning.

While you are observing, there may be periodic blackouts of night vision due to simple fatigue. This is normal and is not a cause for alarm. Night vision will quickly return after the eyes are moved and blinked a few times.

It is more effective to scan from a prone position or a position closer to the ground than the object being observed. This creates a silhouetted view of the object.

When scanning an area, look and listen for the same target indicators as in daylight: movement, sound, and improper camouflage.

Objects in bright moonlight/starlight cast shadows just as in sunlight.

Sound seems louder at night than during daylight.

Types of Illumination

There are two types of illumination that assist engagement during low light or darkness: ambient light and artificial illumination. Both ambient light and artificial illumination can affect target perception (distance and size) and night vision capabilities.

- Ambient light is the light produced by natural means (i.e., the sun, moon, and stars). Variations occur in ambient light due to the time of day, time of year, weather conditions, terrain, and vegetation.
- Artificial illumination is the light produced by a process other than natural means. Artificial light can be used to illuminate an area for target detection

or to illuminate a specific target to pinpoint its position. There are two types of artificial illumination used in combat: air and ground.

Effects of Illumination

In some combat situations, ambient light and artificial illumination may assist a Marine in locating targets. However, this light can affect perception of the target and disrupt night vision. The introduction of artificial light requires the eyes to make a sudden, drastic adjustment to the amount of light received. This can cause a temporary blinding because night vision was abruptly interrupted. Ambient light also can cause a blinding effect; e.g., a Marine may experience temporary blindness or reduced night vision if a bright moon suddenly appears from behind the clouds.

- Light behind a Marine or light between the Marine and a target illuminates the front of the target and makes it appear closer than it is.
- Light beyond the target displays the target in silhouette and makes it appear farther away than it is. If the target is silhouetted, it is easier to see and easier to engage.
- Air illumination devices are in constant motion as they descend to the ground. This movement creates changing shadows on any illuminated target causing a stationary target to appear as if it is moving.

10010. Engaging Targets while Wearing the Field Protective Mask

While engaging targets in a combat environment, a Marine is under considerable stress caused by fear, fatigue, and the noise of battle. His stress is further aggravated by the fear and uncertainty associated with a nuclear, biological, and chemical (NBC) threat. However, a Marine must be able to operate under any battlefield condition, including an NBC environment. If a Marine wears the field protective mask, its bulk and reduced visibility can affect his firing position which in turn affects the rifle's zero and his ability to engage the target. A Marine must make adjustments to his firing position and the application of marksmanship fundamentals to counter the additional gear worn in an NBC environment. Therefore, a Marine should practice wearing his field protective mask when he is not in a combat environment. This allows him to overcome any restrictions caused by the mask, develop confidence in his ability to execute well-aimed shots while wearing the mask, and develop a plan of action. This plan should address how the rifle is presented to the target, how long the mask is worn, and the likelihood of enemy contact. If a Marine expects to wear the mask for an extended period and enemy contact is likely, he should consider adjusting the rifle sights so that his first rounds are on target.

Marksmanship Fundamentals

Wearing the field protective mask requires a Marine to make modifications to his aiming and breath control techniques.

Aiming

Wearing the field protective mask affects the aiming process and the ability to locate targets. The bulk of the mask may require an adjustment to stock weld, eye relief, head position, and placement of the buttstock in the shoulder.

Breath Control

Wearing the field protective mask affects breath control because breathing may be more difficult. Temporary fogging of the lens also may be experienced. If fogging occurs, a Marine should take a deep breath and fire while holding a full breath of air (inhaling clears the fog).

Firing Position

A good firing position provides balance, control, and stability during firing. The field protective mask's added bulk and other restrictions may require a Marine to make changes to his firing position. The adjustments are unique to each Marine and based on his body size and shape and his ability to adapt to the mask. Adjustments should be minor. However, all firing positions will be affected in the following areas:

Stock Weld

Changing the placement of the cheek on the stock may affect the rifle's BZO. Therefore, a Marine should obtain a BZO for the rifle in full mission-oriented protective posture gear. Stock weld will not be as comfortable or feel as solid as it does without the field protective mask. The loss of sensitivity between the cheek and the stock, due to the mask, may cause the cheek to be pressed too firmly against the stock. Pressing the cheek too firmly against the stock can cause the seal of the field protective mask to break. If this occurs, quickly clear the mask and resume a firing position. If the lens of the field protective mask fogs up while in a firing position, this indicates that the mask's seal has been broken. Clear the mask and resume the firing position.

Eye Relief

The added bulk of the field protective mask may increase eye relief because the head is farther back along the stock. If the eye is too far from the rear sight aperture it may be difficult to acquire the target and to maintain sight picture; if the eye is too close, the rear sight can hit the mask, possibly breaking a lens or its seal.

Head Position

The mask's shape and bulk can make sight alignment difficult to achieve. The restrictive vision caused by the mask may force a Marine to roll or tilt his head over the stock to achieve sight alignment. The Marine should keep his head as erect as possible while maintaining sight alignment.

Placement of Buttstock in the Shoulder

Placement of the buttstock in the shoulder pocket may have to be altered due to the mask's added bulk. If the rifle is canted, a Marine may place the buttstock of the rifle just outside the pocket of the shoulder to achieve sight alignment. Holding the rifle straight is the preferred method of obtaining sight alignment. However, if sight alignment cannot be achieved in this position, a Marine may alter the hold of the rifle to bring the aiming eye in line with the sights. Canting the rifle drastically affects the rifle's zero. A Marine should cant the rifle only as much as is needed to obtain a good stock weld and proper sight alignment. If the rifle is canted, the point of impact may not coincide with the point of aim. For example, when wearing the mask, a right-handed Marine's point of impact is usually high and to the left of center mass (for a left-handed Marine, high and to the right of center mass). Therefore a Marine has to offset aim an equal and opposite distance low and to the right. See paragraph 10003 for a discussion on offset aiming and the known strike of the round technique.

Appendix A

Data Book

Note

The principles for recording data in the data book are the same for the Entry Level Rifle (ELR) Program and the Sustainment Level Rifle (SLR) Program. This appendix provides generic information for completing the data book. Specifics pertaining to each program are contained in the ELR and SLR lesson plans.

1. Data Book

Of all the tools that assist the Marine in firing accurately and consistently, the data book, if properly used, is the most valuable asset. It contains a complete record of every shot fired and the weather conditions and their effects on shooting. When used properly it will assist the Marine in establishing and maintaining a battlesight zero (BZO).

2. Recording Data Before Firing

Recording information in the data book prior to firing saves valuable time on the firing line that should be used to prepare for firing. Some information can be recorded before going to the firing line. In the BEFORE FIRING section of the data book, record the following (see fig. A-1):

True Zero

Front Elevation

Enter the front sight post setting by recording the number of clicks up (+) or down (-) under FRONT ELEV.

Rear Elevation

Circle the rear sight elevation knob setting under REAR ELEV for the yard line firing from—

- 200 yards: 8/3-2
- 300 yards: 8/3
- 500 yards: 5

Note

At 500 yards, the rear sight elevation knob setting may be plus or minus one or two clicks off of 5.

Wind

Under the WIND column under TRUE ZERO, the R represents clicks right on the rifle from the initial sight setting and the L represents clicks left on the rifle. Enter the rear sight windage knob setting by recording the number of clicks right (clockwise) or left (counterclockwise) under WIND.

Wind

Prior to firing, check the wind. If wind conditions are present, a sight adjustment will have to be made prior to firing to ensure shots impact the center of the target.

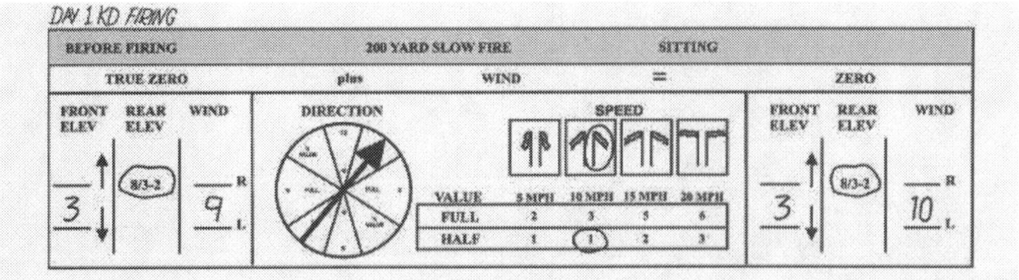

Figure A-1. Recording Data Before Firing.

Direction
Determine the direction of the wind and draw an arrow through the clock indicating the direction the wind is blowing. If the wind is not blowing, a data entry is not needed.

Note

Remember that your position is represented in the center of the clock and the target is at 12 o'clock.

Value
Look at the clock to determine if the wind is full, half, or no value wind. Under VALUE, circle FULL or HALF to indicate the wind value.

Speed
Observe the flag on the range and circle the appropriate flag indicating the wind's velocity (SPEED).

Determining Windage Adjustment
The chart beneath the flag indicates the number of clicks on the rear sight windage knob to offset the effects of the wind. Circle the number of clicks where the wind value and wind speed intersect.

Zero

Determine the zero you will place on your rifle to accommodate wind conditions to begin firing. This ZERO will be the TRUE ZERO plus the rear sight windage setting to compensate for the effects of wind.

Front Elev and Rear Elevation
Elevation adjustments are not affected by wind so the same settings are carried over from TRUE ZERO.

Wind
Wind will affect the strike of the round right or left on the target. Therefore, if wind is a factor, the rear sight windage knob must be adjusted to compensate for the effects of wind.

If the wind is blowing, add the number of clicks circled under the flags to the rear sight windage knob setting from TRUE ZERO.

Once the windage setting is determined, record it in the WIND column.

3. Recording Data During Firing

Recording Data During Slow Fire Stages

The method for calling and plotting slow fire shots in the data book is called "the shot behind method." It allows the Marine to spend less time recording data and more time firing on the target. This is because all the calling and plotting is done while the target is in the pits being marked. This information is recorded in the DURING FIRING portion of the data book page. The proper and most efficient method for recording data during KD slow fire stages is as follows (see fig. A-2):

Fire the First Shot
Fire the first shot. Then immediately check the wind flag to see if the speed or direction of the wind changed. If the wind direction changed, indicate the windage adjustment needed to compensate for it in the WIND row under CALL number 1. This windage adjustment will have to be applied to the rifle prior to firing the second shot.

Call the Shot Accurately
As soon as the shot is fired and the target is pulled into the pits, record the exact location where the tip of the front sight was on the target at the exact instant the shot was fired (assuming sight alignment was maintained). Plot this location on the target provided under number 1 in the block marked CALL.

Prepare to Fire the Second Shot
As soon as you have recorded the call for the first shot, prepare to fire the second shot.

Look at Where the First Shot Hit
As the target reappears out of the pits, look where the first shot hit the target. Remember this location so it can be plotted after firing the second shot.

Fire the Second Shot
Fire the second shot. Then check the wind flag to see if the wind changed speed or direction.

Call the Second Shot and Plot the First Shot
As soon as the second shot is fired and the target is pulled into the pits, record the call of the second shot.

Now plot the precise location of the first shot by writing the numeral 1 on the large target diagram provided in the block marked PLOT.

Prepare to Fire the Third Shot

Repeat steps until all shots have been fired. Indicate each slow fire shot with the appropriate number (e.g., 1, 2, 3, 4, 5).

Make a Sight Adjustment if Required

Sight adjustments should be made off of a shot group, not a single shot. Determine if a sight adjustment is necessary off of the first three shots fired. If the shots form a group (i.e, a group that fits inside the "A" target bull's-eye or the center scoring ring of the "D" target), but are not where they were called, make the necessary sight adjustment.

CAUTION

Generally, major sight adjustments from established sight settings are caused by poor application of the fundamentals, inconsistencies in firing positions, inconsistencies in sight picture at different ranges and different positions, and inconsistent tension on the sling. Every effort should be made to correct shooting errors prior to making a sight adjustment on the rifle.

Note

The plotting targets in the data book for the KD Course of Fire are in inches, requiring you to calculate the number of clicks to center your shot groups.

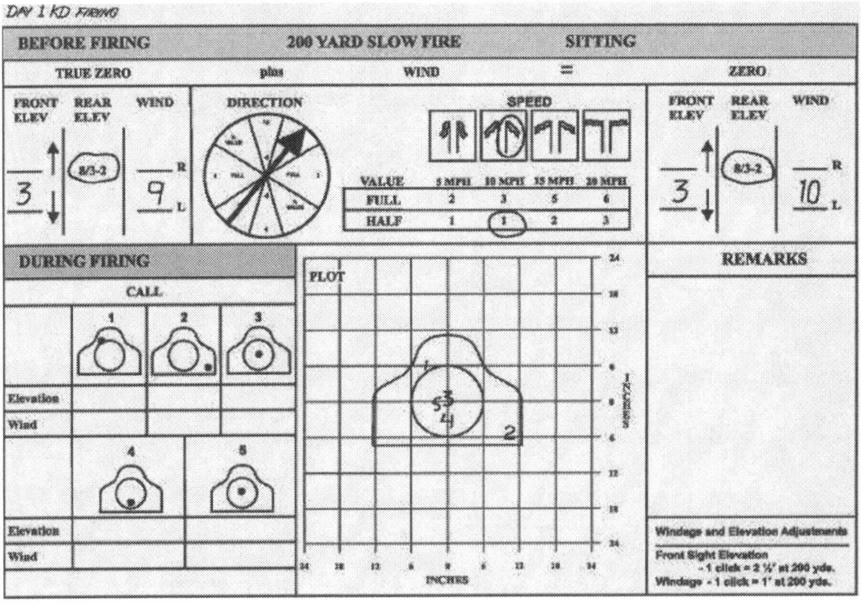

Figure A-2. Recording Slow Fire Data During Firing.

Elevation. Locate the closest horizontal grid line to the center of the plotted shot group. Follow the line across to the numbered vertical scale to determine the number of inches of elevation the shot group is off of target center. Calculate the number of clicks on your front sight post (for 200 and 300 yards) or rear sight elevation knob (for 500 yards) to bring your shot group center.

Windage. Locate the closest vertical grid line to the center of the plotted shot group. Follow the line down to the numbered horizontal scale to determine the number of inches of windage the shot group is off of target center. Calculate the number of clicks on your rear sight windage knob to bring shot group center.

Recording Data During Rapid Fire Stages

In the DURING FIRING section of the data book, record the following (see fig. A-3):

Mentally Call Shots While Firing

While firing the rapid fire string, make a mental note of any shots called out of the group.

Plot the Shot Group

After firing the rapid fire string, and when the target is marked, plot all visible hits with a dot precisely where they appear on the large target diagram in the block marked PLOT.

Make a Sight Adjustment if Required

Locate the center of the shot group. If the shots form a group, make the necessary sight adjustments off of the center of the group. If shots do not form a group and do not just contain a poor shot, do not make a sight adjustment. Determine the sight adjustment by locating the center of the shot group and using the grid lines on the target in the data book. These grid lines represent the number of inches to bring a shot group center. Looking at the shot group:

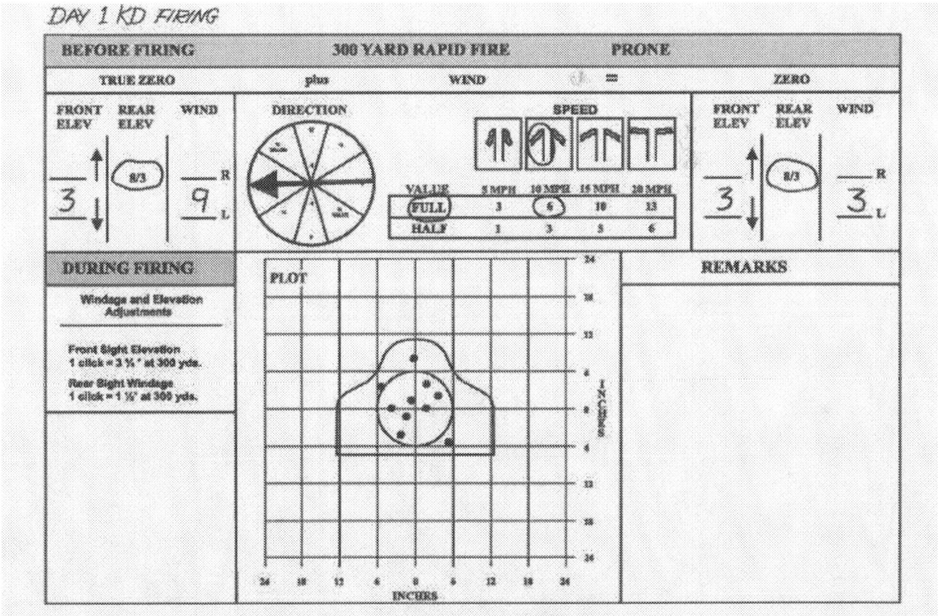

Figure A-3. Recording Rapid Fire Data During Firing.

CAUTION

Generally, major sight adjustments from established sight settings are caused by poor application of the fundamentals, inconsistencies in firing positions, inconsistencies in sight picture at different ranges and different positions, and inconsistent tension on the sling. Every effort should be made to correct shooting errors prior to making a sight adjustment on the rifle.

Note

The plotting targets in the data book for the KD Course of Fire are in inches, requiring you to calculate the number of clicks to center your shot groups.

Elevation. Locate the closest horizontal grid line to the center of the plotted shot group. Follow the line across to the numbered vertical scale to determine the number of inches of elevation the shot group is off of target center. Calculate the number of clicks on your front sight post to bring your shot group center.

Windage. Locate the closest vertical grid line to the center of the plotted shot group. Follow the line down to the numbered horizontal scale to determine the number of inches of windage the shot group is off of target center. Calculate the number of clicks on your rear sight windage knob to bring your shot group center.

Remarks

After firing a stage, record any data or information that can be helpful in improving shooting in the future. Anything done or observed should be recorded. These items will be helpful when analyzing daily shooting performance. Record this information in the REMARKS column. What the Marine fails to record may be the information he will need to improve.

4. Recording Data After Firing

In the AFTER FIRING section of the data book, record the following (see fig. A-4 on page A-6):

Zero

Upon completion of firing, determine the elevation and windage required to center the shot group (if necessary) and record this sight setting in the ZERO block of the AFTER FIRING section. If no adjustments are needed to center the shot group, record the sight settings currently on the rifle.

Front Elevation

Under the column FRONT ELEV, record the final elevation setting made on the front sight post.

Rear Elevation

Under the column REAR ELEV, record the rear sight elevation knob setting for the yard line firing from.

Note

If firing from 500 yards, record the final elevation setting made on the rear sight elevation knob.

Wind

Under the column WIND under ZERO, record final windage setting made on the rear sight windage knob.

Wind

Calculate the prevailing wind during the string of fire.

Direction

Determine the direction of the wind and draw an arrow through the clock indicating the direction the wind is blowing.

Note

Remember that your position is represented in the center of the clock and the target is at 12 o'clock.

Value

Look at the clock to determine if the wind is full, half, or no value wind. Under VALUE, circle FULL or HALF to indicate the wind value.

Speed

Observe the flag on the range and circle the appropriate flag indicating the wind's velocity (SPEED).

Determine Windage Adjustment

The chart beneath the flag indicates the number of clicks on the rear sight windage knob to offset the effects of the wind. Circle the number of clicks where the wind value and wind speed intersect.

True Zero

A true zero is the established zero without the windage adjustments to compensate for the effects of the string of fire's wind. A true zero is calculated because, the next time you fire, the wind conditions will probably be different. Therefore, the rear sight windage knob adjustments made to compensate for the string of fire's wind will not be the correct setting for wind conditions on other strings of fire or other days of firing.

Front Elevation and Rear Elevation

Because elevation adjustments are not affected by wind, the same settings are carried over from ZERO.

Wind

Calculate the windage adjustment to compensate for the string of fire's wind conditions the same way it was calculated in the BEFORE FIRING information of the data book. The only exception is now windage adjustments are being removed from the rifle rather than added to the rifle. Because the windage setting is being removed from the rifle, remove the number of clicks of windage right or left from the ZERO windage setting. Once the windage setting is determined, record it in the WIND column.

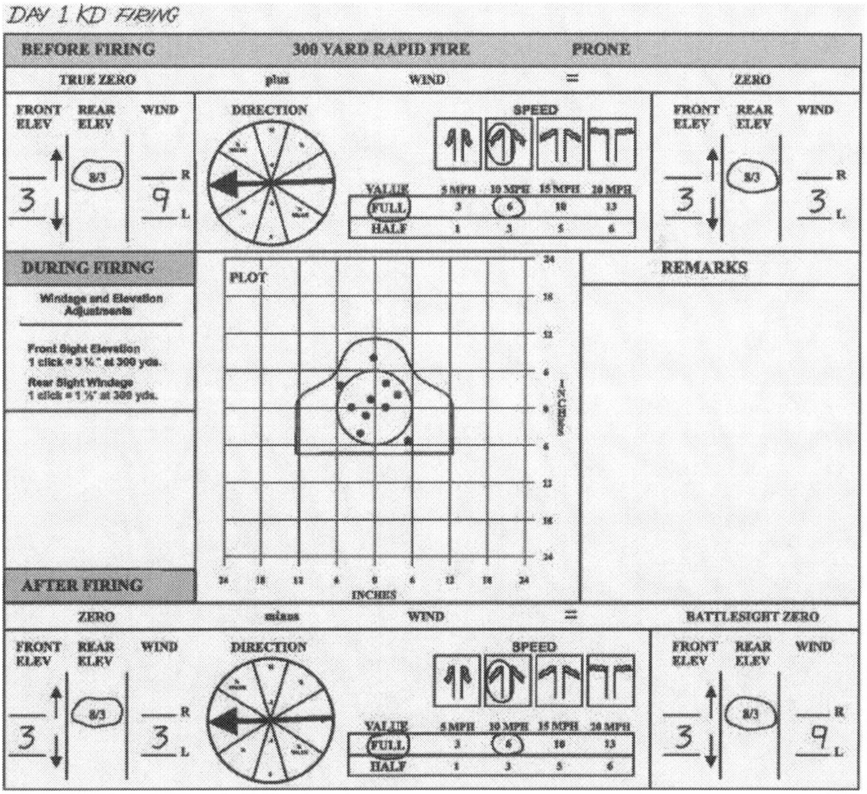

Figure A-4. Recording Data After Firing.

Appendix B

Glossary

Section I. Acronyms

BZO . . . . battlesight zero

CLP . . . . cleaner, lubricant, and preservative

KD . . . . known distance

LAW . . . . .lubricating oil, arctic weapons

NBC. . . . . nuclear, biological, and chemical

ROE. . . . . .rules of engagement

SOP . . . . .standing operating procedures

TM. . . . . technical manual

Section II. Definitions

A

aiming point – The aiming point is the precise point where the tip of the front sight post is placed in relationship to the target.

alibi – Any condition caused by the weapon, ammunition, or range operation that causes the shooter not to have an equal opportunity to complete a string of fire as all other shooters on the range.

B

battlesight zero (BZO) – The elevation and windage settings required to engage point targets from 0-300 yards/meters under ideal weather conditions (i.e., no wind).

bone support – The body's skeletal structure supporting the rifle's weight.

breath control – Procedure used to fire the rifle at the moment of least movement in the body and the rifle.

C

canting – An angular deviation of the weapon to the left or right from a vertical position during firing.

center of mass – A point that is horizontally and vertically centered on the target.

center of mass hold – The placement of the tip of the front sight on the target center of mass prior to the shot breaking.

centerline of the bore – An imaginary straight line beginning at the chamber end of the barrel and proceeding out of the muzzle.

chamber check – Procedure used to determine a weapon's condition.

D

double feed – Attempted simultaneous feeding of two or more rounds from the magazine.

dry fire – Cocking, aiming, and squeezing the trigger of an unloaded rifle in order to practice the fundamentals of marksmanship.

detailed search – Method for conducting a systematic search of an area for specific target indicators.

E

eye relief – The distance between from the rear sight aperture to the aiming eye.

F

function check – Procedure used to ensure the selector lever operates properly.

flag method – Procedure used to determine wind velocity and direction on a Known Distance (KD) range.

field expedient battlesight zeroing – Process used to zero the rifle at 36 yards or 30 meters when a 300-yard/-meter range is not available.

G

gas operated – A self-loading firearm that utilizes the expanding force of the propellant's gases to operate the action of the weapon.

H

hasty search – Method for quickly searching an area for enemy activity.

hasty sight setting – A zero established for distances out to 800 yards/meters.

I

immediate threat target – A target engaged at a short range (50 meters or less) with little or no warning.

initial sight setting – Sight setting placed on a rifle that serves as the starting point from which all sight adjustments are made for the initial zeroing process.

L

limited technical inspection (LTI) – An inspection performed by an armorer on a weapon to determine its operational status (safety and function, not accuracy).

line of sight – An imaginary line extending from the shooter's eye through the rifle's sights and onto an aiming point on a target.

load – Command/Procedure used to take a weapon from Condition 4 to Condition 3 by inserting a magazine with rounds.

M

magazine – A container that holds ammunition in a position to be chambered.

magazine fed – A mechanical, automatic means of supplying a firearm with ammunition to be chambered.

make ready – Command/Procedure used to take a weapon from Condition 3 to Condition 1 by chambering a round.

muscular relaxation – The state of tension required to properly control the rifle. The shooter's muscles are in a relaxed state of control—tightened but not tensed.

N

natural point of aim – The location at which the rifle's sights settle if bone support and muscular relaxation are achieved.

O

observation method – Procedure used to determine wind velocity and direction in a tactical situation.

P

pie technique – A technique used to locate and engage targets from behind cover, while minimizing the Marines exposure to enemy fire.

R

range – **1.** A designated place where live fire weapons training is conducted. **2.** The horizontal distance to which a projectile can be propelled to a specified target. **3.** The horizontal distance between a weapon and target.

recoil management – The ability to manage or control the recoil of the rifle for a given shooting position.

rollout – A technique used to locate and engage targets from behind cover, while minimizing the Marines exposure to enemy fire.

S

semiautomatic – One full cycle of operation. The firing, extraction, and ejection of the spent cartridge, the cocking of the weapon and chambering of the succeeding round of ammunition after the trigger is pulled.

shooter error — Any action induced by the shooter that causes the weapon to fail to operate properly or miss the intended target.

sight alignment – The placement of the tip of the front sight post in the center of the rear sight aperture.

sight picture — The placement of the tip of the front sight post in the center of the target while maintaining sight alignment.

sling – When properly attached to the rifle, the sling provides maximum stability for the weapon and helps reduce the effects of the rifle's recoil. Also used for individual weapon transport.

stability of hold – The ability to acquire a stable position and to hold the rifle steady for any given rifle position.

stock weld – The firm, consistent contact of the cheek with the weapon's buttstock.

stoppage – Any condition that causes the rifle to fail to fire.

T

target indicators – Anything that reveals the enemy's position.

trajectory – The path of a projectile through the air and to a target.

triangulation process – Process used to determine the vertical and horizontal sight adjustments that must be made to center a shot group.

trigger control – The skillful manipulation of the trigger that causes the rifle to fire without disturbing sight alignment or sight picture.

U

unload – Command/Procedure used to take a weapon from any condition to Condition 4.

unload and show clear – Command/Procedure used to take a weapon from any condition to Condition 4 while requiring a second individual to check the weapon to verify that no ammunition is present.

user serviceability inspection – Procedure used to ensure a weapon is in an acceptable operating condition, conducted by the shooter.

V

velocity – Speed at which the projectile travels.

W

weapons carry – Procedure used to effectively handle the rifle while remaining alert to possible threat levels.

weapons condition – Describes a weapon's readiness for live fire.

weapons transport – Procedure used to carry the rifle for long periods of time and when one or both hands are needed for other work.

windage and elevation rules – Rules that define how far the strike of the round will move on the target for each click of front and rear sight elevation or rear sight windage for each 100 yards/meters of range to the target.

Z

zeroing – The process used to adjust the rifle sights that cause it to shoot to point of aim at a desired range.

Made in United States
Orlando, FL
04 November 2026

90203562R10269